John Edgar McFadyen

The anatomy of the horse

A dissection guide

John Edgar McFadyen

The anatomy of the horse
A dissection guide

ISBN/EAN: 9783337257644

Printed in Europe, USA, Canada, Australia, Japan

Cover: Foto ©berggeist007 / pixelio.de

More available books at **www.hansebooks.com**

ANATOMY OF THE HORSE

A

DISSECTION GUIDE

BY

J. M'FADYEAN, M.B., C.M., B.Sc.,

MEMBER OF THE ROYAL COLLEGE OF VETERINARY SURGEONS,
LECTURER ON ANATOMY AT THE ROYAL (DICK'S) VETERINARY COLLEGE, EDINBURGH.

WILLIAM R. JENKINS,
VETERINARY PUBLISHER AND BOOKSELLER,
850 SIXTH AVENUE, NEW YORK.

W. AND A. K. JOHNSTON, PRINTERS, EDINBURGH AND LONDON.

PREFACE.

———o———

THE want of an illustrated topographical treatise on equine anatomy has, in the experience of the author, been a great barrier to the efficient teaching of that all-important branch of veterinary education. In this work the object of the author has been to place in the hands of veterinary students a dissection guide comparable, in some degree, to the text-books at the service of the practical student of human anatomy. The order of dissection laid down is that which the author has found to be most advantageous, and he has attempted to describe with accuracy and moderate fulness the different organs as they present themselves in that order. This description is largely supplemented by the illustrations, which are so complete that almost every organ in the body is delineated. The majority of these illustations are original, being faithful portraits of the author's own dissections. It is hoped that they will prove useful to the student, in the first place, as a plan and a guide in his work, and, secondly, as a means by which he may afterwards summon up a mental picture of his own dissections.

While the book is specially designed for use in the dissecting-room, the author ventures to hope that it may also be serviceable to the veterinary practitioner. Special care has been taken in portraying those regions that possess a surgical interest, and the illustrations furnish a ready means by which the surgeon may refresh his memory regarding the objects to be met in the course of an operation.

The greater number of the original drawings were made in the dissecting-room of the Royal (Dick's) Veterinary College, by Mr J. BAYNE, artist; a few were executed by Mr R. S. REID, artist; and the remainder by Mr R. H. POTTS, veterinary student. A few of the illustrations were directly drawn on stone by Messrs W. & A. K. JOHNSTON. To all of these gentlemen the author is much indebted for the clearness and fidelity with which they have delineated the various objects.

The source of each of the borrowed illustrations is duly acknowledged elsewhere, but the author is constrained to make special mention of those from the systematic text-book of Professor Chauveau, who generously consented to the copying of as many of his figures as might be thought useful for this work.

To insure accuracy, the author has been careful to compare the results of his own dissections with the descriptions of other writers, and more especially with the works of Percivall, Leyh, and Chauveau, to which he begs to express his indebtedness.

To Professor Turner the author is under deep obligation for the revision of the chapters on the brain, the eye, and the ear, and for much-esteemed suggestions regarding other points.

Finally, the author's best thanks are due to his brothers Gavin and Andrew, who have carefully revised the entire proof-sheets, and to Mr. T. Barker, veterinary student, for assistance in making the index.

ROYAL (DICK'S) VETERINARY COLLEGE, EDINBURGH,
October, 1884.

A few words are here necessary in explanation of the system of nomenclature used throughout this work. Although reluctant to add to the confusion already prevailing in the nomenclature of veterinary anatomy, the author has not conformed to any of the systems in general use. The system here employed is based on the principle of naming each object after the homologous object in human anatomy. So far, indeed, as any of the systems in use can be said to follow a principle, it is that just stated; but the violations of the principle are numerous, and, in most cases, appear to have been dictated by the merest caprice. The most vicious form of departure from the principle is that in which terms adopted from human anatomy are employed to designate not the actual homologues, but other parts having, it may be, some faint resemblance in shape or otherwise to the objects bearing these names in the human subject. This method is indefensible, since it tends to produce the greatest confusion, and, if generally adopted, would render a comparison of the anatomy of any two animals an impossibility. Many such terms have long been in use, but it is hoped that they are not ineradicable.

In cases where objects appear to be without homologues in human anatomy, new names must, of course, be found. In only a few of these instances, however, has the author employed terms of his own invention, preferring, in general, to adopt some of those already in use.

The greatest diversity of names, it will be found, exists in the case of muscles, and the following table of synonyms has been compiled for the convenience of those already familiar with the terms employed in some other works.

TABLE OF SYNONYMS OF MUSCLES.

	PERCIVALL.	CHAUVEAU.	LEYH.
Adductor magnus.	Adductor longus.	Great adductor of thigh.	Posterior pubio-femoralis.
Adductor parvus.	Adductor brevis.	Small adductor of thigh.	Middle pubio-femoralis.
Accelerator urinæ.	Idem.	Bulbo-cavernosus.	(part of) Urethral muscle.
Anconeus.	Idem.	Small extensor of fore-arm.	Small humero-olecranius.
Anterior deep pectoral.	Pectoralis parvus.	(part of) Deep pectoral.	Sterno-scapularis.
Anterior superficial pectoral.	(part of) Pectoralis transversus.	(part of) Superficial pectoral.	Small sterno-humeralis.
Aryteno-pharyngeus.	(Not described.)	Idem.	Idem.
Azygos uvulæ.	Circumflexus palati.	Palato-staphyleus.	Palato-staphylinus.
Biceps.	Flexor brachii.	Long flexor of the fore-arm.	Coraco-radialis.
Biceps femoris.	Biceps abductor femoris.	Long vastus.	Sacro-ischio-tibialis ant.
Brachialis anticus.	Humeralis externus.	Short flexor of fore-arm.	Humero-radialis.
Buccinator.	Idem anti caninus.	Alveolo-labialis.	Idem anti molaris.
Caput magnum.	Idem.	Chief extensor of the fore-arm.	Great scapulo-olecranius.
Caput medium.	Idem.	Short extensor of the fore-arm.	External scapulo-olecranius.
Caput parvum.	Idem.	Middle extensor of fore-arm.	Internal humero-olecranius.
Cerato-hyoid.	Hyoideus parvus.	Idem.	Small kerato-hyoid.
Cervico-auriculares.	Retrahentes.	Idem.	Idem.
Complexus.	Complexus major.	Great complexus.	Dorso-occipitalis.
Compressor coccygis.	Idem.	Ischio-coccygeus.	Ischio-coccygeus.
Coraco-humeralis.	Idem.	Idem.	Middle scapulo-humeralis.
Corrugator supercilii.	Levator palpebræ superioris.	Fronto-palpebral.	Superior external palpebral.
Crico-pharyngeus.	Constrictor pharyngis posterior.	Idem.	Idem.
Curvator coccygis.	Idem.	Sacro-coccygeus lateralis.	Sacro-coccygeus lateralis.

TABLE OF SYNONYMS OF MUSCLES—continued.

	PERCIVALL.	CHAUVEAU.	LETH.
Deep flexor of digit (fore limb).	(See flexor pedis perforans.)		
Deep flexor of digit (hind limb).			
Deep gluteus.	Gluteus internus.	Idem.	Small ilio-trochanterius.
Deltoid.	Teres minor.	Long abductor of arm.	Great scapulo-trochanterius.
Depressor labii inferioris.	Idem.	Maxillo-labialis.	Maxillo-labialis inferior.
Depressor labii superioris.	Idem.	Intermediate anterior.	Incisive of upper lip.
Depressor coccygis.	Idem.	Sacro-coccygeus inferior.	Long and short sacro-coccygeus inferior.
Dilatator naris inferior.	(part of) Nasalis brevis labii superioris.	(part of) Small super-maxillo nasalis.	Small super-maxillo nasalis.
Dilatator naris lateralis.	Idem.	Great super-maxillo nasalis.	Great super-maxillo nasalis.
Dilatator naris superior.	(part of) Nasalis brevis labii superioris.	(part of) Small super-maxillo nasalis.	Short muscle of nose.
Dilatator naris transversalis.	Dilatator naris anterior.	Naso transversalis.	Transversus nasi.
Erector clitoridis.	Idem.		
Erector coccygis.	Idem.	Sacro-coccygeus superior.	Sacro-coccygeus superior.
Erector penis.	(Un-named.)	Ischio-cavernosus.	Ischio-penial muscle. (As Chauveau.)
Extensor brevis.	Idem.	Pedal muscle.	
Extensor metacarpi magnus.	Idem.	Anterior extensor of metacarpus.	Humero-metacarpeus anterior.
Extensor metacarpi obliquus.	Idem.	Oblique extensor of metacarpus.	Radio-metacarpeus.
Extensor pedis (fore limb).	Idem.	Anterior extensor of phalanges.	Humero-pre-phalangeus.
Extensor pedis (hind limb).	Idem.	Lateral extensor of phalanges.	Femoro-pre-phalangeus.
Extensor suffraginis.	Idem.	Idem.	Radio-pre-phalangeus.
External oblique of abdomen.	Idem.	Idem.	Costo-abdominalis exterior.
External pterygoid.	Idem.		(part of) Spheno-maxillaris.
Flexor accessorius.	Idem.	Oblique flexor of phalanges.	Small tibio-phalangeus.
Flexor metacarpi externus.	Epitrochlo-carpeus.	External flexor of metacarpus.	Humero-supercarpeus externus.
Flexor metacarpi internus.	Epicondylo-metacarpeus.	Internal flexor of metacarpus.	Humero-metacarpeus internus.
Flexor metacarpi medius.	Epicondylo-carpeus.	Oblique flexor of metacarpus.	Humero-supercarpeus internus.
Flexor metatarsi.	Idem.	Idem.	Tibio-pre-metatarseus.

Flexor pedis per- (fore { ulnar head. / radial head. / forans limb) { humeral head.	Ulnaris accessorius. / Radialis accessorius. / Flexor pedis perforans.	Deep flexor of phalanges. }	Radio-phalangeus.
Flexor pedis perforans (hind limb).	Flexor pedis.	Deep flexor of phalanges.	Great tibio-phalangeus.
Flexor pedis perforatus (fore limb).	Idem.	Superficial flexor of phalanges.	Humero-phalangeus.
Flexor pedis perforatus (hind limb).	Gastrocnemius internus.	Superficial flexor of phalanges.	Femoro-phalangeus.
Gastrocnemius.	Gastrocnemius externus.	Gemelli of leg.	Bi-femoro-calcaneus.
Gemelli.	Gemini.	Gemelli of pelvis.	Gemelli of pelvis.
Genio-glossus.	Genio-hyo-glossus.	Idem.	Idem.
Gracilis.	Idem.	Short adductor of the leg.	Pubio-tibialis.
Great hyo-glossus.	Hyo-glossus brevis.	Idem.	Hyo-glossus.
Hyoideus transversus.	(Not described.)		
Iliacus.	Idem.	Idem.	Idem.
Infraspinatus.	(part of) l'ostea-spinatus.	Iliac psoas.	Great and middle ilio-femoralis.
Internal intercostal.	Idem and sterno-costales externus.	Subspinous.	Posterior spinous.
Internal oblique of abdomen.	Idem.	Idem.	Idem.
Internal pterygoid.	Idem.	Idem.	Ilio-abdominalis.
Interossei.	Lumbrici anteriores.	Idem.	(part of) Spheno-maxillaris.
Intertransversales of neck.	(Not described.)	Idem.	Idem.
Ischio-urethral.	Triangularis penis.	Idem.	Idem.
		Compressor of Cowper's glands.	Part of urethral muscle.
Lateralis sterni.	Idem.	Transverse muscle of ribs.	Transverse muscle of ribs.
Latissimus dorsi.	Idem.	Great dorsal.	Dorso-humeralis.
Levator anguli scapulæ.	Part of serratus magnus.	Angularis scapulæ.	Trachelo-scapularis.
Levatores costarum.	Idem.	Supercostales.	Idem.
Levator labii superioris alæqui nasi.	Idem.	Supernaso-labialis.	Fronto-labialis.
Levator labii superioris proprius.	Nasalis longus labii superioris.	Supermaxillo-labialis.	Maxillo-labialis superior.
Levator menti.	Idem.	Intermediate posterior and mento-labialis.	Incisive of lower lip and mento-labialis.
Levator palati.	Stylo-pharyngeus.	Peristaphyleus internus.	Stylo-staphylinus.
Levator palpebræ superioris.	Levator palpebræ superioris internus.	Idem.	Superior internal palpebral.
Longissimus dorsi.	Idem and spinalis dorsi.	Ilio-spinalis.	Ilio-spinalis.
Longus colli.	Idem.	Idem.	Dorso-atloideus.
Lumbricales.	Lumbrici posteriores.	Idem.	Idem.
Masseter.	Idem.	Idem.	Zygomato-maxillaris.

TABLE OF SYNONYMS OF MUSCLES—continued.

	PERCIVALL.	CHAUVEAU.	LEYH.
Mastoido-auricularis.	(Not described.)	Tympano-auricularis.	Idem.
Mastoido-humeralis.	Levator humeri.	Idem.	Common muscle of the arm, neck, and head.
Middle gluteus.	Gluteus maximus.	Idem.	Great and middle ilio-trochanterius.
Middle hyo-glossus.	(Not described.)	(Not described.)	Kerato-glossus internus.
Mylo-hyoideus.	Idem.	Idem.	Idem and mylo-glossus.
Obliquus capitis inferior.	Idem.	Great oblique of head.	Axoido-atloideus.
Obliquus capitis superior.	Idem.	Small oblique of head.	Atloido-occipitalis lateralis.
Occipito-styloid.	Stylo-hyoideus.	Idem.	Stylo-hyoideus.
Orbicularis oris.	Idem.	Labialis.	Orbicularis of the lips.
Palato-glossus.	(Not described.)	Pharyngo-glossus.	Pharyngo-glossus.
Palato-pharyngeus.	Idem. (in part).	Pharyngo-staphylinus.	Staphylinus communis.
Parieto-auricularis externus.	Attolens maximus.	Temporo-auricularis externus.	Common muscle of ear.
Parieto-auricularis internus.	Attolens posticus.	Temporo-auricularis internus.	Parieto-auricularis.
Parotido-auricularis.	Abducens vel deprimens aurem.	Idem.	Idem.
Pectineus.	Idem.	Idem.	Pubio-femoralis anterior.
Peroneus.	Idem.	Lateral extensor of phalanges.	Tibio-præ-phalangeus.
Popliteus.	Idem.	Idem.	Oblique femoro-tibialis.
Posterior deep pectoral.	Pectoralis magnus.	(part of) Deep pectoral.	Great sterno-humeralis.
Posterior superficial pectoral.	(part of) Pectoralis transversus.	(part of) Superficial pectoral.	Sterno-radialis.
Psoas magnus.	Idem.	Idem.	Lumbo-femoralis.
Psoas parvus.	Idem.	Idem.	Lumbo-ilialis.
Pterygo-pharyngeus.	(part of) Palato-pharyngeus.	Idem.	Idem.
Pyriformis.	Idem.	(part of) Obturator internus.	Sacro-trochanterius.
Quadratus femoris.	(Not described.)	Quadrate crural.	Small ischio-femoralis.
Quadratus lumborum.	Sacro-lumbalis.	Idem.	Idem.
Rectus abdominis.	Idem.	Idem.	Sterno-pubialis.
Rectus capitis anticus major.	Idem.	Idem.	Trachelo-occipitalis.
Rectus capitis anticus minor.	Idem.	Idem.	Atloido-occipitalis inferior.

Rectus capitis lateralis.	Obliquus capitis anticus.	Idem.	Atloido-styloideus.
Rectus capitis posticus major.	Idem and complexus minor.	Idem.	Atloido-occipitalis longus and brevis.
Rectus capitis posticus minor.	Idem.	Idem.	Atloido-occipitalis superior.
Rectus femoris.	Rectus.	Anterior straight of thigh.	Ilio-rotuleus anterior.
Rectus parvus.	Cruræus.	Anterior gracilis.	Small ilio-femoralis.
Retractor costæ.	(part of) Obliquus abdominis internus.		Lumbo-costalis.
Rhomboideus.	Idem.	Idem.	Cervico-subscapularis and dorso scapularis.
Sartorius.	Idem.	Long adductor of the leg.	Ilio-rotuleus internus.
Scalenus.	Idem.	Idem.	Costo-trachelius.
Scapulo-humeralis gracilis.	Idem.	Small scapulo-humeralis.	Idem.
Scapulo-ulnaris.	(part of) Caput magnum.	Long extensor of the fore-arm.	Long scapulo-olecranius.
Scuto-auricularis externus.	Anterior conchæ.	Idem.	Idem.
Scuto-auricularis internus.	Posterior conchæ.	Idem.	Idem.
Semimembranosus.	Adductor magnus.	Idem.	Great ischio-femoralis.
Semispinalis (back and loins).	Idem.	Transverse spinous.	Transverse spinous.
Semispinalis (colli).	Spinalis colli.	Transverse spinous.	
Semitendinosus.	Adductor tibialis.	Idem.	Sacro-ischio-tibialis post.
Serratus anticus.	(part of) Superficialis costarum.	Anterior small serratus.	Idem.
Serratus magnus.	Idem and levator anguli scapulæ.	Idem.	Costo-scapularis.
Serratus posticus.	(part of) Superficialis costarum.	Posterior small serratus.	Idem.
Small hyo-glossus.	(Not described.)	Idem.	(Not described.)
Small stylo-pharyngeus.	(Not described.)	Idem.	Kerato-pharyngeus inferior.
Soleus.	Plantaris.	Idem.	Peroneo-calcaneus.
Sterno-thyro-hyoideus.	Idem.	Sterno-thyroid and sterno-hyoid.	Sterno-thyroid and sterno-hyoil.
Stylo-glossus.	Hyo-glossus longus.		Kerato-glossus externus.
Stylo-hyoideus.	Hyoideus magnus.	(part of) Digastricus	Great kerato-hyoid.
Stylo-maxillaris.	Idem.	Idem.	Idem.
Stylo-pharyngeus.	Hyo-pharyngeus.	Idem.	Kerato-pharyngeus superior.
Subscapulo-hyoid.	Idem.		Scapulo-hyoid.
Superficial flexor of digit (fore limb).	See flexor pedis perforatus.		
Superficial flexor of digit (hind limb).			
Superficial gluteus.	Gluteus externus.	Idem.	Ilio-trochanterius externus.
Supraspinatus.	Antea-spinatus.	Supraspinous.	Anterior spinous.
Temporalis.	Idem.	Idem.	Temporo-maxillaris.
Tensor palati.	Idem.	Peristaphyleus externus.	Peristaphylinus externus.

xiv

TABLE OF SYNONYMS OF MUSCLES—*continued.*

	PERCIVALL.	CHAUVEAU.	LETH.
Tensor vaginæ femoris.	*Idem.*	Muscle of fascia lata	Ilio-rotuleus externus.
Teres major.	*Idem.*	Adductor of arm.	Great scapulo-humeralis.
Teres minor.	(part of) Infraspinatus.	Short abductor of arm.	Middle and small scapulo-trochiterius.
Trachelo-mastoid.	*Idem.*	Small complexus.	Dorso-mastoideus.
Transversalis abdominis.	*Idem.*	*Idem.*	Costo-abdominalis internus.
Transversalis costarum.	*Idem.*	Common intercostal.	Common intercostal.
Transversus perinæi.	(Not described.)	*Idem.*	*Idem.*
Trapezius.	*Idem.*	*Idem.*	Cervico-acromialis.
Triangularis sterni.	Sterno-costales interni.	*Idem.*	Sterno-costalis.
Vastus externus.	*Idem.*	*Idem.*	Femoro-tibialis externus.
Vastus internus.	*Idem.*	*Idem.*	Femoro-tibialis internus.
Wilson's muscle.	(part of) Triangularis penis.	Sphincter urethræ.	Prostatic muscle.
Zygomatico-auricularis.	Atolens anterior.	*Idem.*	Temporo-auricularis and fronto-auricularis.
Zygomaticus.	*Idem.*	Zygomatico-labialis.	Zygomatico-labialis.

PLATES.

CONTENTS.

———:o:———

CHAPTER I.

DISSECTION OF THE ANTERIOR LIMB.

CHAPTER II.

DISSECTION OF THE POSTERIOR LIMB.

CHAPTER III.

DISSECTION OF THE BACK AND THORAX.

xviii

CONTENTS.

CHAPTER IV.

DISSECTION OF THE HEAD AND NECK.

CHAPTER V.

DISSECTION OF THE LARYNX.

CHAPTER VI.

DISSECTION OF THE BRAIN.

CHAPTER VII.

DISSECTION OF THE EYEBALL.

CHAPTER VIII.

DISSECTION OF THE EAR.

CHAPTER IX.

DISSECTION OF THE PERINÆUM IN THE MALE.

CHAPTER I.

DISSECTION OF THE ANTERIOR LIMB.

THE PECTORAL REGION AND THE AXILLA.

As the first step in the examination of the fore limb, the student should dissect the structures which pass between the trunk and the ventral aspect of the limb.

Position.—The subject should be placed on the middle line of its back, and its limbs should be forcibly drawn upwards and outwards by ropes running over pulleys fixed to the ceiling. If only one side is being dissected, the subject may be inclined as in Plate 1. This will put the muscles and other structures on the stretch, and thus facilitate their dissection.

Surface-marking.—In the fore part of the pectoral region the student will notice the well-marked prominence formed by the anterior superficial pectoral muscle. Between this muscle and the lower edge of the mastoido-humeralis there is a groove in which will afterwards be dissected the cephalic vein and a branch of the inferior cervical artery. Extending inwards from the point of the elbow is a prominent fold of skin over the hinder edge of the posterior superficial pectoral muscle.

Directions.—An incision through the skin, but not deeper, is to be made along the middle line of the sternum, from the ensiform cartilage as far forwards as the cariniform cartilage. From the middle of this incision another is to be carried transversely outwards, and terminated a little beyond the elbow-joint. Where this second incision stops, another is to be made across the inner face of the fore-arm. Beginning at the point where these incisions meet, the student should raise and turn outwards the two flaps of skin, so as to denude the superficial pectoral muscle. In doing this, it may be noticed that here, as in other unexposed situations, the skin is comparatively thin. Beneath the skin is the subcutaneous fascia, and search is to be made in it for the cutaneous nerves of this region.

CUTANEOUS NERVES. A nerve of considerable size, derived from the 6th cervical nerve (Plate 1), crosses the groove between the mastoido-humeralis and the anterior superficial pectoral, and distributes branches to the skin over the latter muscle and part of the posterior superficial pectoral. Other small cutaneous twigs, which are branches of the inter-

B

costal nerves, appear near the middle line, and are directed transversely
outwards.

Directions.—The surface of the superficial pectoral muscles should
now be carefully cleaned by the removal of the subcutaneous fascia;
and this operation should be conducted by beginning at the anterior or
posterior border of the muscle and working parallel to the direction of
the muscular fibres. When this has been effected, a line will be seen
on the surface of the muscle; and by dissecting carefully down on this
line, the student will be able to separate the anterior from the posterior
part of the muscle. Search is to be made, in the groove already mentioned, for the cephalic vein, and the fat is to be carefully removed from
the vein and its accompanying arterial branch.

Superficial Pectoral Muscle (*Pectoralis transversus* of Percivall).—This
muscle is divided, though not very distinctly, into two portions, which
may be distinguished as the anterior superficial pectoral and the posterior superficial pectoral.

The ANTERIOR SUPERFICIAL PECTORAL (Plate 1) *arises* from the first
two or three inches of the inferior border of the sternum, its posterior
fibres overlapping the anterior part of the next muscle. It is *inserted*
into the external lip of the musculo-spiral groove.

The POSTERIOR SUPERFICIAL PECTORAL (Plate 1) *arises* from the inferior
border of the sternum from within an inch of its anterior end as far
back as a point behind the 6th costal cartilage, and from a fibrous cord
which joins the muscle along the middle line to its fellow of the opposite
side. It is *inserted* into the superficial fascia which descends on the
inner face of the fore-arm; and a few of its anterior fibres, forming
a band about one inch in breadth, are *inserted* along with the preceding
muscle into the external lip of the musculo-spiral groove. At the
elbow-joint the muscle covers the posterior radial vessels and the median
nerve, but these are not to be exposed at present.

Action.—The superficial pectoral muscle is an adductor of the limb at
the shoulder, and the posterior division of the muscle is also a tensor of
the fascia of the fore-arm.

Directions.—Both divisions of the muscle are now to be cut across near
their origin, and dissected carefully from the subjacent deep pectoral;
and while this is being done, search is to be made for their nerves, which
come from the brachial plexus by passing between the two divisions of
the deep pectoral muscle. In reflecting the muscle, the dissector will
cut many small branches of the external or internal thoracic vessels.
The reflected muscles are now to be fastened outwards with chain and
hooks, and the dissection of the deep pectoral is to be undertaken after
the cephalic vein has been examined.

The CEPHALIC VEIN (Plate 1). This is the upward continuation
of one of the divisions of the internal subcutaneous vein of the fore-arm.

It ascends in the groove between the anterior superficial pectoral and the mastoido-humeralis. In the inner third of this groove it lies on the anterior deep pectoral, in company with a branch of the inferior cervical artery. It empties itself into the jugular about two inches from the lower end of that vessel.

Deep Pectoral Muscle.—This consists of two distinct divisions, which may be distinguished as the anterior deep pectoral and the posterior deep pectoral.

The POSTERIOR DEEP PECTORAL (*Pectoralis magnus* of Percivall) (Plate 2) is a muscle of large size ; and its posterior part, being subcutaneous, was visible before reflection of the superficial pectoral. It *arises* from the abdominal tunic covering the external oblique and the straight muscles of the abdomen ; from the tips of the cartilages of the 5th, 6th, 7th, and 8th ribs, and from the immediately subjacent lateral surface of the sternum. It is *inserted* into the inner tuberosity of the humerus, into the tendon of origin of the biceps, and into the fascia which retains that muscle in the bicipital groove. By its deep face the muscle serves to bound the axillary space ; while its upper border is closely united to the panniculus, and bordered by the subcutaneous thoracic nerve and vessels.

The ANTERIOR DEEP PECTORAL (*Pectoralis parvus* of Percivall) (Plate 2) *arises* from the cartilages of the first four ribs, and from the immediately subjacent lateral surface of the sternum ; and, being carried upwards in front of the supraspinatus nearly as far as the cervical angle of the scapula, it is somewhat loosely *inserted* into the fascia covering the last-named muscle. This insertion is concealed by the mastoido-humeralis, and will be better seen in the dissection of the muscles on the outer surface of the scapula (Plate 4). The deep face of the muscle forms part of the inferior boundary of the axilla.

Action.—The two divisions of the deep pectoral have the same action, which is to pull the shoulder-joint, and thus the whole limb, backwards. When the limbs are fixed, the muscle may to some extent act as a muscle of inspiration.

Directions.—The deep pectoral muscles are now to be severed carefully about midway between their origin and insertion, and the cut portions are to be turned outwards and inwards. Their nerves, which come from the brachial plexus, will be found entering their deep face ; and care is to be taken of the external thoracic and inferior cervical arteries. By the reflection of these muscles, the axilla is exposed. Owing to the limited power of abduction at the shoulder-joint of the horse, the dissection of the space is attended with much greater difficulty than in man. The best method of procedure is as follows:—All the pectoral muscles having been cut across, the limb is to be forcibly separated from the chest-wall ; and, to permit this to a sufficient extent, it may be necessary to cut the mastoido-humeralis in

front of the shoulder.   On looking into the space, the dissector will now
see it occupied by a large amount of loose, areolar connective-tissue,
which envelops its contents, and facilitates the play of the shoulder on
the wall of the thorax.   This areolar tissue must be cleaned away from
the axillary vessels and the brachial plexus of nerves, but most of the
branches of these will be more conveniently followed after separation of
the limb from the trunk.

The AXILLA corresponds to the arm-pit of the human subject, and is
the important space across which the large vessels and nerves for the
supply of the fore limb are transmitted.

Boundaries of the space.—In the natural movements of the limb, and
before dissection, the space can hardly be said to have any existence
except at its lower part; but in the dissected condition it may be
observed to have the following boundaries.   On its *outer side* are the
subscapularis, teres major, and (in part) latissimus dorsi muscles.   The
inner side of the space is formed by the anterior part of the chest-wall
covered by the serratus magnus, lateralis sterni, and intercostal muscles.
Inferiorly the space is enclosed by the deep pectoral muscles, and there
the space is most extensive.   *Superiorly* the outer and inner boundaries
meet at the insertion of the serratus magnus into the scapula.   The
anterior limit of the space may be taken as formed by the mastoido-
humeralis and the reflected portion of the anterior deep pectoral ; while
posteriorly the space is closed by the panniculus carnosus and skin where
these are carried from the wall of the thorax over the outer aspect of
the shoulder.

The AXILLARY ARTERY (Plates 3 and 5) begins within the thorax.   On the
left side it arises as one of the terminal branches of the anterior aorta; while
on the right it is a branch of the arteria innominata.   It leaves the chest
and reaches the axilla by turning round the anterior border of the first
rib, below the inferior insertion of the scalenus.   It crosses the axillary
space, inclining downwards and backwards ; and at the anterior border
of the teres major tendon it is directly continued as the brachial artery.
In this course it gives off four vessels, viz., inferior cervical, external
thoracic, suprascapular, and subscapular ; but only the first two are to
be followed at present.

The INFERIOR CERVICAL ARTERY (Plates 1 and 2) arises from the front
of the axillary where that vessel turns round the first rib.   After a course
of about two inches it bifurcates, its superior branch passing between
the mastoido-humeralis and the subscapulo-hyoideus, while the inferior
division passes into the groove between the mastoido-humeralis and the
anterior superficial pectoral, where it has already been seen in company
with the cephalic vein.

The EXTERNAL THORACIC ARTERY (Plate 3) arises about the same point
as the preceding, but from the opposite side of the parent vessel; and

passing backwards in relation to the axillary surface of the deep pectorals, it distributes branches to these, and also to the superficial pectorals. A slender branch from it accompanies the subcutaneous thoracic vein to the panniculus carnosus.

The AXILLARY VEIN is the upward continuation of the brachial vein, and is, at its lower part, posterior to the artery; but at the anterior border of the first rib it is below the artery, and it here joins the jugulars and the axillary vein of the opposite side, thus forming the anterior vena cava.

Directions.—The axillary vessels may now be cut as they turn round the first rib, and the limb may be further abducted to facilitate the dissection of the brachial plexus, which, in its first step, should be undertaken by the dissectors of the limb and of the neck conjointly.

The BRACHIAL PLEXUS (Plate 3) is composed of the nerves for the supply of the fore limb. It is formed by the inferior primary branches of the last three cervical (6th, 7th, and 8th) and first two dorsal nerves. These, however, do not enter into it in equal proportions. The 6th cervical sends only a very slender branch to it, while the 7th and 8th, after detaching a communicating filament to the sympathetic, are wholly expended in it. The 1st dorsal is, with the exception of a similar communicating filament and a slender intercostal branch, also entirely expended in it, but the 2nd dorsal gives off, besides the usual communicating branch, a considerable intercostal nerve before joining the plexus.

These roots of the plexus converge towards each other, and come out as a flat fasciculus between the upper and lower portions of the scalenus. In descending to this point, the dorsal roots of the plexus turn round the anterior border of the first rib, leaving on it a smooth impression near its upper end. The several roots anastomose in an intricate fashion, contributing to the formation of the various branches of the plexus, in proportions that the student will not be able to trace accurately in the course of an ordinary dissection.

The manner in which the several roots of the plexus comport themselves is liable to slight variation, but the following is probably as common as any other :—

I. The root from the 6th cervical nerve is a slender branch detached from the division which that nerve sends to aid in the formation of the phrenic. Passing obliquely backwards on the scalenus muscle, it resolves itself into three divisions—or rather its fibres are traceable in three groups, viz.,—1. To the suprascapular nerve ; 2. to the anterior root of the median ; 3. to join branches from all the other roots of the plexus in forming a broad, flat fasciculus from which arise the subscapular, circumflex, and musculo-spiral nerves.

II. The root from the 7th cervical nerve gives a branch to the nerve for the serratus magnus, and then divides its fibres in three directions, viz.,—1. To the above-mentioned fasciculus giving off the subscapular, circumflex, and musculo-spiral nerves ; 2. to the suprascapular nerve ; 3. to the phrenic, anterior root of the median, and nerve for the anterior deep pectoral muscle.

III. The root from the 8th cervical nerve gives a branch to the nerve for the serratus magnus, and then sends its fibres in three directions, viz.,—1. To the before-mentioned flat fasciculus giving off the subscapular, etc.; 2. to join the cord from which arise the posterior root of the median, the ulnar, and the subcutaneous thoracic nerve; 3. to the anterior root of the median and the nerve for the anterior deep pectoral muscle.

IV. The roots from the 1st and 2nd dorsal nerves unite to form a common cord which divides its fibres in two directions, viz.,—1. To join the above-mentioned cord giving off the posterior root of the median, etc.; 2. to join the broad fasciculus from which arise the subscapular, etc.

The following is a list of the branches of the plexus :—

1. The phrenic or diaphragmatic nerve (in part).
2. The suprascapular nerve.
3. Nerves to the pectoral muscles.
4. The nerve to the subscapularis.
5. Nerves to the serratus magnus and levator scapulæ (cervical portion of the serratus), the latter only in part.
6. The circumflex nerve.
7. Nerves to the teres major and latissimus dorsi.
8. The musculo-spiral nerve.
9. The median nerve (two roots). •
10. The ulnar nerve.
11. The subcutaneous thoracic nerve.

The PHRENIC NERVE. This nerve is formed by the union of two, or sometimes three, branches. The inconstant branch comes from the 5th cervical; the other two come from the 6th and 7th respectively. The root from the 6th nerve gives off a branch to the brachial plexus, and then unites on the scalenus with the root from the 5th—when that is present. The single cord resulting passes obliquely backwards and downwards, and at the lower edge of the scalenus it joins with the root from the 7th nerve. This last comes from the fore part of the brachial plexus. The trunk of the nerve, as thus formed, passes backwards between the axillary artery and its inferior cervical branch, and enters the thorax. It is the motor nerve to the diaphragm.

The NERVES to the LEVATOR ANGULI SCAPULÆ and RHOMBOIDEUS. In Plate 3 two nerves are seen at the upper edge of the scalenus. They are not, strictly speaking, branches of the brachial plexus; but come from the inferior primary branch of the 6th nerve, and pierce the muscle either together or separately. They are distributed to the levator anguli scapulæ, and the posterior of the two is continued in that muscle to reach the rhomboideus.

The NERVE to the SERRATUS MAGNUS is formed by the union of two branches, which pierce the upper division of the scalenus before uniting. These are branches of the 7th and 8th nerves respectively. By their fusion there is formed a broad nerve, which passes backwards on the surface of the serratus, distributing its filaments upwards and downwards. Before fusion, the branch from the 7th gives off a nerve which is distributed to both the levator and the serratus.

The SUBCUTANEOUS THORACIC NERVE (Plates 1 and 3) derives its fibres from the dorsal roots of the plexus and from the 8th cervical, but principally from the former. It accompanies the spur vein to near the flank, being distributed with perforating intercostal branches on the deep face of the panniculus carnosus. A branch from it unites with perforating branches from the 2nd and 3rd intercostal nerves, and turns round behind the limb, to be distributed to the panniculus over the shoulder and arm.

The NERVES to the PECTORAL MUSCLES have already been referred to. The nerve to the anterior deep pectoral leaves the fore part of the plexus, deriving its fibres from the 7th and 8th cervical nerves. The nerve to the superficial pectoral muscle (both divisions) derives its fibres from both roots of the median. In general, there are two nerves to the posterior deep pectoral. The first—to the anterior part of the muscle, comes off with the posterior root of the median, the other—to the posterior part of the muscle, comes off in common with the subcutaneous thoracic.

Directions.—The remaining nerves of the brachial plexus can be more satisfactorily followed after separation of the limb from the trunk, and the dissector should therefore now proceed as follows :—Pass a cord round the nerves of the plexus as they emerge from between the two divisions of the scalenus, and then cut the roots of the plexus as near their points of origin as possible. Cut also the axillary artery and vein at the first rib. This will allow the limb to be carried well out from the trunk, so as to expose the serratus magnus and levator anguli scapulæ, which are now to be cleaned.

Serratus Magnus and *Levator Anguli Scapulæ.*—These muscles are, in the horse, not very distinctly marked off from each other, and have therefore been frequently described as one muscle under the first name.

The SERRATUS MAGNUS (Plate 4) *arises* from the outer surfaces of the eight (or nine) anterior ribs, its eight slips of origin forming a curved, serrated line which gives to the muscle its name. The posterior four of these slips inter-digitate with slips of origin of the external oblique muscle of the abdomen (Plate 39), and are overspread by the abdominal tunic. It is *inserted* into a triangular area on the ventral surface of the scapula near its dorsal angle, and, in common with the next muscle, into another triangular area at the cervical angle.

Action.—It pulls the dorsal angle of the scapula downwards and backwards on the chest-wall, causing the shoulder-joint at the same time to move upwards and forwards ; but when the limbs are fixed, it can become a muscle of inspiration, pulling the ribs upwards and forwards. In the standing posture of the animal at rest, the chest is, in a manner, slung on the fore limbs by means of the right and left serratus muscles.

The Levator Anguli Scapulæ (Plate 4) *arises* from the transverse processes of the last four cervical vertebræ; and its fibres converge to be *inserted* into the triangular area on the ventral surface of the scapula near its cervical angle, in common with the anterior fibres of the preceding muscle, from which it is not distinct. The two muscles, taken together, have a well-marked fan-like arrangement, having an extensive convex border where they take origin, while they converge to a comparatively narrow point at their insertion.

Action.—The levator anguli scapulæ carries the articular angle of the scapula backwards by pulling the cervical angle forwards; but when the scapula is fixed, the right and left muscles, acting together, can raise the cervical portion of the spinal column, or the single muscle can incline it to one side.

The Subscapulo-hyoid. This muscle, which arises from the subscapular fascia, is described with the dissection of the neck.

THE OUTER SCAPULAR REGION.

Position.—The muscles which pass between the shoulder and the trunk, on the outer aspect of the former, must next be dissected; and, to permit this, the subject must be placed in an entirely new position. The standing posture of the animal is the best for this purpose; and it may be imitated by suspending the subject to a stout iron rod provided with chains and hooks, and capable of being raised or lowered by means of a system of pulleys or a small windlass.

Surface-marking.—About the centre of the region to be dissected the student will feel the spine of the scapula, the most prominent part of which is its tubercle. In a well-nourished, sound horse the spine should not be very distinctly visible, but in an emaciated animal, or in one whose scapular muscles are atrophied as an accompaniment of joint-disease, it forms a very prominent ridge.

Directions.—An incision through the skin is to be made along the spine of the scapula from the withers to the middle of the arm, where a transverse incision is to be made from the anterior to the posterior border of the limb. Another incision is to be carried along the middle line of the back, and prolonged forwards along the neck by the dissector of that region, and backwards to the lumbar region by the dissector of the back. The dissectors of the three regions should here work together, the skin being turned down as a single flap from the neck and anterior half of the shoulder, and as another flap from the back and posterior half of the shoulder. The skin, it will be observed, is thicker than in the pectoral region, and it has the panniculus carnosus attached to its inner surface. Care must be taken not to remove this panniculus with the skin.

The Panniculus Carnosus is the muscle which enables the horse to twitch its skin, and thus remove offending insects. It is most extensive over the thorax and abdomen, but it is here carried over the muscles covering the scapula and humerus. Before the muscle passes on to the limb, it sends an aponeurotic layer inwards between the limb and the chest-wall. At its upper border this layer is provided with a small tendon, which becomes inserted into the inner tuberosity of the humerus, and which will be seen when the limb is dissected from the trunk. A nerve will be seen ramifying in the scapulo-humeral part of the panniculus. This turns round the posterior border of the limb; and, as already seen, it is formed by the union of the subcutaneous thoracic with some perforating intercostal nerves.

Directions.—The panniculus is now to be dissected away from the limb; and in doing this in front, care is to be taken of the thin cervical trapezius muscle, which might be mistaken for a portion of the panniculus.

The *Trapezius* in the horse has its muscular substance interrupted by a tendinous portion, and is therefore sometimes described as two separate muscles, distinguished as the cervical and the dorsal trapezius.

The CERVICAL TRAPEZIUS (Plate 4) *arises* from the funicular portion of the ligamentum nuchæ; and it is *inserted* into the tubercle on the spine of the scapula, while its most anterior fibres are continuous with an aponeurosis covering the scapular muscles. Both the deep and the superficial face of the muscle have a thin, adherent, fibrous covering, the direction of whose fibres is at right angles to that of the muscular fibres.

Action.—It draws the scapula forwards and upwards.

The DORSAL TRAPEZIUS (Plate 4) is continuous with the preceding by the aponeurotic centre already mentioned. It *arises* from the summits of a few of the anterior dorsal spines, and is *inserted* into the tubercle on the scapular spine.

Action.—It pulls the scapula backwards and upwards.

Directions.—Both divisions of the trapezius are now to be severed close to their origin, and reflected downwards; and while this is being done, search is to be made for the branches of the 11th, or spinal accessory, nerve, which enter their deep face. The muscles which were covered, wholly or in part, by the trapezius, will now be exposed. These are : the splenius, the levator anguli scapulæ, the supraspinatus, the infraspinatus, the anterior deep pectoral, the latissimus dorsi, and the rhomboideus.

It will be remembered that in the dissection of the pectoral region the anterior deep pectoral could not be followed to its termination. The reflected portion of the muscle is here seen (Plate 4), but is partly covered by the insertion of the mastoido-humeralis.

The LATISSIMUS DORSI (Plate 4). Though neither the origin nor the insertion of the muscle is found here, attention should be given to it as it is being exposed by the dissector of the back. It *arises* by an aponeurotic tendon from the series of vertebral spines, beginning about the 4th dorsal, and extending backwards to the last lumbar. This tendon is succeeded by a thick muscular portion, which contracts and passes in between the limb and the trunk, where it will afterwards be followed to its *insertion* into the internal tubercle of the humerus. Its anterior fibres will be noticed to play over the dorsal angle and cartilage of prolongation of the scapula.

Action.—It is a flexor and an inward-rotator of the shoulder-joint.

The RHOMBOIDEUS (Plate 4), like the trapezius, comprises a cervical and a dorsal portion. The *cervical* part is an elongated, narrow muscle, which extends as far forward as the axis, and *arises* from the funicular part of the ligamentum nuchæ. Its fibres take a very oblique direction downwards and backwards, and are *inserted* into the anterior part of the cartilage of prolongation on its inner surface, being there confounded with the insertion of the levator anguli scapulæ. The *dorsal* portion consists of fibres which *arise* from the anterior dorsal spines, and pass in a nearly vertical direction to be *inserted* into the inner surface of the cartilage of prolongation, behind the fibres of the cervical division. It will be recollected that the nerve to these muscles passes from the 6th cervical nerve, and reaches its destination by traversing the levator anguli scapulæ.

Action.—To pull the scapula upwards and forwards on the chest-wall.

The MASTOIDO-HUMERALIS, or LEVATOR HUMERI (Plate 4). This muscle, in the greatest part of its extent, is found in the head and neck, where it takes its *origin* from the mastoid crest and the transverse processes of the first four cervical vertebræ; but attention must here be given to its *insertion*, which is into the external lip of the musculo-spiral groove, after covering the shoulder-joint. It receives here some branches from the circumflex nerve.

Action.—It is an extensor and inward-rotator of the shoulder-joint. When the limb is fixed, it bends the neck laterally.

Directions.—The limb may now be detached from the trunk by severing the attachment of the rhomboideus, serratus magnus, levator anguli scapulæ, mastoido-humeralis, and latissimus dorsi, the last being cut where it plays over the angle of the scapula. Pieces of clean cloth saturated with some preservative solution should be placed on the outer aspect of the shoulder where the skin has been removed, while the dissector proceeds to examine the structures over the inner surface of the scapula and humerus.

INNER ASPECT OF THE SHOULDER AND ARM.

Directions.—The dissector should now identify the terminal portions of the muscles already dissected, and cut them off within an inch or two of their insertion, except in the case of the latissimus dorsi, which is to be left at its present length until its nerve and artery have been followed. The posterior superficial pectoral should be cut away on a level with the olecranon, but care is to be taken not to disturb the vessels and nerves which it covers. The aponeurosis which the panniculus sends within the shoulder will now be observed, and, at its upper border, a small glistening band passing to be inserted into the internal tuberosity of the humerus.

The next step is to dissect out the axillary and brachial vessels, and the remaining branches of the brachial plexus ; and this is an operation demanding time and care. While an assistant holds the nerves on the stretch, the fat and areolar connective-tissue which surround them and the vessels, are to be cleaned away piecemeal, always proceeding from the main trunks to the branches. In doing this, the dissector will meet two groups of lymphatic glands.

BRACHIAL LYMPHATIC GLANDS. The upper group consists of a cluster placed behind the brachial vessels, on a level with the middle of the humerus. The lower group consists of one or two glands in relation to the vessels, just above the elbow-joint.

The AXILLARY ARTERY (Plates 5 and 6). This vessel has already been seen passing in a curved direction from the anterior border of the 1st rib, across the inner aspect of the shoulder-joint, where it rests above the terminal insertion of the posterior deep pectoral, and on the tendon of the subscapularis. It passes on to the teres major, and is continued as the brachial artery. In this course it gives off four vessels, viz., the inferior cervical, external thoracic, suprascapular, and subscapular. The first two have already been dissected in the axilla.

The SUPRASCAPULAR ARTERY (Plate 5) is a small, tortuous vessel springing from the upper surface of the axillary artery about the middle of its extra-thoracic course. It passes upwards for a short distance, and then divides into branches, the longest of which passes over the subscapularis to reach the anterior deep pectoral. A branch passes in between the subscapularis and the supraspinatus, while smaller branches are expended in the tendons about the shoulder.

The SUBSCAPULAR ARTERY (Plates 5 and 6) is a comparatively large vessel, and beyond its origin the parent trunk is much reduced in calibre. It arises at the interstice between the subscapularis and teres major muscles ; and, disappearing between these muscles, it ascends behind the glenoid border of the scapula, as far as its dorsal angle. It gives off a considerable number of vessels that cannot at this stage be completely followed, but near its origin it will be seen to throw off a branch which runs upwards and backwards on the latissimus dorsi (Plate 5).

The BRACHIAL ARTERY (Plates 5 and 6) is the direct continuation of the

axillary, which changes its name when it passes on to the teres major.
It descends in a nearly vertical direction to the lower extremity of the
humerus, where, above the inner condyle, it divides to form the anterior
and posterior radial arteries.* In its course it crosses the direction of
the humerus obliquely, and rests successively on the tendons of the teres
major and latissimus dorsi, the small head of the triceps, and the bone.
In front of it is first the coraco-humeralis, and then the biceps; but
these are separated from it by the median nerve, which is in close
contact with the vessel. Behind the artery is the satellite vein,
posterior to which is the ulnar nerve. Its collateral branches are: the
pre-humeral, the deep humeral, the ulnar, the nutrient artery of the
humerus (sometimes), and innominate muscular branches.

The PRE-HUMERAL or ANTERIOR CIRCUMFLEX ARTERY (Plate 6) arises
at the tendon of the teres major, and passes in front of the humerus,
between the upper and lower insertions of the coraco-humeralis, to
terminate in the biceps or the mastoido-humeralis. Some of its fine
twigs may anastomose with divisions of the posterior circumflex.

The DEEP HUMERAL ARTERY (Plates 5 and 6) arises at the lower border of
the latissimus dorsi tendon, and soon splits into three or four branches,
the larger of which perforate the large head of the triceps extensor
cubiti, while the smaller supply the small and medium heads of the
same muscle. A branch is continued round behind the humerus, in
company with the musculo-spiral nerve, to the front of the elbow-joint,
where it anastomoses with branches of the anterior radial. This
branch will not be followed at present.

MUSCULAR BRANCHES of the Brachial. The largest and most constant
of these is a vessel of considerable size which penetrates the lower part
of the biceps (Plate 6).

The Ulnar artery and the two terminal branches of the brachial will
be followed in the dissection of the fore-arm.

The BRACHIAL VEIN is a large vessel which ascends behind the
artery, and receives branches that for the most part correspond to those
of the artery. It receives also the subcutaneous thoracic or spur vein.

Directions.—As the brachial vein generally contains a large quantity
of blood which exudes from the smaller cut branches, it will contribute
to the neatness and cleanness of the dissection if the dissector will
carefully remove the vein and all its branches before he proceeds to
follow the nerves.

The BRACHIAL PLEXUS. The mode of formation of the plexus has
already been explained, and the student will recollect that he has
already followed branches from it to the levator anguli scapulæ,
serratus magnus, and pectoral muscles, as well as the subcutaneous

* In Plate 6 the termination of the brachial artery has been pulled slightly forwards in order to
show the origin of the anterior radial artery.

thoracic nerve, and the filament furnished by the plexus to the phrenic nerve. He can now easily identify and trace the following branches :—

The NERVE to the LATISSIMUS DORSI (Plate 5) derives its fibres from the 8th cervical and the dorsal roots of the plexus.

The NERVE to the TERES MAJOR (Plate 5)—one or more filaments, generally deriving fibres, in common with the circumflex nerve, from the 7th and 8th cervical roots (with possibly some fibres from the 6th).

The NERVE to the SUBSCAPULARIS (Plate 5) derives its fibres from all the cervical roots of the plexus.

The CIRCUMFLEX NERVE (Plates 5 and 6). Its fibres come from the 7th and 8th cervical roots, and possibly also from the 6th. It turns round behind the shoulder-joint in company with the posterior circumflex artery ; and on the outside of the joint it supplies branches to the teres minor, deltoid, mastoido-humeralis, and skin (Plate 7). It gives a twig to the small scapulo-humeral muscle.

The SUPRASCAPULAR NERVE (Plate 5), deriving its fibres from the 6th, 7th, and 8th cervical roots, passes into the interstice between the subscapularis and the supraspinatus. It then turns round the anterior border of the scapula ; and gaining its dorsal surface, is expended in the supraspinatus and subspinatus muscles (Plate 8).

The MUSCULO-SPIRAL NERVE (or radial nerve) (Plates 5 and 6) is, at its origin, the thickest of the nerves of the brachial plexus. Deriving its fibres from the 7th and 8th cervical, and from the dorsal roots of the plexus, it passes downwards and backwards on the subscapularis and teres major muscles, and some little distance behind the axillary vessels, from which it is separated by the ulnar nerve. On reaching the deep humeral artery, it disappears in front of the large head of the triceps, and is continued round the humerus in the musculo-spiral groove, where it rests on the brachialis anticus (humeralis externus), and, after-wards, at the posterior or outer border of that muscle. It reaches the front of the elbow-joint, being here deeply placed between the brachialis anticus inwardly, and the origin of the great extensor of the metacarpus outwardly. Before the nerve disappears behind the humerus, it gives branches to the great and small heads of the triceps, and a long branch which passes backward to divide under the scapulo-ulnaris for the supply of that muscle. Behind the limb it supplies the medium head of the triceps and the anconeus, and furnishes a few cutaneous branches which perforate the caput medium, or emerge at its lower part, to be distributed to the skin of the outer side of the fore-arm, below the elbow. The termination of the nerve will afterwards be followed in the fore-arm, where it supplies the extensor muscles and the flexor metacarpi externus.

The ULNAR NERVE (Plates 5 and 6) derives its fibres from the dorsal roots of the brachial plexus. At first it lies close behind the main vessels; but as it passes downwards, it recedes from them, and passing under cover of the scapulo-ulnaris, it reaches the space between the olecranon and the inner condyle. Thence it descends to the back of the fore-arm, where it will subsequently be dissected. At present it is seen to give off only one branch, which disappears within the superficial pectoral muscle, and afterwards becomes distributed to the skin of the fore-arm (Plate 5).

The MEDIAN NERVE (Plates 5 and 6) is formed by the union of two roots. The anterior of these comes from the 6th, 7th, and 8th cervical, while the posterior is derived from the 8th cervical and the 1st dorsal. These roots gives off some pectoral twigs, and then unite by forming a loop in which the axillary artery rests. The nerve then descends in front of the axillary artery and its brachial continuation, and will afterwards be seen to accompany the posterior radial artery. The following branches of the nerve may be found at present :—

The *Nerve to the Biceps* and *Coraco-humeralis* comes off close below the union of the two roots of the median, or from the anterior root above the point of union. It passes between the upper and lower insertions of the coraco-humeralis, supplying that muscle and terminating in the biceps.

Musculo-cutaneous branch.—This is given off from the median about the middle of the humerus; and passing underneath the biceps, it divides into a muscular branch for the brachialis anticus, and a cutaneous branch for the front of the fore-arm.

Directions.—The muscles of this region should now be examined in the order of their description.

The LATISSIMUS DORSI (Plates 5 and 6). The *insertion* of this muscle into the inner tubercle of the humerus is here seen. About an inch or two from its termination the tendon gets a twist which alters the direction of its surfaces, and brings it to be inserted in front of the termination of the teres major on the same tubercle.

Action.—The muscle is a flexor and an inward-rotator of the shoulder-joint.

The TERES MAJOR (Plate 5). It *arises* from the dorsal angle of the scapula, and from an aponeurosis between it and t' subscapularis. It is *inserted* into the internal tubercle of the humerus, its terminal tendon resting in the twist formed by the tendon of the latissimus dorsi muscle.

Action.—It is a flexor and an inward-rotator of the shoulder.

The SCAPULO-ULNARIS (Plate 5). This is a thin, flat muscle which rests on the inner surface of the triceps, and is provided, in front and above, with a thin, transparent tendon. It *arises* from the posterior border of the scapula, and is *inserted* into the posterior border of the olecranon, and into the fascia of the fore-arm. At its lower extremity the muscle covers the ulnar vessels and nerves.

Action.—To extend the elbow-joint, and tense the fascia of the forearm.

The TRICEPS EXTENSOR CUBITI. This is an immense muscular mass which, with the preceding, fills up the angle formed behind the shoulder-joint. It has three divisions or heads, which may be distinguished as the caput magnum, the caput medium, and the caput parvum.

The *Caput Magnum*, or large head (Plate 5), forms a great mass which is seen on both the outside and the inside of the limb. It *arises* from the dorsal angle and glenoid (posterior) border of the scapula; and it is *inserted* into the olecranon, there being a synovial bursa between the summit of that eminence and the tendon.

The *Caput Parvum*, or small head (Plate 5), is, when compared with the preceding, a very small muscle. It *arises* from the shaft of the humerus below and behind the internal tubercle, and it is *inserted* into the olecranon.

The *Caput Medium*, which is not now visible, will be dissected with the outside of the shoulder.

Action of the triceps. It is an extensor of the elbow-joint, and acts as a lever of the first order, the joint, which represents the fulcrum, being between the power and the weight. The large head is also a flexor of the shoulder.

The SUBSCAPULARIS (Plates 5 and 6). This muscle is lodged in the fossa of the same name on the ventral surface of the scapula, and it *arises* from the whole extent of that fossa. It is *inserted* into the inner tuberosity of the humerus, a small synovial bursa being interposed between the tendon and the bone. The tendon is crossed by the origin of the coraco-humeralis, and another small bursa is here interposed between the tendons. Above its insertion it is closely related to the capsular ligament of the joint. The muscle is partly united in front with the supraspinatus, and behind with the teres major.

Action.—It is an adductor of the shoulder.

The CORACO-HUMERALIS (or coraco-brachialis) (Plates 5 and 6). This, which is rather a small muscle, *arises* from a small tubercle on the inner side of the coracoid process of the scapula. It has two *insertions*, the first into the inner surface of the shaft of the humerus above the internal tubercle, the second into a line which begins on a level with the internal tubercle, and runs down the anterior surface of the shaft near its inner border. Between these two insertions, the pre-humeral artery and the nerve to the biceps pass. The tendon of origin of the muscle comes out between the supraspinatus and subscapularis muscles, and the posterior border of the muscle is related to the brachial artery.

Action.—To adduct and flex the shoulder.

The BICEPS (Plates 5 and 6). This muscle receives its name in the human subject from its having two heads of origin. It is also known as the *flexor brachii* or *coraco-radialis.* It *arises* from the whole of the coracoid process of the scapula with the exception of the tubercle on its inner side, which is for the coraco-humeralis. Its strong tendon of origin emerges from between the outer and inner tendons of the supraspinatus, and passes over the shoulder-joint, a pad of fat separating its deep face from the capsular ligament of the joint. The tendon, which is of fibro-cartilaginous consistency, then plays over the bicipital groove of the humerus, on which its deep face is moulded, and a synovial bursa facilitates the movements of the tendon in the groove. The central portion of the muscle, which is thick and fusiform, has numerous tendinous intersections, and is traversed throughout by a fibrous cord. It rests on the anterior face of the humerus, and at its lower end terminates by a tendon which, passing over the anterior ligament of the elbow-joint (to which it is adherent), is *inserted* into the bicipital tuberosity of the radius. The tendon is partly covered by the internal lateral ligament of the elbow. The muscle has a second insertion, in the shape of a strong fibrous band, detached from the main tendon to blend with the sheath of the extensor metacarpi magnus, and deep fascia on the front of the fore-arm.

Action.—To flex the elbow-joint, and make tense the fascia of the fore-arm. In the first of these actions it is a good example of a lever of the third order, where the power is applied between the fulcrum—represented by the elbow-joint, and the weight—represented by the distal portion of the limb. The fibrous cord which traverses the muscle is a mechanical extensor of the shoulder-joint, as long as the elbow is kept extended by the triceps extensor cubiti.

Directions.—The teres major from the shoulder upwards should now be removed, in order to follow more thoroughly the course of the sub-scapular artery with its branches, and to expose the small scapulo-humeral muscle, which lies on the capsular ligament behind the joint; but care should be taken, in dissecting the tendons in the neighbour-hood of the joint, to preserve the capsular ligament intact.

The SUBSCAPULAR ARTERY (Plate 6) springs from the axillary trunk at the interstice between the subscapularis and teres major muscles, and disappearing from view, runs upwards at the posterior border of the scapula. It gives off as its most important branches :—

1. A *Muscular branch* of considerable volume which passes backwards and upwards on the deep face of the latissimus dorsi.

2. The *Posterior circumflex* artery, which turns round behind the shoulder, passing through a triangular space bounded by the teres major, caput magnum, and scapulo-humeralis gracilis. At the outer side of the joint (Plate 7) it appears between the caput magnum, caput medium, and teres minor, and is covered by the deltoid. It splits into branches

which are distributed to these muscles and the supraspinatus (Plate 7). It is accompanied by the circumflex nerve.

3. Other branches of the subscapular are as follows :—A few inches above the origin of the posterior circumflex, a vessel is detached which passes backwards, and divides to supply the caput magnum. A number of smaller branches come off from the anterior aspect of the vessel, and are distributed on both surfaces of the scapula. One of these supplies the *nutrient artery* of the scapula.

The SCAPULO-HUMERALIS GRACILIS is a very slender muscle. It *arises* from the scapula above the rim of its glenoid cavity ; and passing over the capsular ligament of the shoulder, on which some of its fibres seem to terminate, it insinuates itself between the fibres of the brachialis anticus (humeralis externus), and is *inserted* into the posterior surface of the shaft of the humerus. It is supplied by a small nerve from the circumflex.

Action.—The muscle is too inconsiderable in size to exercise any appreciable action on the joint over which it passes, and, probably, its function is to raise the capsular ligament and prevent its injury during flexion of the joint.

OUTER ASPECT OF THE SHOULDER AND ARM.

Directions.—The limb is now to be turned over, and the muscles and other structures on the outer side of the scapula and humerus are to be dissected.

SCAPULAR FASCIA.—This is a strong, glistening, fibrous covering which is spread over the muscles on the dorsum of the scapula, affording by its inner surface an origin to many of their fibres. When traced upwards, it is seen to be inserted into the scapula or its cartilage of prolongation; while before, behind, and inferiorly, it becomes less fibrous, and is continuous with the fascia covering the muscles on the inner surface of the scapula and the outer aspect of the arm. It furnishes septa to pass between the subjacent muscles, and it is adherent to the tubercle on the scapular spine. If an attempt be made to dissect it off these muscles, they will be exposed with a rough surface, showing that they there take origin from the inner aspect of the fascia.

The DELTOID MUSCLE (scapular portion) (Plates 4 and 7). This muscle was by Percivall erroneously termed the teres minor. It is not the homologue of either of the teres muscles of human anatomy, but is, most clearly, the representative of that part of the deltoid muscle which in man takes origin from the scapula. A linear depression which traverses the muscle corresponds to an imperfect division of it into an anterior and a posterior portion. It *arises* by its anterior portion from the

c

scapular fascia, and by its posterior portion from the dorsal angle of
the scapula. It is *inserted* into the deltoid (external) tubercle of the
humerus.

Action.—To abduct the humerus, and rotate it outwards. Acting
with the teres major, it is also a flexor of the shoulder.

Directions.—The last-mentioned muscle should be carefully cut at
the level of the shoulder, and reflected upwards and downwards. This
will expose the divisions of the circumflex vessels and nerve, branches
of which will be seen entering the muscle, and it will at the same time
bring into view the next muscle.

The TERES MINOR (Plates 7 and 8). ¡This small muscle *arises* from the
posterior border of the scapula, from the rough lines at the lower part
of the infraspinous fossa, and from the small tubercle on the outer rim
of the glenoid cavity. Its tendon, which is crossed by a glistening
band of fascia, is *inserted* into the lower half of the ridge running
upwards from the deltoid tubercle to the external tuberosity.

Action.—The same as the preceding muscle.

The INFRASPINATUS (subspinatus, or postea-spinatus) (Plates 7 and 8)
occupies the greater part of the fossa of the same name. It *arises* from
the whole extent of the fossa, and from the inner surface of the scapular
fascia. It possesses two tendons of insertion, the outer of which passes
over the convexity of the external tuberosity, a synovial bursa being
interposed, and is *inserted* into the upper half of the ridge connecting
that tuberosity to the deltoid tubercle. If this tendon be cut where it
plays over the convexity, the synovial bursa will be opened, and,
at the same time, the inner *insertion* of the muscle into the inside
of the convexity will be exposed. This inner tendon is more fleshy
than the outer, and is in contact with the capsular ligament of the
shoulder.

Action.—It abducts the humerus, and rotates it outwards.

The SUPRASPINATUS (antea-spinatus) (Plates 7 and 8) fills the whole of
the fossa of the same name, and takes *origin* from it as well as from
the scapular fascia. It is bifid inferiorly, having an inner tendon
inserted into the internal tuberosity at its highest point, and an outer
tendon *inserted* into the corresponding point of the external tuberosity.
These two tendons are in contact with the capsular ligament of the
joint, and the tendon of origin of the biceps emerges from between them.

Action.—It is an extensor of the shoulder-joint.

Directions.—The outer aspect of the triceps extensor cubiti is here
seen; and when its surface has been cleaned, a line will be observed
running from the shoulder to the point of the elbow. Careful dissection
downwards into the mass, along this line, will separate the caput magnum (already described) from the caput medium, which lies below it.
While the surface of the muscle is being cleaned, some small cutaneous

nerves from the musculo-spiral will be found to pierce the muscle, or emerge at its lower edge, and become distributed to the outer side of the fore-arm.   These should, as far as possible, be preserved.

The CAPUT MEDIUM (Plates 7 and 8) *arises*, by a short aponeurotic ten-don, from a curved line beginning on the deltoid tubercle and continued upwards to the external tuberosity.   It is *inserted* into the olecranon.

Action.—Like the other divisions of the triceps, this muscle is an extensor of the elbow-joint.

Directions.—By raising the lower edge of the last muscle and dissect-ing upwards, the anconeus will be partly exposed; but to effect a com-plete and natural separation of the two muscles, is a matter of some difficulty.

The ANCONEUS (Plates 7 and 8) is a small muscle which lies above the olecranon fossa, and there covers the synovial membrane of the joint, a pad of fat being interposed.   It *arises* from the margin of the fossa, and is *inserted* into the olecranon on its outer and anterior aspect.

Action.—To assist in extending the elbow, and at the same time to raise the synovial membrane and prevent its injury between the bones.

Directions.—If the caput medium be now severed at its origin, and turned backwards, the musculo-spiral nerve and some branches of the deep humeral artery will, as already described, be found turning round the humerus in the musculo-spiral groove, which is mainly filled by the brachialis anticus muscle.

The BRACHIALIS ANTICUS muscle (Plate 8), also known as the humeralis obliquus or externus, is lodged in the furrow of torsion on the shaft of the humerus.   The muscle has its *origin* on the posterior aspect of the shaft of the humerus below its articular head.   Its tendon, which cannot be followed at present, passes in front of the elbow-joint, and is afterwards reflected under the internal lateral ligament of the joint, to be *inserted* into the radius and ulna.

Action.—To flex the elbow-joint.

THE FORE-ARM.

Surface-marking.—At the elbow-joint the olecranon process of the ulna is distinctly seen; but the shafts of the bones of the fore-arm are clothed with muscles, except at the lower third of the inner border of the radius, where the bone is subcutaneous.   On the outer side of the front of the elbow-joint a large muscular mass is formed by the extensor metacarpi magnus and the anterior extensor of the digit (extensor pedis).   In the living animal (in which it is preferable to study these surface-markings) this is more distinctly visible, and the tendons of these muscles and that of the lateral extensor (extensor suffraginis) may be distinctly traced.   On the inner side of the elbow-joint one may feel the tendon of insertion of the biceps; and just behind the tendon the posterior

radial vessels and the median nerve may be felt as they lie on the bone
under cover of the posterior superficial pectoral, and they may be made
to roll under the finger.   This should be practised, as the posterior
radial artery is a convenient vessel at which to feel the pulse.   The
internal subcutaneous vein crosses the inner face of the fore-arm
obliquely upwards and forwards; and in the living animal, pressure at
the upper part will distend the vessel and bring it into view.   At the
outer side of the carpus the prominence formed by the pisiform bone
may be seen and felt.   On the inner surface of the fore-arm, at its lower
third, the skin presents an oval-shaped, horny callosity, vulgarly termed
the *chestnut.*   This is largest in coarse-bred animals.

Directions.—The skin is now to be carefully removed from the
fore-arm and carpus, and the cutaneous nerves and vessels are to be
sought.

CUTANEOUS NERVES.   (1) At the front of the elbow-joint (Plate 8) the
cutaneous division of the musculo-cutaneous branch of the median appears
from beneath the biceps, and splits into two branches, one accompanying
the anterior, the other the internal, subcutaneous vein; (2) a little way
below the elbow, on its inner aspect, the cutaneous branch of the ulnar
(Plate 5) appears from beneath the insertion of the posterior superficial
pectoral, and divides for the supply of the skin of the back of the fore-arm
on both its outer and its inner side; (3) perforating the caput medium,
or emerging at its lower edge, are some twigs from the musculo-spiral
nerve, which are distributed to the skin of the outer side of the fore-arm
beneath the elbow; (4) on the outer side of the carpus (Plate 8) are the
ramifications of a cutaneous branch of the ulnar, which comes out between the tendons of the external and oblique flexors of the metacarpus.

SUBCUTANEOUS VEINS.—1. The *Median* or *Internal subcutaneous vein*
begins at the inner side of the carpus, where it continues upwards the
internal metacarpal vein.   It crosses the fore-arm obliquely upwards
and forwards, in company with a cutaneous nerve already described,
and divides into the cephalic and basilic veins.   The *Cephalic vein* has
already been seen ascending in the groove between the mastoido-
humeralis and the anterior superficial pectoral to terminate in the
jugular.   The *Basilic vein* pierces the posterior superficial pectoral
to concur in forming the brachial vein.

2. The *Anterior subcutaneous* or *radial vein* is much smaller than the
preceding vessel.   It begins at the front of the carpus, and, ascending
on the middle line of the fore-arm, it empties itself into the cephalic or
the median vein.

Directions.—The thin superficial fascia in which these nerves and
vessels are distributed should be removed to show the deep fascia.

DEEP FASCIA of the fore-arm.—This is spread in the form of a close-
fitting fibrous envelope around the fore-arm.   Above it receives an

insertion from the biceps, and another from the scapulo-ulnaris ; below
it is continued over the carpus to form sheaths for the tendons ; while
by its deep face it furnishes septa to pass between the muscles of the
fore-arm.

Directions.—The dissection of the back of the fore-arm is now to be
undertaken. The before-mentioned fascia is to be incised along the
lines of separation of the muscles, and these are to be cleaned and
isolated. The remaining portion of the posterior superficial pectoral
muscle, which covers the posterior radial vessels and the median nerve
at the inner side of the elbow, is to be removed ; and care is to be
taken of the ulnar vessels and nerve, which are placed beneath the
deep fascia, on the middle line at the back of the limb.

The ULNAR ARTERY (Plates 6 and 7) is a collateral branch of the brachial,
from which it comes off at the lower border of the caput parvum. It
descends parallel to the lower border of that muscle, to the space between
the olecranon and the inner condyle, where it is covered by the scapulo-
ulnaris. It here places itself in company with the ulnar nerve ; and,
crossing beneath the ulnar origin of the middle flexor of the metacarpus,
it descends to the carpus by following the tendon of the ulnar portion of
the deep flexor (ulnaris accessorius), being placed between the external
and oblique flexors of the metacarpus. At the upper limit of the carpus
it concurs in the formation of the supracarpal arch, by joining a branch
detached from the large metacarpal artery. In this course it gives off
—(1) the nutrient artery to the humerus (sometimes); (2) articular
branches to the elbow-joint ; (3) muscular branches in the neighbourhood of the joint, to the scapulo-ulnaris, caput parvum, and posterior
superficial pectoral ; (4) cutaneous branches to the skin on the inner
side of the fore-arm.

The ULNAR VEIN accompanies the artery and nerve, and at the elbow
concurs in the formation of the brachial vein.

The ULNAR NERVE (Plates 6 and 8) has already been partly described in
the dissection of the arm. At the lower part of that region it crosses
the ulnar artery, with which it places itself in company between the olecranon and the inner condyle. It here gives off branches to the following
muscles :—(1) the anterior head of the middle flexor of the metacarpus ;
(2) the ulnar head of the same muscle ; (3) the superficial flexor of the
digit (perforatus); (4) the ulnar origin of the deep flexor (ulnaris
accessorius). In the fore-arm it descends in close company with the
vessels of the same name, and at the carpus it gives off the cutaneous
branch already described (page 20). At the upper border of the pisiform
bone, and beneath the tendon of the middle flexor,* it joins a branch
from the median to form the external plantar nerve.

* In Plates 6 and 9 the termination of the nerve has been pulled slightly forwards to show its
junction with the branch from the median.

The POSTERIOR RADIAL ARTERY (Plate 6) is one of the terminal branches of the brachial. It is so much larger than the other terminal branch (the anterior radial), that it might be described as the direct continuation of the brachial, whose direction it prolongs. Beginning above the inner condyle, it descends on the bone, and then lies over the internal lateral ligament of the elbow-joint, and posterior to the tendon of insertion of the biceps. It is here covered by the posterior superficial pectoral, and is related to the median nerve, which lies close behind it, and to its satellite veins. At this point it is favourably placed for taking the pulse, and its situation and relations should be carefully noted. After crossing the elbow, it inclines forwards and disappears with the median nerve between the radius and the internal flexor of the meta-carpus. In this position it descends to within a short distance of the carpus, where it divides into two terminal branches of unequal size—the large and small metacarpal arteries. It gives off the following collateral branches :—

1. *Articular Branches* to the elbow-joint.

2. The *Interosseous Artery* of the fore-arm, which reaches the outside of the limb by passing through the radio-ulnar arch. It then descends along the outer side of the line of junction of the radius and ulna (Plate 7), where it will be followed in the dissection of the front of the fore-arm.

3. *Muscular Branches* to the flexors of the metacarpus and digit.

4. *Cutaneous Branches.*

The POSTERIOR RADIAL VEINS. The artery is accompanied by three or four satellite veins, which surround it and the nerve, and anastomose freely with each other. They begin at the carpus, where they anasto-mose with the metacarpal veins, and at the elbow-joint they unite with the basilic and ulnar veins to form the brachial vein. They receive branches corresponding more or less exactly to those of the artery.

The MEDIAN NERVE in the fore-arm (Plate 6). This nerve has already been followed in the dissection of the arm, where it was seen descending in front of the brachial artery. It preserves the same relationship to the first few inches of the posterior radial artery, but at the elbow it crosses the artery superficially to take up a posterior position. Below the joint it again changes its position by mounting on the surface of the artery, or it may even again place itself in front. At a variable point in the fore-arm it terminates by dividing into two branches, one of which is continued as the internal plantar nerve, while the other joins the ulnar to form the external plantar. In the subject from which Plate 6 was taken, the division took place considerably above the middle of the fore-arm, but more frequently it occurs in the lower third. Immediately below the elbow the nerve furnishes a branch to the internal flexor of the metacarpus, and branches to the deep flexor of the digit (humeral and radial heads).

Directions.—The muscles on the back of the fore-arm must now be learnt. These consist of the three flexors of the metacarpus, and the two flexors of the digit.

The FLEXOR METACARPI INTERNUS (Plate 6). This muscle lies along the inner edge of the posterior surface of the radius, where it conceals the posterior radial vessels and the median nerve. It *arises* from the inner condyle of the humerus, just behind the point of origin of the internal lateral ligament, where it is confounded with the origin of the middle flexor. It terminates inferiorly in a long, slender tendon, which, after passing through a synovial sheath at the inner side of the carpus, is *inserted* into the head of the inner small metacarpal bone.

Action.—It is a flexor at the carpal articulations—*i.e.*, it flexes the manus on the fore-arm.

The FLEXOR METACARPI MEDIUS (Plate 6). This muscle descends in contact with the posterior edge of the internal flexor. It has two heads of origin—an anterior and a posterior. It *arises* by its anterior head just behind the origin of the preceding muscle, and by its posterior head from the upper part of the posterior edge of the olecranon. After a course of three or four inches these two heads unite, and the single inferior tendon is *inserted* into the upper border of the pisiform bone. The ulnar nerve and vessels pass beneath the posterior or ulnar head of the muscle.

Action.—The same as the preceding muscle.

The FLEXOR METACARPI EXTERNUS (Plates 7 and 8) is situated at the outer side of the back of the fore-arm, having the lateral extensor of the digit (extensor suffraginis) in front of it, while behind it is separated from the last-described muscle by the ulnar division of the deep flexor of the digit (ulnaris accessorius). It *arises* from the lowest point of the outer ridge bounding the olecranon fossa. At its lower end it has two *insertions*, viz., (1) into the upper border of the pisiform bone, where it is confounded with the insertion of the middle flexor; (2) by a cord-like tendon which, after descending in a synovial sheath formed inwardly by the oblique groove on the outer surface of the pisiform bone, is inserted into the head of the external small metacarpal bone.

Action.—Like the preceding two muscles.

Directions.—The three flexors of the metacarpus surround the flexors of the digit, and they should be cut about their middle and reflected to bring these latter into view.

The SUPERFICIAL FLEXOR of the DIGIT (flexor pedis perforatus) (Plate 6) *arises*, by a tendon common to it and the deep flexor, from the lower extremity of the ridge bounding the olecranon fossa on the inside. Its muscular belly contains much tendinous tissue, and cannot without difficulty be separated from the deep flexor, on which it rests. At the lower part of the radius its muscular portion is succeeded by a tendon, which, after being reinforced by a fibrous band from the back of the radius, passes through the carpal sheath behind the carpus, and is ultimately *inserted* by a bifid tendon into the second phalanx. The examination of this and the succeeding muscle, from the carpus downwards, must be postponed till the dissection of the metacarpus and digit is undertaken.

Action.—The muscle flexes successively the pastern, fetlock, and carpal joints.

The DEEP FLEXOR of the DIGIT (flexor pedis perforans) (Plate 6). This muscle is situated in contact with the posterior surface of the radius, and consists of three divisions, which may be distinguished as the humeral, the radial, and the ulnar portions. The *humeral* or main division *arises*, in common with the preceding muscle, from the lower extremity of the ridge bounding the olecranon fossa on the inside. The *radial* portion, or *radialis accessorius*, is deeply placed, and *arises* from the back of the radius. The *ulnar* division, or *ulnaris accessorius*, is placed beneath the deep fascia of the fore-arm, where it lies between the external and oblique flexors of the metacarpus, and is accompanied by the ulnar nerve and vessels. It *arises* from the summit and posterior border of the olecranon. These three divisions unite above the carpus, and have a common tendon which passes through the carpal sheath, and is ultimately *inserted* into the os pedis.

Action.—It flexes successively from below upwards the inter-phalangeal joints, the fetlock, and the carpus.

Directions.—The front of the fore-arm must now be dissected; and here it will be convenient to turn attention in the first place to muscles; but while these are being isolated, care is to be taken of the interosseous vessels, which descend along the lateral extensor at the outer side of the region, and of the tendon of the oblique extensor where it crosses over the tendon of the extensor metacarpi magnus above the carpus.

The EXTENSOR METACARPI MAGNUS (Plates 7 and 8) corresponds to the long and short radial extensors of the wrist in the human subject. It is a powerful muscle, having at its upper end a massive muscular belly, which tapers downwards, and terminates a few inches above the carpus in a tendon. It *arises* from the anterior and upper part of the outer ridge of the olecranon fossa (the outer condyloid ridge), where this ridge bounds the musculo-spiral groove ; and by a second tendon, in common with the extensor pedis, from a depression which is placed external to the coronoid fossa. Its inferior tendon lies in the largest and most internal of the vertical grooves at the lower end of the radius ; and after gliding over the front of the carpus in a synovial sheath, it is *inserted* into a special tubercle on the upper end of the large metacarpal bone at its inner side.

Action.—It extends the manus on the fore-arm.

The EXTENSOR METACARPI OBLIQUUS (Plates 8 and 9). This is the representative of the extensor muscles of the thumb in man. It *arises* from the outer side of the radius; and its tendon, after passing obliquely downwards and inwards over that of the great extensor, is *inserted* into the head of the inner small metacarpal bone. It lies in an oblique groove at the lower end of the radius, where the play of its tendon is facilitated by a small synovial bursa.

Action.—Like the preceding muscle.

The EXTENSOR PEDIS, or anterior extensor of the digit (Plate 7), represents the extensor communis digitorum of man. At its origin it lies immediately to the outer side of the extensor metacarpi magnus, but at the lower part of the fore-arm the extensor metacarpi obliquus emerges from between the two muscles. It *arises*, by a tendon common to it and the extensor metacarpi magnus, from a depression external to the coronoid fossa; also from the external lateral ligament of the elbow, and the external tuberosity at the upper end of the radius. It consists of two parallel portions of unequal size, and these are succeeded by two tendons which lie close together, but are distinct from each other. These tendons pass in common through a vertical groove at the lower end of the radius, and over the front of the carpus, where they are provided with a synovial sheath. In the dissection of the metacarpus and digit, the tendons will be pursued to their insertion, the outer and smaller * joining the tendon of the extensor suffraginis, while the inner and main tendon becomes *inserted* into the pyramidal process of the os pedis.

Action.—This muscle extends in succession the interphalangeal joints, the fetlock, and the carpus.

The EXTENSOR SUFFRAGINIS, or lateral extensor of the digit (Plates 7 and 8), is a smaller muscle than the extensor pedis, to the outer side of which it lies. It is the homologue of the extensor of the little finger in man. It *arises* from the external lateral ligament of the elbow, from the external tuberosity at the upper end of the radius, from the line of junction of the radius and ulna, and from the outer border of the radius. Its tendon passes first through a vertical groove on the external tuberosity at the lower end of the radius, then through a synovial sheath at the outer side of the carpus, and it will subsequently be followed to its *insertion* into the first phalanx.

Action.—It is an extensor of the fetlock and of the carpus.

Directions.—The nerves and bloodvessels on the front of the fore-arm must next be sought, and in order to fully expose them, some of the foregoing muscles must be cut. The biceps is to be cut about its

---

' This is sometimes termed the *muscle of Phillips*. Occasionally there occurs, to the inner side of the preceding, another and smaller fasciculus, with a slender tendon which joins the main tendon before reaching the carpus. This is the *muscle of Thiernesse*.

middle in order to follow the anterior radial artery; and by dissecting deeply down in front of the elbow, between the brachialis anticus and the extensor metacarpi magnus, the artery will be found to meet the musculo-spiral nerve. The extensor metacarpi magnus is to be cut about its middle and carefully reflected in order to follow the artery, which lies in relation to the deep face of the muscle; and the extensor pedis is to be similarly reflected to trace the termination of the musculo-spiral nerve.

The ANTERIOR RADIAL ARTERY (Plate 8) is the smaller terminal branch of the brachial. It separates at an acute angle from the posterior radial, and passes forwards beneath the biceps and then beneath the brachialis anticus. It meets the musculo-spiral nerve in the interspace between the brachialis anticus and the extensor metacarpi magnus, and afterwards descends on the anterior surface of the radius, where it is covered by the last-mentioned muscle. It terminates at the carpus by anastomosing inwardly with branches from the posterior radial, and outwardly with the interosseous artery of the fore-arm. It supplies articular branches to the elbow, and muscular branches to the muscles on the front of the fore-arm.

The INTEROSSEOUS ARTERY of the fore-arm (Plate 7) is a branch given off by the median at the back of the fore-arm. It comes outwards through the radio-ulnar arch, and descends along the extensor suffraginis, terminating in slender branches in front of the carpus. It supplies articular branches to the elbow; the nutrient artery of the radius; and muscular twigs to the extensor suffraginis, extensor pedis, and extensor metacarpi obliquus.

The anterior radial and interosseous arteries are, generally, comparatively slender vessels, but they are liable to some variation in size and distribution, and the one may partly supplant the other.

VEINS. Satellite veins of the same names run in company with the foregoing arteries.

The MUSCULO-SPIRAL NERVE in the fore-arm (Plate 8). In the dissection of the axilla and arm, this nerve has already been seen as a large trunk descending from the brachial plexus, and taking a spiral course behind the humerus. It reaches the front of the elbow, where it meets the radial artery in the interspace between the brachialis anticus inwardly, and the origin of the extensor metacarpi magnus outwardly. It here gives off branches to the extensor metacarpi magnus, extensor pedis, extensor suffraginis, and flexor metacarpi externus; and, much reduced in size, it descends between the shaft of the radius and the extensor pedis, and terminates in the extensor metacarpi obliquus. The nerve to the flexor metacarpi externus is furnished after the branches to the extensor pedis, and passing outwards between the latter muscle and the bone, it penetrates its muscle at the radio-ulnar arch.

Directions.—In this stage of the dissection the student will be better able to trace the musculo-cutaneous branch of the median nerve, and the insertions of the biceps and brachialis anticus muscles (see pages 16 and 19). When these have been examined, he may, as the next step, either dissect the articulations of the shoulder and elbow (pages 41 and 43), or he may saturate the parts already dissected with some preservative solution, and postpone the examination of these joints till after the dissection of the metacarpus and digit.

THE METACARPUS AND DIGIT.

The distal portion of the horse's fore limb, beyond the lower extremity of the radius, is technically termed the manus, as it corresponds to the hand of man. The carpus, or, as it is commonly but erroneously termed, the knee, of the horse corresponds to the wrist of the human subject. The portion of the limb between the carpus and the fetlock, representing the palmar portion of man's hand, is called the metacarpus ; while the rest of the limb, beyond the fetlock, is the digit, and is the homologue of man's middle-finger.

Surface-marking.—By flexing the carpal and fetlock joints, the *splint* bones may be felt at the back of the metacarpus. Behind the bones in the same region lie the flexor tendons, the subcarpal ligament, and the suspensory ligament. These, whose edges may be more or less distinctly seen in a well-bred animal, have the relation to each other shown in Plate 7. Behind the fetlock-joint is a tuft of hair in which will be found a horny spur or *ergot*, which is largest in coarse-bred animals. By manipulation, the flexible lateral cartilages may be felt above the hoof, in the region of the heels.

Directions.—The entire remaining portion of skin should now be carefully removed from the limb. Should it be intended to study from the same preparation the parts contained within the hoof, this must, before the removal of the skin, be detached by force in the manner described on page 35. The various structures are now to be defined by dissection in the order of the following description ; and while the vessels and nerves are being cleaned, care must be taken of the small *lumbricales* muscles, which lie on the tendon of the deep flexor above the fetlock. The palmar arterial arches cannot be fully exposed at this stage of the dissection, but it is convenient to describe them here, from their relationship to the vessels of the region. The same applies to the large metacarpal artery and the plantar nerves behind the carpus, all of which can be fully traced in the examination of the carpal sheath (page 33).

The LARGE METACARPAL ARTERY (Plate 9). This is the largest artery in the part of the limb now exposed, and is, by means of its ter-

minal branches, the main vessel of supply to the digit. It has already
been seen at its origin, as the larger of the two terminal branches of the
posterior radial artery ; and, indeed, from its volume and direction, it
might be described as the direct continuation of that vessel. From its
point of origin at the lower end of the radius, it descends in company
with the flexor tendons, by passing behind the carpus and beneath the
carpal arch. Emerging from beneath the last-named structure, it continues to descend on the inner side of the flexor tendons until a little
above the fetlock, where it sinks slightly inwards to bifurcate into the
digital arteries. From the carpus downwards the artery is related to
the internal metacarpal vein, which ascends in front of it, and to the
internal plantar nerve, which is in contact with it posteriorly. The
relative position of the three structures should be carefully noted in
reference to the higher operation of neurotomy. Only two of its
collateral branches are of sufficient size to merit description, and both
are somewhat irregular in their origin. The first of these comes
off near the origin of the parent vessel, and may come from the
posterior radial itself. It crosses behind the lower extremity of the
radius, and anastomoses with the termination of the ulnar artery to
form the supracarpal or superficial palmar arch. The second is an
un-named vessel which springs from the large metacarpal at or near its
point of bifurcation, and divides into branches that ascend to anastomose
with the interosseous metacarpal arteries.

The SUPRACARPAL or SUPERFICIAL PALMAR ARCH is formed behind the
lower extremity of the radius, by the junction of the above-mentioned
branch of the large metacarpal artery with the termination of the ulnar.
The convexity of the arch is turned downwards, and from it there arise
several branches. The largest and most regular of these descends
within the carpal arch, and joins the small metacarpal artery to form
the subcarpal or deep palmar arch, which will be dissected at a later
stage.

The DIGITAL ARTERIES (Plates 9 and 10) are the terminal branches
of the large metacarpal artery. They separate at an acute angle, the
outer one passing above the fetlock, between the deep flexor and the
suspensory ligament. Each passes over the side of the fetlock-joint, and
descends at the edge of the flexor tendons as far as the inner face of
the basilar process, where it bifurcates to form the plantar and preplantar arteries. Each artery is related in front to the vein of the
same name, and behind to the posterior branch of the plantar nerve.
The anterior branch of the same nerve crosses the vessel at the
fetlock ; while other twigs cross over the artery and form the
middle branch, which will be found between the artery and vein, or
resting on the former. Crossing these vessels and nerves obliquely,
is a small glistening ligamentous cord (Plate 9) which stretches

downwards and forwards from the horny spur behind the fetlock, becoming attached within the wing of the os pedis. A knowledge of these relationships is of importance for the performance of the lower operation of neurotomy. The collateral branches of the digital arteries are :—

1. At different levels numerous small branches for the skin, tendons, or articulations. Among these may be included the *rameaux échelonnés* of Bouley (Plate 10). These branches, some of them of considerable size, spring from the posterior aspect of the artery, and anastomose across the back of the digit with corresponding branches from the opposite side, forming arches arranged like the steps of a ladder.

2. The *Perpendicular Artery*, which comes off at a right angle about the middle of the first phalanx, and divides almost immediately into an ascending and a descending set of branches, both of which are distributed on the front of the first phalanx. Branches from each of these sets anastomose with corresponding vessels from the opposite side.

3. The *Artery of the Plantar Cushion.*

4. Vessels forming the *Coronary Circle.*

The last two, as well as the terminal branches of the digital arteries, will be described in connection with the foot.

The SMALL METACARPAL ARTERY (Plate 9). This, the smaller terminal branch of the posterior radial artery, descends behind the knee and towards its inner side. It is superficially placed to the fibrous band completing the carpal arch, while the large metacarpal lies beneath that structure. In company with it is the first part of the median vein. At the level of the head of the inner metacarpal bone it crosses to the outer side by passing between the suspensory ligament and the subcarpal ligament, or check-band furnished from the back of the carpus to the tendon of the deep flexor. It here anastomoses with a branch already described as descending from the supracarpal arch. In this way the subcarpal arch is formed.

The SUBCARPAL or DEEP PALMAR ARCH gives off the following two pairs of arteries :—

1. The *Anterior* or *Dorsal Interosseous Metacarpal Arteries.*—These are small vessels (Plate 9), one on each side of the limb, which turn forward round the heads of the small metacarpal bones, and descend in the grooves between these bones and the large metacarpal. They supply the skin and subjacent structures on the front of the metacarpus, and anastomose above the fetlock with divisions of the artery springing from the large metacarpal at its point of bifurcation.

2. The *Posterior* or *Palmar Interosseous Metacarpal Arteries.*—These descend on the edge of the suspensory ligament, each being internally placed to the small metacarpal bone of its own side. They anastomose like the preceding, and supply small branches to the suspensory ligament

and flexor tendons. One of them gives off the *nutrient artery* of the large metacarpal bone. They are of unequal size, the outer being the larger.

The DIGITAL VEINS (Plate 9). These are the satellites of the digital arteries, in front of which they ascend. They drain away the blood from the venous plexuses within the hoof, and, uniting with one another above the fetlock, they form an arch between the deep flexor and the suspensory ligament. From this arch spring the metacarpal veins.

The METACARPAL VEINS are three in number :—

1. The *Internal Metacarpal Vein* (Plate 9), which is the largest of the three, ascends in front of the large metacarpal artery, on the inner edge of the flexor tendons. At the inner side of the back of the carpus it is continued as the median vein.

2. The *External Metacarpal Vein* is similarly disposed on the outside of the flexor tendons, in company with the external plantar nerve. At the carpus it divides into several anastomosing branches, which are continued as the ulnar and posterior radial veins.

3. The *Interosseous* or *Deep Metacarpal Vein* is an irregular vessel ascending between the suspensory ligament and the inner splint bone. At the back of the carpus it breaks up into branches that anastomose with the external and internal metacarpal veins.

The *Plantar Nerves* (*metacarpal nerves* of Percivall).—These are the nerves which confer sensibility on the digit, and which, in their main trunks, or in one of their terminal branches, are cut in the operation of neurotomy. They must therefore be dissected with great care, and the student must make himself thoroughly acquainted with their situation and relations.

The INTERNAL PLANTAR NERVE (Plate 7). This is one of the terminal branches of the median nerve. Beginning at a variable point above the carpus, it passes within the carpal arch, in close company with the large metacarpal artery, both resting on the side of the deep flexor tendon. Here the nerve crosses beneath the artery, to place itself behind it. Throughout the metacarpal region the same relationship is preserved, the nerve lying immediately behind the artery, in front of which is the internal metacarpal vein. Just above the fetlock the artery sinks in somewhat more deeply than the vein and nerve, and thereby allows these to approach each other. In the higher operation of neurotomy the nerve is cut a little way above the fetlock, and before it divides. About the middle of the metacarpus it gives off a considerable branch which winds obliquely downwards and outwards behind the flexor tendons, to join the external plantar nerve an inch or more above the *button* of the splint bone. At the level of the sesamoid bones the trunk of the nerve divides into three digital branches, which are distinguished as anterior, middle, and posterior. These are of very unequal size, the posterior being much the largest, and also the most important, as it is the nerve which is cut in the lower operation of neurotomy when performed for navicular arthritis. The middle is the smallest and most irregular, and all three branches are in close relationship with the digital vessels.

The *Anterior branch* descends in front of the vein, distributes cutaneous branches to the front of the digit, and terminates in the coronary cushion.

The *Middle branch*, which is small and irregular, descends between the artery and vein. It is generally, as in Plate 9, formed by the union of several smaller branches which cross forwards over the artery before uniting, and it terminates in the sensitive laminæ and coronary cushion.

The *Posterior branch* lies close behind the artery, except at the fetlock, where the nerve is almost superposed to the artery. It accompanies the digital artery into the hoof, and passes with the preplantar branch of that vessel to be distributed to the os pedis and the sensitive laminæ. Within the hoof it gives off several branches, which for the most part accompany the arteries.

The EXTERNAL PLANTAR NERVE (Plate 9). This is formed by the fusion of the termination of the ulnar nerve with one of the terminal branches of the median. These two branches unite at the upper border of the pisiform bone, beneath the middle flexor of the metacarpus. Behind the carpus the nerve inclines downwards and outwards, in the texture of the annular ligament that completes the carpal sheath. In the metacarpal region it occupies, on the outside of the limb, a position on the flexor tendons analagous to that of the internal plantar nerve on the inside. Unlike the latter nerve, however, it is accompanied by only a single vessel—the external metacarpal vein, which lies in front of it. An inch or more above the *button* of the splint bone it is joined by the oblique branch from the internal nerve. In the higher operation of neurotomy it is cut at the same point as the inner nerve. At the level of the sesamoid bones it divides into three digital branches, exactly similar to those of the internal nerve already described.

The plantar nerves give filaments to the lumbricales and interossei muscles, and to the suspensory ligament.

Directions.—The student must now pursue the dissection of the following muscles which have already been dissected in the fore-arm, viz., the extensor pedis and extensor suffraginis on the front of the limb, and the superficial and deep flexors behind. In addition to these, there are the lumbricales and interossei muscles, which entirely belong to this region ; and, as they are of small size, and might easily be overlooked, their dissection must be first undertaken.

The LUMBRICALES MUSCLES (Plate 9) receive their name in the human hand from their resemblance to a common earthworm. In the

---

horse they are of small but very variable size. Frequently they contain but little muscular tissue, but now and again a subject is met in which they are very distinct. They are two in number, one being placed on each side of the deep flexor tendon, above the fetlock. The fibres of the small muscular belly *arise* from the side of the deep flexor, and terminate in a small tendon which is lost in the tissue beneath the horny spur of the fetlock.

The INTEROSSEI MUSCLES (Plate 9). These are the representatives of the muscles which, in the human hand, fill up the interspaces of the metacarpal bones, and give lateral movement to the fingers. In the horse they are two in number, and are extremely rudimentary. Each is to be sought to the inner side of the small metacarpal bone of its own side, between that bone and the edge of the suspensory ligament. Each has at its upper end a small muscular belly taking *origin* from the neighbourhood of the head of the small metacarpal bone. It is succeeded by a long, slender, nerve-like tendon, which at the fetlock blends with the band sent from the suspensory ligament to the extensor pedis tendon, or with the connective-tissue on the side of the joint.

The interossei and lumbricales muscles are of great interest to the comparative anatomist, but, from their small size, they can have no appreciable effect on the movements of the digit.

The *Tendon* of the EXTENSOR SUFFRAGINIS (Plate 7) is to be followed from the point below the carpus to which it has already been dissected.

The flat tendon, after crossing the carpus, descends to the outer side of the anterior surface of the large metacarpal bone. As it passes over the fetlock-joint, it becomes somewhat broader, and its play over the anterior ligament of the joint is facilitated by means of a small synovial bursa. Immediately below the joint it is *inserted* into the fore part of the upper end of the first phalanx. In the region of the metacarpus the tendon receives on each side a reinforcing band. The outer band comes from the external side of the carpus; the inner is detached from the extensor pedis tendon.

Action.—The muscle is primarily an extensor of the digit on the metacarpus. When contraction is carried beyond this, it extends the metacarpus on the fore-arm.

The *Tendon* of the EXTENSOR PEDIS (Plate 7). This tendon, after throwing off the slip to the extensor suffraginis, descends over the front of the metacarpus and digit, and lies on the middle line. Its play over the anterior ligament of the fetlock is facilitated by a small synovial bursa; while, over the front of the interphalangeal joints, the synovial membrane is directly supported by the deep face of the tendon, there being no anterior ligament for these joints. At the middle of the first phalanx the tendon is joined on each side by a strong band which descends obliquely over the side of the fetlock from the suspensory ligament.

The tendon is finally *inserted* into the pyramidal process of the os pedis.

Action.—The first action of the muscle is to extend the third phalanx on the second, and then the second on the first. When contraction is continued, it produces successively extension of the fetlock and of the carpus.

Directions.—The tendons on the back of the metacarpus and digit must next be dissected; and as a preliminary step, the carpal and metacarpo-phalangeal sheaths formed in connection with these tendons should be examined.

The CARPAL SHEATH (Fig. 1) is the tubular passage through which the flexors of the digit are transmitted behind the carpus. It is formed in front by the back of the carpus covered by the posterior common ligament of that joint. Behind it is bounded in its outer third by the pisiform bone, and in its inner two-thirds by a strong fibrous band representing the anterior *annular ligament* of the human wrist. This band stretches like an arch from the pisiform bone to the inner side of the carpus. It is continuous above with the deep fascia on the back of the fore-arm, of which it may be considered a thickened portion; and below it becomes thinner, and is continued as the fascia on the back of the metacarpus (*palmar fascia* of man). The carpal sheath is provided with

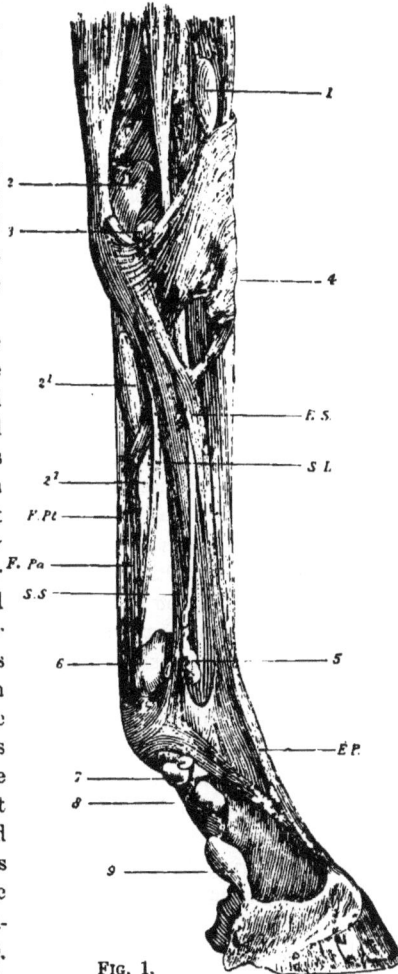

FIG. 1.

DISSECTION OF THE METACARPUS AND DIGIT, SHOWING THE TENDONS AND THEIR SYNOVIAL SHEATHS (*Chauveau*).

1. Synovial bursa of the extensor metacarpi magnus; 2. Superior *cul-de-sac*, or pouch, of the synovial membrane of the carpal sheath; 2¹, 2². Inferior part of the same; 3. Pouch of the radio-carpal synovial membrane, appearing as a hernia between the posterior common ligament and the outermost radio-carpal ligament; 4. Synovial bursa of the extensor pedis; 5. Protrusion of the synovial membrane of the fetlock-joint; 6, 7, 8. Superior, middle, and inferior pouches of the synovial membrane of the metacarpo-phalangeal sheath; 9. Inferior extremity of the same, exposed by the removal of the reinforcing sheath of the perforans tendon; E. S. Extensor suffraginis; S. L. Subcarpal ligament; E. P. Extensor pedis; S. S. Superior sesamoidean (suspensory) ligament; F. Pa. Flexor perforans; F. Pt. Flexor perforatus.

D

a synovial membrane, which lines it, and is reflected over the flexor tendons to facilitate their gliding. If the fibrous band just described be cut, and a probe be passed upwards and downwards within the sheath, an idea of the extent of the synovial sac will be gained. It will be found to extend upwards for two or three inches above the carpus, and downwards as far as the middle of the metacarpus.

Directions.—The fibrous band should be entirely removed in order to permit the examination of the tendons, and of the nerves and bloodvessels which accompany these within the sheath.

The METACARPO-PHALANGEAL or GREAT SESAMOID SHEATH (Fig. 1). This is a second synovial apparatus developed in connection with the flexor tendons. If a vertical incision be made through the superficial flexor just above the fetlock, and a probe passed into the incision, it will enter the synovial cavity, and may be pushed upwards for two or three inches above the fetlock, and downwards as far as the middle of the second phalanx. The synovial membrane lubricates the pulley-like surface formed by the sesamoid bones and the inter-sesamoid ligament, and is reflected on to the tendons. It is supported laterally by a fibrous expansion which, adhering to the superficial flexor behind, is inserted in front by three slips on each side, the highest insertion being into the sesamoid, and the other two into the first phalanx. At its lower extremity this synovial membrane meets that of the navicular sheath, and in front of the same point it is separated from the synovial capsule of the coffin-joint by a kind of partition of yellow fibrous tissue connecting the front of the perforans tendon to the back of the os coronæ (Plate 10, fig. 2).

The SUPERFICIAL FLEXOR tendon (Plates 5, 9, 10, and 11). The tendon succeeds the fleshy portion of the muscle at the lower part of the fore-arm, and it is there reinforced by a fibrous band which springs from the back of the radius and is sometimes termed the *superior carpal ligament*, in contradistinction to the band which reinforces the tendon of the deep flexor below the carpus. The tendon passes through the carpal sheath in company with and behind the deep flexor, and then descends behind the metacarpus. Having arrived at the fetlock, there is formed in it a remarkable ring, through which the tendon of the deep flexor plays. It is in consequence of this arrangement that the superficial muscle is termed *perforatus*, and the deep one *perforans*. As already seen, the tendons are here enveloped by the synovial membrane of the meta-carpo-phalangeal sheath. At its extremity the tendon is bifid, and each slip is *inserted* into the upper extremity of the second phalanx on its lateral aspect.

Action.—The muscle flexes successively the pastern, fetlock, and carpal joints.

The DEEP FLEXOR tendon (Plates 5, 9, 10, and 11) is, through-out its course, closely related to the preceding, in front of which it lies.

After descending through the carpal sheath, it is joined by a very strong fibrous band—the *inferior carpal ligament*, which is the downward continuation of the posterior common ligament of the carpus. This fuses with the tendon about the middle of the metacarpus, and it is of considerable importance, being frequently involved in what is commonly termed "sprain of the back tendons." In that condition it may be very distinctly felt by manipulating in front of the flexor tendons, just below the carpus. The tendon, as thus reinforced, descends between the suspensory ligament in front, and the perforatus tendon behind ; and at the fetlock it glides over the sesamoid pulley, and passes through the ring of the superficial flexor. It then passes between the terminal branches of the last-mentioned muscle, glides over the smooth surface on the back of the second phalanx, plays over the navicular bone, and finally becomes *inserted* into the semilunar crest of the os pedis. The terminal portion of the muscle, as well as the navicular sheath developed in connection with it, will be examined with the parts contained within the hoof.

Action.—The muscle flexes successively the interphalangeal joints, the fetlock, and the carpus.

THE FOOT.

Directions.—By the term foot, as here applied, is meant the hoof and the parts contained within it. If it is intended to study this in a limb the whole of which is to be successively dissected, the student must proceed in the following manner. When the dissection of the fore-arm has been completed, and before the removal of the skin from the metacarpus and digit, the hoof must be forcibly removed by the aid of a shoeing-smith's hammer, toe-knife, and pincers. To facilitate this, the hoof may be heated in a fire, the skin of the digit being swathed in a wet cloth to prevent charring. This is the speediest method of removing the hoof, but it has the double disadvantage of destroying in great measure the hoof itself, and also the injection of the vessels, provided that has been executed. The following is a preferable method of procedure :—Procure a foot severed a few inches above the fetlock, and inject the arteries and veins from the metacarpal vessels. When the injection has solidified, roll the foot in a piece of wet cloth, and bury it in a fermenting heap of stable manure. Decomposition will speedily set in, and after a week the preparation should be examined at intervals of two or three days, the metacarpal bone being fixed in a vice while forcible attempts are made to pull off the hoof. When this has been effected, the foot and removed hoof should be immersed for a day in a saturated solution of carbolic acid in water, to which a little methylated spirit may be added. This will speedily remove all odour of decomposition, and dissection may then be proceeded with.

The Hoof (Plate 10, figs. 4 and 6). This is made up of the *wall*, the *bars*, the *sole*, and the *frog*.

The WALL is that part of the hoof which is exposed when the foot rests in its natural position on a flat surface. It is divided, though not by any well-defined boundaries, into *toe*, *quarters*, and *heels*. The *toe* includes an area on each side of the middle line of the wall in front; and it passes on each side into the *quarter*, which comprises the lateral region of the wall. Posteriorly the wall changes its direction, and disappears from view, forming an angular part, which is termed the *heel*. In reality, the wall does not stop at the heel, and it is this concealed continuation that is termed the *bar*. In a well-formed hoof the wall in the region of the toe slopes at an angle of about 50°.

The *External Surface* of the wall is, in a state of nature, covered by a kind of epithelial varnish termed the *periople*, which is thickest at the top of the wall, just under the hair. This, which is a natural varnish provided to check evaporation and consequent cracking of the subjacent horn, is generally rasped away by the shoeing-smith. The *internal sur-face* of the wall is traversed in a vertical direction by the series of *horny laminæ*. These number about five or six hundred; and before separa-tion of the hoof, they were interleaved with the sensitive laminæ to be presently described. The *superior border* of the wall shows a kind of gutter, termed the *cutigeral groove*, which is the mould left by the coronary cushion. The floor of this groove has a closely punctated appearance, each minute perforation being the upper end of one of the horn tubes of the wall, and lodging, in the natural state, one of the papillæ of the coronary cushion. The *inferior border* embraces the sole, and in the unshod animal comes into contact with the ground.

The wall is thicker at the toe than at the quarters or heels; and in each of these areas, it is thicker on the outside than in the correspond-ing area on the inside.

The BARS. These are the reflected terminations of the wall behind the heels; and if the foot be turned up, the continuity will be distinctly seen.

The *Outer Surface* of the bar, which is here seen, slopes towards the frog, and bounds outwardly the lateral lacuna of that body. It shows an *inferior border*, which runs towards the centre of the sole, but stops a little behind the point of the frog. The bars are also seen in the interior of the hoof, where they show an *internal surface* bearing horny laminæ like those of the wall. The *superior border* of the bars is included between the frog and the sole, and blended with them.

The SOLE presents an *inferior face*, which is vaulted, and this inde-pendently of any paring to which the foot may have been subjected, as the horn of which it is composed exfoliates so as to give it this con-figuration naturally. The *superior face* is somewhat convex, and has a punctated appearance similar to that already seen in the cutigeral groove. The minute holes lodge the papillæ of the so-called sensitive sole, which is the horn secreting structure of this region. *Anteriorly* the sole presents a convex border, which unites it intimately to the lower border of the wall, a line of whitish horn marking the junction of the two structures. *Posteriorly* it has a deep V shaped indentation, into the central point of which the frog penetrates, while behind that on each side it is related to the bar.

The sole of the hind hoof is distinguished from that of the fore by being more vaulted, and by being more pointed (less circular) at the toe, this latter difference affecting also the form of the wall in the same region. The outer edge of the sole is more convex than the inner, which enables one to readily distinguish between a right and a left hoof.

The FROG. This is a distinctly elastic mass of horn which, in a state of nature, projects sufficiently to come into contact with the ground, and thus give the animal a secure foothold. Its *inferior surface* shows posteriorly a shallow cleft, or depression, termed the *median lacuna*. The *lateral lacunæ* lie at the sides of the frog, the outer boundary of each lacuna being formed by the bar. The *superior surface* shows, vertically over the median lacuna, a projection termed the *frog-stay*. On each side of the frog-stay this surface is depressed, and the whole is moulded on the plantar cushion. This surface is punctated, and the papillæ of the plantar cushion are received into the minute apertures The *posterior extremity*, or *base*, of the frog consists of two rounded eminences—the *bulbs*, or *glomes*—separated from each other by the median lacuna. The *anterior extremity*, or *point*, is wedged into the centre of the sole. The *lateral borders* bring the frog into relation with the bars and the sole, and there is an intimate union with each of these at the point of contact.

MINUTE STRUCTURE of the hoof. The entire hoof is an aggregation of modified epithelial cells, which here represent the horny layer of the epidermis. When a thin section across the wall, sole, or frog is examined, the horn substance is seen to be arranged in the form of tubes, cemented together by an intertubular substance, and containing within their lumen a quantity of intratubular material. All of these—tubular, intertubular, and intratubular—are composed of modified epithelial cells, differing in the three situations in the direction of the cells, their state of aggregation, or the presence or absence of contained pigment. The tubes of the wall are straight, and extend parallel to the surface, from the coronary to the inferior edge of the wall. The tubes of the sole have the same disposition, but those of the frog are slightly flexuous. The upper end of each tube is occupied by an elongated vascular papilla, which belongs, in the case of the wall, to the coronary cushion ; in the periople, to the perioplic ring ; and in the sole and frog, to the sensitive structures of the same names. In the growing hoof the bond of connection between these papillated surfaces (which represent the corium of the skin) and the corresponding part of the hoof, is a stratum of soft protoplasmic epithelial cells by whose growth and multiplication the hoof-horn is formed. This stratum of cells represents the deepest cells of the rete mucosum in the skin, and it is by its ready decomposition that the bond of connection between the sensitive and insensitive structures is destroyed, permitting the extremity of the digit to be extracted from its horny investment.

Directions.—The student should next turn his attention to the extremity of the digit as exposed by the removal of the hoof, and he will find it to present a configuration not unlike the exterior of the hoof itself (Plate 10, figs. 1 and 5). And in the first place, let him examine that part which he will easily recognise as having been separated from the inner surface of the wall. This is traversed by a series of leaves which, in contradistinction to those already seen on the inner surface of the wall, are termed the sensitive laminæ, and sometimes the podophyllous tissue.

The SENSITIVE LAMINÆ. Each lamina is fixed by one of its borders to the periosteum of the os pedis, and extends in a vertical direction from near the coronary cushion to the sharp edge of the bone, where it terminates in five or six long papillæ. In the natural state the sensitive and the horny laminæ are interleaved, and the former here represent the corium, or true skin. The laminæ, it will be noticed, become progressively shorter as they are traced backwards ; and at the end of the series on each side, and adjacent to the plantar cushion, there is a number of small leaves that were interleaved with the horny laminæ of the bars.

The CORONARY CUSHION. This is a projecting, cornice-like structure, placed above the laminæ and below the limits of the skin of the digit. It fits into the cutigeral groove at the upper border of the wall, and its surface is closely set with long papillæ which were received into the apertures found in that groove. These papillæ give the coronary cushion a velvety pile, which may be rendered very evident by immersing the foot in water. If the coronary cushion be traced backwards, it will be seen to pass into the plantar cushion. Above the cushion is a narrow groove separating it from the periopolic ring. Below the cushion there is a narrow smooth space which runs between the cushion and the sensitive laminæ. The coronary cushion is a modified portion of the corium, and through the agency of the cells which cover the surface of its papillæ, the wall of the hoof is formed.

The PERIOPLIC RING. This ring is composed of papillæ like those of the coronary cushion, but smaller in size ; and it is by its agency that the periople which covers the exterior of the wall is formed.

The PLANTAR CUSHION. This is a fibro-elastic pad interposed between the horny frog and the terminal part of the perforans tendon. It possesses two faces, two borders, a base, and an apex. The *lower face* looks backwards as well as downwards when the foot rests on a flat

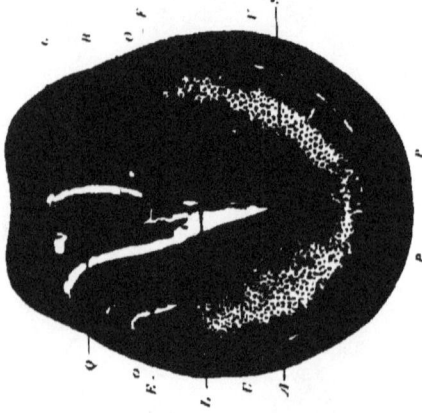

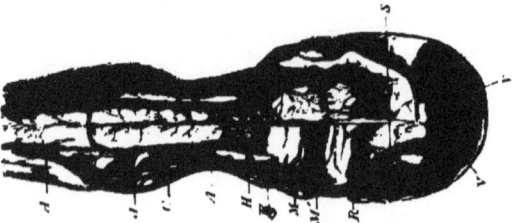

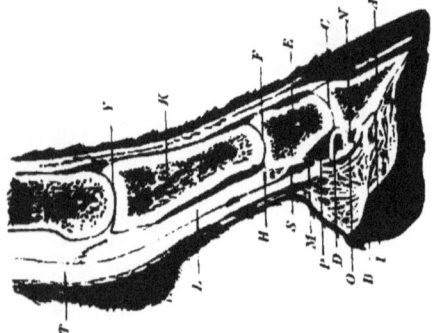

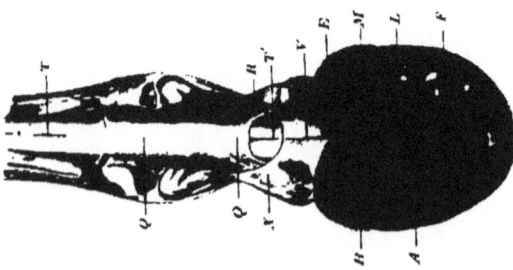

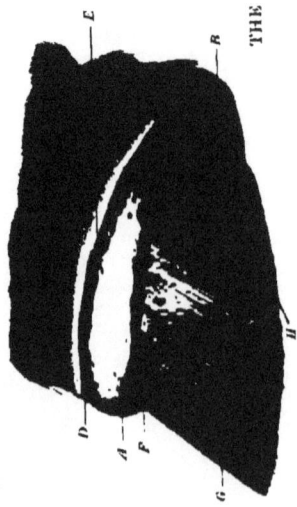

PLATE X

Fig. IV

Fig. VI

Fig. III

Fig. II

Fig. I

Fig. V

THE FOOT (Botley)

Drawn & Printed by W. & A.K. Johnston, Edinburgh & London

PLATE X.

Fig. I.—THE DIGIT WITH THE HOOF REMOVED, FLEXED AND VIEWED FROM BEHIND.

A. Sensitive sole; B. Sensitive laminæ that were interleaved with the horny laminæ of the bar; F. The pyramidal body, or sensitive frog; L. Lateral lacuna of the same; M. Median lacuna of the same; Q. Q. Fibrous sheath uniting the two branches of the perforatus; R. Branches of the perforatus passing to be inserted into the os coronæ; T. Tendon of the perforatus; T'. Tendon of the perforans in its passage between the branches of the perforatus; V. Reinforcing sheath of the plantar aponeurosis; X. Attachment of the same to the side of the os suffraginis.

Fig. II.—VERTICAL MESIAL SECTION OF THE DIGIT.

A. Os pedis; B. Coronary cushion; C. Coffin-joint; D. Navicular bone; E. Os coronæ; F. Pastern-joint; H. Branch of the perforatus at its insertion into the lateral aspect of the os coronæ; I. Insertion of the plantar aponeurosis into the semilunar crest; K. Os suffraginis; L. The perforatus tendon; M. Ligament of yellow fibrous tissue which unites the anterior face of the perforans to the posterior face of the os coronæ, and separates the inferior *cul-de-sac* of the great sesamoid sheath from that of the synovial membrane of the coffin-joint; N. Protrusion of the synovial membrane of the corono-pedal joint between the navicular bone and the os pedis; O. Small sesamoid sheath; P. Synovial membrane of the coffin-joint in contact superiorly with the great sesamoid sheath, from which it is separated by the yellow transverse ligament M.; T. Tendon of the perforans; Y. Fetlock-joint.

Fig. III.—ARTERIES OF THE DIGIT.

A. A. Digital artery; C. Perpendicular artery at its **origin**; H. One of the posterior branches (*rameaux échelonnés*), for the perforans tendon; J. **Another of the same**; K. Origin of the artery of the plantar cushion; M. Origin of anterior branch of **coronary circle**; M.' Posterior branch of the same circle; R. Origin of preplantar artery; S. Plantar artery in the plantar groove and in the os pedis, forming with the opposite artery the semilunar anastomosis; V. V. Descending branches from the semilunar anastomosis.

Fig. IV.—THE HOOF—PLANTAR ASPECT.

P. P. Region of the toe; S. Sole; L. Frog; A. Line indicating the junction of wall and sole; B. Angle of inflexion of the wall, showing the continuity of the wall and the bar; E. Inferior edge of the bar; F. Lateral lacuna of the frog; G. Bulbs of the frog; Q. Median lacuna of the frog; U. Regions of the quarters; O. Regions of the heels.

Fig. V.—EXTREMITY OF THE DIGIT WITH THE HOOF REMOVED—VIEWED FROM THE SIDE.

A. B. Plantar cushion with its villosities; D. Groove between the plantar cushion and the perioplic ring; E. Perioplic ring; F. Inferior border of the plantar cushion; G. Sensitive laminæ, or podophyllous tissue; H. Villosities which terminate the laminæ.

Fig. VI.—ANTERO-POSTERIOR MESIAL SECTION OF THE HOOF—SHOWING ITS INTERIOR.

M. Series of horny laminæ; O. Section of the wall; P. Section of the sole; S. Upper edge of the periople above the cutigeral groove; T. Section of the frog; X. Cutigeral groove.

surface, and it is moulded on the upper face of the horny frog, to which
it has a close resemblance in form. The central portion of the cushion
is therefore sometimes termed the sensitive frog, and it is also known as
the pyramidal body. It shows in front a single ridge, which posteriorly
becomes divided into two by a deep median cleft for the reception of
the frog-stay. This surface has a villous aspect, the papillæ being
imbedded in the foramina seen on the upper surface of the horny frog.
The horny frog is formed by the agency of the cells covering these
papillæ. The *upper face* looks forwards as well as upwards, and is
applied to the reinforcing sheath of the deep flexor tendon. The
borders, which are right and left, bring the plantar cushion into relation
with the inner surface of the lateral cartilages. The *apex* lies in front
of the semilunar crest of the os pedis, with whose periosteum the tissue
of the cushion is intimately blended. The *base* of the cushion consists
of two thick rounded masses termed the *bulbs* of the plantar cushion.
These are continuous in front with the ridges of the pyramidal body,
and they present the same velvety aspect; while, on each side, the
villous tissue joins the extremities of the coronary cushion.

The SENSITIVE SOLE. The student should next examine that part of
the foot which, before separation of the hoof, came into contact with
the upper surface of the horny sole, and which for that reason is termed
the sensitive sole. It is of a roughly crescentic form, being penetrated
by the pyramidal body behind; and it is co-extensive with the plantar
surface of the os pedis. Its connective-tissue basis is firmly adherent to
the periosteum of the bone, while its free surface bears long papillæ
which penetrate the horn tubes of the sole. The horny sole is formed
by the agency of the cells which clothe the papillæ of the sensitive sole.

Directions.—On manipulating the bulbs of the plantar cushion, the
student will feel the lateral cartilages of the foot; and one of these is
to be exposed and defined by removing one half of the plantar cushion.

The LATERAL CARTILAGES. These are in the main composed of
hyaline cartilage, though often erroneously termed the *fibro-cartilages* of
the foot. As is common with fibro-cartilage in many other regions, it
shows a transitional structure at its periphery, where its matrix becomes
more or less fibrous. Each plate of cartilage possesses two faces, and
four borders separated by four angles. The *external face* is convex and
covered by a plexus of veins, some of which penetrate the plate and
connect the plexus with another lying beneath it. The *internal face* is
concave. Behind it is united to the plantar cushion, while anteriorly
it protects the corono-pedal articulation ; and a *cul-de-sac* of the synovial
membrane of the joint lies in direct contact with the cartilage, a fact
which it is important to remember in connection with operations for
"quittor." The *superior border* is thin and flexible, and may be felt in
the living animal. The digital vessels cross this border in passing into

the foot. The *inferior border* is supported by the wing of the os pedis in front, while posteriorly it blends with the plantar cushion. The *anterior border* slopes downwards and backwards, and is blended with the antero-lateral ligament of the corono-pedal joint. The *posterior border* is parallel to the anterior, and is covered by the plantar cushion. The four borders meet at four angles, of which the postero-superior one and the one diago-nally opposite are obtuse, while the other two are acute.

In the disease termed "Side-bones," the lateral cartilages lose their mobility, in consequence of their conversion into bone.

The *Bloodvessels of the Foot* (Plate 10, fig. 3). These should be studied in an injected limb from which the hoof has been removed by the method of decomposition described at page 35. The arteries of the foot are derived from the digital artery, which has already been dissected in its descent towards the foot, where, within the wing of the os pedis, it divides into the plantar and preplantar arteries. Some of the collateral branches of the digital artery have already been described at page 29; but there remain for examination the artery of the plantar cushion and the coronary circle, as well as the plantar and preplantar terminal branches.

The ARTERY of the PLANTAR CUSHION arises from the digital, just as that vessel passes within the upper border of the lateral cartilage, and it passes obliquely downwards and backwards to its destination. Besides supplying the plantar cushion, it gives off a branch which turns forwards to concur in the formation of the circumflex artery of the coronary cushion.

The CORONARY CIRCLE. Where each digital artery lies under cover of the lateral cartilage, it gives off an anterior and a posterior branch which inosculate on the middle line before and behind with the corre-sponding branches of the opposite side, and thus form an arterial circle. This circle closely embraces the os coronæ; and among the largest branches furnished by it, are two which emanate from its anterior half, and descend, one at each border of the extensor tendon, to aid in form-ing the circumflex artery of the coronary cushion.

The CIRCUMFLEX ARTERY of the CORONARY CUSHION (*Chauveau*). This is a slender vascular arch placed immediately above the coronary cushion, to which its branches are distributed. It is fed in front by the two above-mentioned vessels from the coronary circle, and behind, on each side, by the before-mentioned branch from the artery of the plantar cushion.

The PREPLANTAR ARTERY is the smaller of the two terminal branches of the digital. It passes forwards through the notch in the wing of the os pedis, and then along the preplantar groove on the laminal surface of that bone, where its branches are expended in the sensitive laminæ.

The PLANTAR ARTERY passes along the plantar groove to enter the foramen of the same name. Within the os pedis it inosculates with the corresponding vessel of the opposite side, forming the *plantar arch*, or *semilunar anastomosis.* From this intra-osseous arch a great number of branches proceed. An ascending (*anterior laminal*) set of these leave the os pedis by the numerous small foramina which cribble its laminal surface. A descending (*inferior communicating*) set escape from the bone by the series of larger foramina which open on the sharp edge separating its laminal and plantar surfaces. These inferior communicating arteries anastomose right and left with each other, and thus form the *circumflex artery of the toe.* From the concavity of this artery branches pass backwards, and supply the tissue of the sole.

The VEINS of the FOOT.—*Intra-osseous vessels.* Within the os pedis the arterial branches are accompanied by satellite veins. There is thus a semilunar venous anastomosis, to which small veins converge from the laminal surface of the bone. The blood from this sinus is drained away by a larger vessel which passes out by the plantar foramen in company with the plantar artery, and joins the posterior part of the coronary plexus. *Extra-osseous vessels.* The foot is richly provided with a superficial system of vessels, which are arranged in the form of a close-meshed network having little or no communication with the deep set. This venous envelope of the foot is divided into a *solar*, a *laminal* (podophyllous), and a *coronary* plexus. Where the solar and laminal plexuses meet, a composite venous vessel runs in company with the circumflex artery of the toe. These two plexuses communicate freely with each other, and with the coronary plexus. This last consists of a central part, which underlies the coronary cushion, and of two lateral parts, which on each side ramify on both surfaces of the lateral cartilage. By the convergence of branches belonging to this cartilaginous division of the coronary plexus, the digital veins are formed; and these drain away the blood from both the intra-osseous and extra-osseous systems of vessels.

Directions.—The terminal portion of the deep flexor tendon, and the synovial apparatus developed in connection with it, should now be examined.

The DEEP FLEXOR tendon (Plates 10 and 11), when it reaches the upper border of the navicular bone, widens out to form what is called the *plantar aponeurosis.* This plantar aponeurosis plays over the navicular bone by means of the navicular sheath, and is covered posteriorly by a fibrous layer which ultimately blends with it. It becomes *inserted* into the semilunar crest of the os pedis, and into the bone behind that crest. The above-mentioned fibrous layer was first described by Bouley, and designated by him the *reinforcing sheath of the perforans.* This expansion is attached on each side by a slip to the lower half of the first phalanx, and it serves to maintain the plantar aponeurosis against the navicular bone.

The NAVICULAR or SMALL SESAMOID SHEATH (Plate 10, fig. 2). This is a synovial apparatus developed in connection with the perforans tendon where it plays over the navicular bone. It lines the deep face of the tendon, and is reflected on to the navicular bone and interosseous ligament. It also extends above the navicular bone, where it is in contact with the synovial membrane of the coffin-joint and that of the metacarpophalangeal sheath.

THE SHOULDER-JOINT.

This joint is formed between the glenoid fossa of the scapula and the head of the humerus. It is enclosed by a single capsular ligament lined internally by the synovial membrane. The absence of lateral or other retaining ligaments in connection with the joint, is compensated for by the numerous tendons which pass from one bone to the other in close relation to the capsular ligament. These muscles are as follows :— the supraspinatus, infraspinatus, teres minor, biceps, and small scapulo-humeralis. The last passes over the joint behind, where some of its fibres seem to be inserted into the ligament. In front of the joint the tendon of the biceps is separated from the ligament by a pad of fat.

MOVEMENTS.—The joint belongs to the class of enarthrodial or ball-and-socket joints, and the amount of its mobility should be proved by manipulation before the removal of the muscles. If the scapula be kept fixed, it will be found that the humerus can be carried backwards so as to diminish the angle formed by the meeting of the bones. This is a movement of *flexion*. Or the humerus can be carried forward in the same plane as in the preceding movement, but increasing the angle. This is *extension*. Or again, the humerus may be moved in a lateral direction either outwards or inwards. When, in the living animal, it is carried inwards, the limb is thrown towards the middle plane of the body, and is said to be *adducted*. The opposite movement, by which the limb is carried outwards from the middle plane, is termed *abduction*. Two other movements are permitted in the joint, viz., rotation and circumduction. In *rotation* the humerus, without change of place as a whole, turns round its own axis. In *circumduction* the shaft of the humerus moves so as to describe the surface of a cone.

(These different terms having been here defined at length, their application in the case of the other joints of the body will be readily understood).

The shoulder-joint of the horse is thus possessed of considerable freedom of movement; but still, the range of its mobility, owing to the absence of a clavicle, and to the different disposition of the pectoral muscles, is much more restricted than in the human arm.

Directions.—The muscles which surround the joint must now be removed, care being taken not to cut the capsular ligament.

The CAPSULAR LIGAMENT loosely surrounds the articular ends of the bones, and may be conceived as having the form of a double-mouthed sack, one mouth being attached around the rim of the glenoid cavity, and the other at the periphery of the head of the humerus. The wall of this sack is comparatively thin, but in front it is strengthened by accessory fibres that pass in a divergent manner from the coracoid process to the outer and inner tuberosities. These correspond to the coraco-humeral ligament of man.

Directions.—If, in the removal of the muscles, the ligament has been preserved perfectly intact, it will be noticed that though a considerable force be exerted to pull the articular surfaces from each other, they still remain in contact. If, however, an incision be made in the ligament, the air will be heard to rush into the joint, while the bones separate to the extent of half an inch or more. In the shoulder then, as in other joints, atmospheric pressure is to be included among the agents keeping the articular surfaces in contact. The capsular ligament is to be slit up so as to expose the smooth and glistening aspect of the synovial membrane, and the articular surfaces of the bones covered by articular cartilage.

The SYNOVIAL MEMBRANE lines the inner surface of the capsular ligament. It secretes the synovia, or joint oil, some of which will be seen escaping from the joint.

THE ELBOW-JOINT (PLATE 11, fig. 1).

This joint is formed by the lower extremity of the humerus and the upper extremities of the bones of the fore-arm. It possesses two lateral ligaments, and an anterior ligament which supports the synovial membrane in front; but behind, there being no ligament, the synovial sac is directly supported by muscles.

MOVEMENTS.—This is a ginglymoid joint, the only movements being *flexion* and *extension*. In *flexion*, while the humerus remains fixed, the bones of the fore-arm are carried forwards until the movement is arrested by the coronoid process passing into the fossa of the same name. In this movement the bones of the fore-arm do not move in the plane in which the humerus lies, but deviate a little outwards. The opposite movement is *extension*, in which the radius and ulna are carried backwards until they are arrested by the tension of the lateral ligaments, and by the passage of the beak of the olecranon into the fossa of the same name.

Directions.—The anterior and lateral ligaments are to be exposed and defined by removing the muscles from the front of the joint, but on the posterior aspect of the joint the muscles should not be removed at present.

The EXTERNAL LATERAL LIGAMENT is a cord-like band which is fixed superiorly to a depression on the outer side of the lower extremity of the humerus, and to the ridge which forms the lower boundary of the musculo-spiral groove; while inferiorly it passes to be inserted into the external tuberosity at the upper end of the radius.

The INTERNAL LATERAL LIGAMENT forms a longer but more slender cord than the preceding, and passes from a small eminence on the outer side of the lower extremity of the humerus to be inserted into the shaft of the radius below the bicipital tuberosity. Some of the anterior fibres join the tendon of the biceps or the anterior ligament, while some of the posterior join the arciform fibres connecting the radius and ulna.

The ANTERIOR LIGAMENT is of a membranous form. Its upper border is fixed to the humerus, its lower border to the radius, while its lateral borders blend with the lateral ligaments.

Directions.—The anterior and lateral ligaments should now be cut transversely about their middle in order to expose the interior of the joint.

The SYNOVIAL MEMBRANE will be seen to line the inner face of the anterior and lateral ligaments, but at the back part of the joint there is no ligament and the membrane is supported by the muscles. If the finger be passed backwards and upwards, it will enter a process of the synovial capsule which extends upwards into the olecranon fossa, where a pad of fat intervenes between it and the anconeus muscle. Just behind the external lateral ligament the membrane lines the origin of the flexor metacarpi externus. On the inner side of the joint, behind the internal lateral ligament, the membrane lines the tendons of origin of the middle and internal flexors of the metacarpus, and of the superficial and deep flexors of the digit. This disposition of the synovial capsule will be rendered more evident by cutting the above-mentioned muscles a few inches below the joint, and then turning their tendons of origin upwards.

Directions.—The humerus being now completely severed from the radius and ulna, the mode of union of these latter bones should be examined.

The RADIO-ULNAR ARTICULATION.—In the adult animal the bones of the fore-arm are fused together below the radio-ulnar arch, by ossification of the interosseous fibres which in the young animal are interposed between the two bones. Above the arch, however, the fibres interposed between the bones do not ossify except in a very old animal, but persist as an *interosseous ligament.* The union of the two bones is further maintained by *arciform fibres* passing on each side from the one bone to the other, and blending with the lateral ligaments of the elbow. At the upper part of their opposed surfaces, the two bones respond to each other by two small synovial facets, which, however, have no special synovial membrane, but are lubricated by processes from the synovial capsule of the elbow-joint.

Movements.—These are inappreciable, the limb of the horse being fixed in a condition of pronation.

46   DISSECTION OF THE ANTERIOR LIMB.

THE KNEE, OR CARPUS (PLATE 11, figs. 2 and 3).

This is not a simple, but a composite, joint, and entering into its formation there are the carpal bones, the lower extremity of the radius, and the upper extremities of the bones of the metacarpus. The carpal bones are arranged in two rows, or tiers, and the bones of each row are firmly bound together and converted into a single piece by ligaments passing between the adjacent bones. A transverse joint is then formed between the upper and the lower tier. This may be called the *inter-carpal* joint, and it is secured by special ligaments passing between the two rows. Another transverse joint is formed between the lower row and the heads of the metacarpal bones; and this, which has also got special ligaments, is termed the *carpo-metacarpal* articulation. A third transverse joint is formed between the lower end of the radius and the upper row. This, which is the *radio-carpal* joint, is also provided with special ligaments. Lastly, there are four ligaments which do not belong specially to any of these articulations, but secure the stability of the entire composite joint, and are therefore termed *common*.

MOVEMENTS.—The movements which take place at the carpus are *flexion* and *extension*, and each of the transverse joints above-mentioned is a ginglymus. When these movements are executed, however, the three joints do not participate in them in an equal degree. The largest share of the movement occurs at the radio-carpal articulation, and the smallest between the carpus and the metacarpus; while, as regards the amount of movement, the inter-carpal transverse joint occupies an intermediate position. When the limb is *flexed* at the carpus, it will be noticed that the metacarpus and digit deviate a little outwards from the plane of the fore-arm. When the limb is fully *extended* the lateral ligaments are tightly stretched, and resist any attempts to produce *abduction* or *adduction ;* but these movements can be produced when the limb is fully *flexed*, in which position the lateral ligaments are relaxed. Lateral movement, however, is not executed at this joint in any appreciable degree in the living animal. The *gliding* movement permitted between adjacent bones in each row is of importance, as tending to distribute pressure, and obviate the bad effects which would have been likely to result from concussion had each row been a single rigid mass.

Directions.—The tendons which pass in relation to the joint before and behind should be removed, and the ligaments should be studied in the order of the following description.

There are four ligaments common to the whole joint, viz., two lateral, an anterior, and a posterior.

The EXTERNAL LATERAL LIGAMENT is a cord-like band composed of a deep and a superficial set of fibres, which slightly cross each other. It is fixed superiorly to the external tuberosity at the lower end of the

PLATE XI

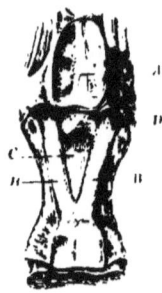

Fig. VII

Fig. VI

Fig. V

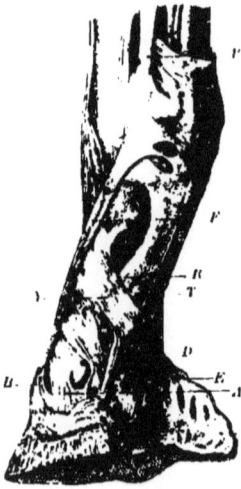

Fig. IV

Fig. I

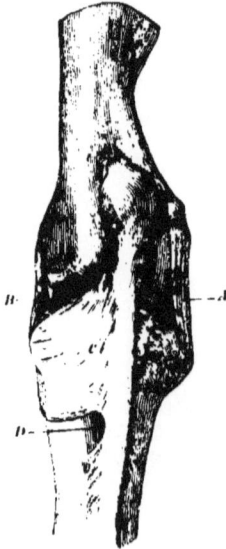

Fig. III

Fig. II

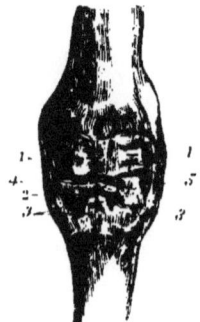

JOINTS AND LIGAMENTS OF FORE LIMB

PLATE XI.

Fig. I.—LIGAMENTS OF THE ELBOW, SEEN FROM BEHIND (*Leyh*).

A. Ext. lateral ligament; B. Int. lateral ligament; C. C. C. Arciform ligaments; D. Radlo-ulnar arch.

Fig. II.—LIGAMENTS OF THE CARPUS, FRONT VIEW (*Chauveau*).

1. 1. Ant. ligaments of upper row; 2. An ant. ligament of lower row; 3. 3. Ant. carpo-metacarpal ligaments; 4. Int. lateral ligament; 5. Ext. lateral ligament.

Fig. III.—LIGAMENTS OF THE CARPUS, VIEWED FROM THE OUTER SIDE (*Chauveau*).

1. 1. 1. Ant. ligaments of upper row; 2. An ant. ligament of the lower row; 3. 3. Ant. carpo-metacarpal ligaments; 4. An intercarpal ligament; 5. Ext. lateral ligament; 6. A radio-carpal ligament.

Fig. IV.—LIGAMENTS OF THE FETLOCK, PASTERN, AND COFFIN-JOINTS; SIDE VIEW (*Chauveau*).

1. Superficial fasciculus of the ext. lateral ligament of the fetlock; 2. 3. Sesamoid and phalangeal slips of the deep fasciculus of the same ligament; 4. 5. 6. Upper, middle, and lower fibrous slips attaching the glenoidal fibro-cartilage to the os suffraginis; 7. Lateral ligament of the pastern-joint; 8. Antero-lateral ligament of the coffin-joint; 9. Postero-lateral ligament of the same joint.

Fig. V.—BACK OF THE DIGIT DISSECTED TO SHOW THE TENDONS AND LIGAMENTS (*Bouley*).

A. Antero-lateral ligament of the coffin-joint; B. Insertion of extensor pedis tendon; D. Postero-lateral ligament of the coffin-joint; E. Divergent fibres of the same ligament passing to be attached to the wing of the os pedis and inner surface of the lateral cartilage; F. Slip sent from suspensory ligament to extensor tendon; P. Branch of bifurcation of the suspensory ligament; R. Branch of perforatus; T. Perforans emerging from between the branches of the perforatus; Y. Attachment of the reinforcing sheath of the perforans tendon to the side of the os suffraginis.

Fig. VI.—BACK OF THE DIGIT DISSECTED TO SHOW THE TENDONS AND LIGAMENTS (*Bouley*).

A. Superficial inferior sesamoidean ligament; B. Highest slip attaching the glenoidal fibro-cartilage of the pastern-joint to the first phalanax; O. Branch of perforatus; P. Middle inferior sesamoidean ligament; S. Insertion of plantar aponeurosis into semilunar crest; T. Reinforcing sheath of the plantar aponeurosis; X. Perforans tendon.

Fig. VII.—BACK OF THE FETLOCK-JOINT (Modified from *Bouley*).

A. Intersesamoid ligament; B. B. Lateral bands of the middle inferior sesamoidean ligament; C. Middle band of the same ligament, its upper attachment cut away to show D. the deep inferior sesamoidean ligament.

radius; and passing over the outside of the carpus, it furnishes slips to
the cuneiform and unciform bones, and terminates on the head of the
external small metacarpal bone. The ligament is perforated by a thecal
canal in which the tendon of the extensor suffraginis plays.

The INTERNAL LATERAL LIGAMENT is fixed superiorly to the internal
tuberosity of the radius, and inferiorly to the heads of the large and
inner small metacarpal bones, furnishing slips, as it passes over the
carpus, to the scaphoid, magnum, and trapezoid bones.

The ANTERIOR COMMON LIGAMENT has a flattened, four-sided form. It
is fixed superiorly to the radius, and inferiorly to the large metacarpal
bone, while its lateral borders are united to the lateral ligaments. Its
deep face is partly adherent to the carpal bones or their anterior
ligaments, and partly it is lined by synovial membrane. The tendons
of the extensor pedis and the extensors of the metacarpus play over its
superficial face, where they are provided with synovial bursæ. The
ligament is somewhat loose when the joint is extended, and is put on
the stretch during flexion.

The POSTERIOR COMMON LIGAMENT is a much stronger ligament than
the preceding. It is fixed above to the radius, and below to the large
metacarpal bone. Its internal border mixes its fibres with the internal
lateral ligament, while its outer border is blended in the same way with
the most external of the intercarpal ligaments. Its anterior or deep
face is very intimately united to the carpal bones, and its posterior face
is smooth and lined by the synovial membrane of the carpal sheath.
The *subcarpal ligament,* or fibrous band which reinforces the perforans
tendon below the carpus, takes origin from the posterior common ligament, or may be described as the downward continuation of that
ligament.

Directions.—The anterior and lateral ligaments just described are to
be carefully dissected away, and in removing the first of these, care is
to be taken of the anterior bands connecting the bones in each row.

RADIO-CARPAL LIGAMENTS.—There are three of these. The strongest
of them is a thick cord that stretches obliquely downwards and inwards
behind the carpus, and connects the radius and scaphoid. It will be
seen, without removing the posterior common ligament, which covers it,
by strongly flexing the joint and looking into it from the front. The
second is a very slender ligament which is fixed to the radius beneath
the preceding, and passes downwards to be attached to the pisiform and
the interosseous ligament uniting the cuneiform and semilunar bones.
The third is situated at the outside of the carpus, where it connects
the radius and the upper border of the pisiform bone, and is partly
covered by the lateral ligament.

The INTER-CARPAL LIGAMENTS are also three in number. Two of
them are situated behind the joint, under cover of the posterior common

ligament, and will be seen without further dissection on flexing the joint and looking into it from the front. One of these connects the scaphoid to the magnum and trapezoid, the other joins the cuneiform and magnum. The third is a strong ligament situated at the outer side of the joint, where it is blended with the lateral ligament in front, and with the posterior common ligament behind. Its fibres are fixed superiorly to the pisiform bone, and inferiorly to the unciform and head of the external small metacarpal bone.

The CARPO-METACARPAL LIGAMENTS are four in number—two anterior and two interosseous. One of the anterior ligaments is composed of two separate slips which connect the os magnum and large metacarpal bone. The other passes from the unciform to the head of the external small metacarpal bone, under cover of the lateral ligament. The two interosseous pass, one on each side, from the point of articulation of the large and small metacarpal bones, to join the interosseous ligaments connecting the bones of the lower row.

Directions.—Attention may at this stage be given to the disposition of the synovial membranes of the carpus, which are three in number.

SYNOVIAL MEMBRANES.—1. The *radio-carpal* synovial membrane not only facilitates the movements between the radius and the bones of the upper row, but also descends between the latter bones as far as their interosseous ligaments. 2. The *inter-carpal* synovial membrane, in the same way, belongs to the intercarpal transverse joint; but it is also insinuated above, between the bones of the upper row as far as their interosseous ligaments, and descends in the same way below, between the adjacent bones of the lower row. It communicates with the next. 3. The *carpo-metacarpal* synovial membrane facilitates the movements between the lower row and the heads of the metacarpal bones, ascends between the adjacent bones of the lower row as far as their interosseous ligaments, and dips down to supply the articulations between the large and small metacarpals.

Directions.—The radio-carpal, inter-carpal, and posterior common ligaments should now be cut transversely. The upper row will thus be isolated as a single piece for the examination of its special ligaments.

The LIGAMENTS of the UPPER Row are three anterior, and three interosseous; and they are extremely simple. The anterior ligaments are flattened bands connecting the adjacent bones in front, while the interosseous bands are very short and connect the contiguous surfaces of the bones.

The LIGAMENTS of the LOWER Row are two anterior, and two interosseous; and they are disposed like those of the upper row. In examining these, the lower tier of bones must not be separated from the metacarpus, as that would involve the destruction, in part, of the suspensory ligament of the fetlock.

E

50THE ANATOMY OF THE HORSE.

The INTER-METACARPAL ARTICULATIONS. The head of the large metacarpal bone responds to one of the small metacarpals on each side by a
small synovial joint lubricated by a process from the carpo-metacarpal
synovial membrane. Below that point the union of the bones is maintained by short *interosseous fibres*, which, in adult animals, are very
frequently ossified. The lower extremities of the splint bones, however,
for a short distance above the little knob that terminates them, remain
freely movable, as may be felt by manipulation in the living animal.
In addition to the interosseous fibres, the ligaments of the carpus which
get inserted in common into the heads of both large and small metacarpal
bones, contribute to the union of these bones.

THE FETLOCK-JOINT (PLATE 11, FIGS. 4-7).

This, which is technically termed the *metacarpo-phalangeal articulation*, is a ginglymoid joint; and its articular surfaces are furnished by
the lower extremity of the large metacarpal bone, the upper extremity
of the first phalanx, and the two sesamoid bones. It corresponds to the
joint at the knuckles in the human hand.

MOVEMENTS.—*Flexion* and *extension* are, in the natural state, the only
movements executed at the joint; but by manipulation, slight lateral
movements may be produced when the joint is fully flexed. In complete *extension* the digit is carried beyond the point at which it lies in a
straight line with the metacarpus (*over-extension*), until the movement is
arrested by tension of the suspensory ligament.

Directions.—The tendons which pass in relation to the joint before
and behind having been carefully removed, the ligaments should be
dissected and studied in the order of their description.

The SUPERIOR SESAMOIDEAN or SUSPENSORY LIGAMENT.—The main portion of this ligament is lodged in the channel formed by the three metacarpal bones, where it is related by its posterior face to the perforans
tendon and its reinforcing band (subcarpal ligament). It has a double
origin behind the carpus, viz., (1) by a superficial layer from the lower
row of carpal bones, and (2) by a deeper layer from the upper end of
the large metacarpal bone. (In the hind limb it has a similar origin from
the tarsus and metatarsus). These two portions blend, and descend behind the metacarpus as a flattened band which bifurcates a few inches
above the sesamoid bones. Each branch passes to the sesamoid bone of
its own side, where a considerable proportion of its fibres become inserted;
while the rest is continued in the form of a band which crosses obliquely
downwards and forwards over the side of the fetlock to join the extensor
tendon on the front of the digit, and be continued with it to the os pedis.
The ligament is composed of white fibrous tissue with a constant admixture of striped muscular tissue. The presence of muscular tissue here,
points to the conclusion (strengthened by other considerations) that the

suspensory ligament is a muscle which, in the evolution of the horse, has undergone retrogressive changes, and lost its original function.*

The INFERIOR SESAMOIDEAN LIGAMENTS. These are three in number, and may be distinguished as superficial, middle, and deep. The *superficial* ligament is fixed below to the glenoidal fibro-cartilage developed behind the superior articular surface of the second phalanx. It ascends as a flattened band behind the os suffraginis, where it is placed between the middle ligament and the tendon of the deep flexor; and, widening a little, it is inserted into the base of the sesamoids and the intersesamoid ligament. By cutting the ligament about its middle, and reflecting it upwards and downwards, the middle ligament will be brought into view. The *middle* ligament consists of a median and two lateral bands. Each is fixed to the back of the os suffraginis, and ascends to be inserted into the base of the sesamoids. This should be cut and reflected like the preceding ligament, in order to expose the next. The *deep* ligament consists of a few short fibres disposed like the letter X, and fixed, on the one hand, to the upper part of the posterior surface of the os suffraginis, and, on the other, into the base of the sesamoid bones. This ligament supports the synovial membrane of the joint.

The LATERAL LIGAMENTS of the fetlock-joint. Each comprises (1) a superficial fasciculus connecting the lower extremity of the large metacarpal bone to the upper extremity of the first phalanx; and (2) a deep fasciculus attached, on the one hand, to the large metacarpal beneath the preceding, and, on the other, to the sesamoid and upper extremity of the first phalanx.

The ANTERIOR LIGAMENT has a membranous, four-sided form. It covers the joint in front, and supports the synovial membrane by its deep face; while the extensor pedis tendon passes over its superficial aspect, a synovial bursa being interposed. It is fixed above to the large metacarpal, below to the first phalanx, and on each side to the lateral ligament.

Directions.—On one side of the joint the lateral ligament and the slip sent from the suspensory ligament to the extensor tendon must be removed to expose the next ligament.

The LATERAL SESAMOIDEAN LIGAMENTS. These are not to be confounded with the lateral ligaments of the joint, by which they are partly covered. Each fixes the sesamoid of its own side to the upper extremity of the first phalanx.

The INTERSESAMOID LIGAMENT is the name given to the fibro-cartilaginous tissue which unites the two sesamoids, and with them forms a pulley-like surface for the passage of the deep flexor tendon.

The SYNOVIAL MEMBRANE is supported in front by the anterior

* According to Professor D. J. Cunningham (*Reports of the Challenger Expedition, Vol. V.*), the ligament is the altered flexor brevis of the middle digit, the corresponding muscle in the human subject being the 1st plantar interosseous muscle.

ligament, and on each side by the lateral ligament. Behind the joint it is supported below the sesamoids by the deep inferior sesamoidean ligament, but above these bones it is unsupported; and when the synovial sac is distended, it bulges upwards between the branches of the suspensory ligament (Fig. 1, page 33).

THE PASTERN-JOINT (PLATE 11).

This joint, which is technically termed the *first interphalangeal articulation*, is formed between the distal end of the os suffraginis and the proximal end of the os coronæ. It is a ginglymus, or hinge joint, and corresponds to the second joint of the human finger.

MOVEMENTS.—As with the joint last described, the only natural movements are *flexion* and *extension*.

Directions.—The tendon of the extensor pedis, which passes over the front of the joint, should be cut and reflected downwards. This will show that the tendon completes the joint in front, where it plays the part of an anterior ligament, and supports the synovial membrane. The lateral ligaments are next to be defined, and after these, the supplementary cartilaginous apparatus placed behind the joint.

The LATERAL LIGAMENTS. Each of these stretches from the lower extremity of the first phalanx on its lateral aspect, to be inserted into the side of the os coronæ, and beyond that point some of its fibres are continued downwards and backwards as the postero-lateral ligament of the second interphalangeal joint.   .

The GLENOIDAL FIBRO-CARTILAGE. This is a piece of fibro-cartilage fixed at the posterior edge of the upper articular surface of the os coronæ. It serves to increase that surface, and its anterior face is moulded on the lower articular surface of the first phalanx, while its posterior face is smooth for the passage of the perforans tendon. Three fibrous slips pass from it on each side, and are attached to the first phalanx. The superficial inferior sesamoidean ligament is inserted into it, and the terminal insertion of the perforatus tendon is blended with it on each side.

SYNOVIAL MEMBRANE. This is supported in front by the extensor tendon, and on each side by the lateral ligament. Posteriorly it lines the glenoidal fibro-cartilage, and is prolonged upwards as a pouch behind the lower extremity of the first phalanx (Plate 10, fig. 2).

THE COFFIN-JOINT (PLATE 11).

This, the *second interphalangeal joint*, has three bones entering into its formation, viz., the os coronæ, the os pedis, and the navicular bone. It is a ginglymus, and corresponds to the first joint of the human finger.

MOVEMENTS.—*Flexion and extension.*

It possesses an interosseous ligament, and two pairs of lateral ligaments. The INTEROSSEOUS LIGAMENT is composed of short fibres passing from the inferior border of the navicular bone to the os pedis behind its articular surface.

The ANTERO-LATERAL LIGAMENTS. Each of these passes from the side of the os coronæ to be inserted into the excavation at the side of the pyramidal process of the os pedis.

The POSTERO-LATERAL LIGAMENTS. These seem to be the downward continuations of the lateral ligaments of the pastern-joint. Passing from the side of the os coronæ, each is inserted into the upper border of the navicular bone, and sends slips to the wing of the os pedis and inner surface of the lateral cartilage.

SYNOVIAL MEMBRANE. This is supported in front by the extensor tendon, and laterally by the lateral ligaments. A protrusion of it passes on each side between the antero-lateral and postero-lateral ligaments, and lies in relation to the deep face of the lateral cartilage. A third protrusion passes upwards posteriorly, between the navicular bone and the back of the os coronæ (Plate 10, fig. 2).

TABULAR VIEW OF THE MUSCLES OF THE FORE LIMB IN THEIR ACTION ON THE DIFFERENT JOINTS.

SHOULDER.

Flexors	Deltoid. Coraco-humeralis. Latissimus dorsi. Teres major. Teres minor. Scapulo-humeralis gracilis (?) Large head of triceps.	Adductors	Superficial pectoral. Subscapularis. Coraco-humeralis.
Extensors	Supraspinatus. Mastoido-humeralis.	Rotators outwards	Deltoid Teres minor Infraspinatus
Abductors	Deltoid. Teres minor. Infraspinatus.	Rotators inwards	Mastoido-humeralis. Latissimus dorsi. Teres major.

ELBOW.

Flexors	Flexor brachi. Brachialis anticus.	Extensors	Triceps extensor cubiti. Anconeus. Scapulo-ulnaris.

CARPUS.

Flexors	Flexor metacarpi externus. Flexor metacarpi medius. Flexor metacarpi internus. Flexor perforans. Flexor perforatus.	Extensors	Extensor metacarpi magnus. Extensor metacarpi obliquus. Extensor pedis. Extensor suffraginis.

FETLOCK.

Flexors	Flexor perforans. Flexor perforatus.	Extensors	Extensor pedis. Extensor suffraginis.

PASTERN.

Flexors	Flexor perforans. Flexor perforatus.	Extensor —Extensor pedis.

COFFIN-JOINT.

Flexor—Flexor perforans. | Extensor—Extensor pedis.

Name of Muscle.	Origin.	Insertion.	Source of Nerve.
Anterior superficial pectoral	Sternum, first 2 or 3 inches of inferior border	Humerus, outer lip of musculo-spiral groove	From brachial plexus.
Posterior superficial pectoral	Sternum, inferior border; and median fibrous cord	Humerus, (with the preceding); and superficial fascia of fore-arm	From brachial plexus.
Anterior deep pectoral	Costal cartilages, 1st four; and sternum, lateral surface	Fascia covering supraspinatus	From brachial plexus.
Posterior deep pectoral	Abdominal tunic; side of sternum; and tips of costal cartilages, 5th, 6th, 7th, and 8th.	Humerus, inner tuberosity; biceps tendon and its retaining fascia	From brachial plexus (2 branches).
Serratus magnus	Ribs, 1st eight or nine	Scapula, two triangular areas on ventral surface	From brachial plexus.
Levator anguli scapulæ	Cervical vertebræ, last four, transverse processes	Scapula, triangular area on ventral surface, at cervical angle	From 6th and 7th cervical nerves.
Trapezius (cervical)	Ligamentum nuchæ, funicular portion	Scapula, tubercle of spine; and aponeurosis over outer scapular muscles	Spinal accessory.
Trapezius (dorsal)	Anterior dorsal spines (or supraspinous ligament)	Scapula, tubercle of spine	Spinal accessory.
Rhomboideus (cervical)	Ligamentum nuchæ, funicular portion	Scapular cartilage of prolongation, inner surface	6th cervical.
Rhomboideus (dorsal)	Anterior dorsal spines (or supraspinous ligament)	Scapular cartilage (behind the preceding)	6th cervical.
Latissimus dorsi	Vertebral spines, 4th dorsal to last lumbar (or supraspinous ligament)	Humerus, inner tubercle	From brachial plexus.
Mastoido-humeralis	Mastoid process and crest; and cervical vertebræ, 1st four, transverse processes	Humerus, outer lip of musculo-spiral groove	Cervical nerves and circumflex.
Teres major	Scapula, dorsal angle; and intermuscular septum (between it and subscapularis)	Humerus, inner tubercle	Brachial plexus.
Subscapularis	Scapula, fossa of same name	Humerus, inner tuberosity	Brachial plexus.

Muscle	Origin	Insertion	Nerve
Scapulo-ulnaris	Scapula, posterior border	Ulna, olecranon process; and fascia of fore-arm	Musculo-spiral.
Triceps extensor cubiti. { caput mag. caput parv. caput med.	Scapula, dorsal angle and posterior border / Humerus, shaft / Humerus, shaft	Ulna, olecranon process .	Musculo-spiral.
Anconeus	Humerus, margin of olecranon fossa	.	.
Scapulo-humeralis gracilis	Scapula, above and behind rim of glenoid cavity	Humerus, shaft	Circumflex.
Coraco-humeralis	Scapula, coracoid process	Humerus, shaft (two insertions)	Median.
Biceps	Scapula, coracoid process	Radius, bicipital tuberosity; and fascia of fore-arm	Median (or its anterior root).
Deltoil	Scapula, dorsal angle; and scapular fascia	Humerus, deltoid (outer) tubercle .	Circumflex.
Teres minor	Scapula, posterior border, lower part of infra-spinous fossa, and tubercle on glenoid rim	Humerus, ridge between outer tubercle and tuberosity .	Circumflex.
Infraspinatus	Scapula, infraspinous fossa; and scapular fascia	Humerus, outer tuberosity and ridge below it (two insertions)	Suprascapular.
Supraspinatus	Scapula, supraspinous fossa; and scapular fascia	Humerus, outer and inner tuberosities (two insertions)	Suprascapular.
Brachialis anticus	Humerus, shaft	Radius and ulna, inner side	{ Musculo-cutaneous (of median). Median.
Flexor metacarpi internus	Humerus, inner condyle	Inner splint bone, head	Ulnar.
Flexor metacarpi medius	Humerus, inner condyle; and ulna, olecranon process (two heads)	Pisiform, upper edge	
Flexor metacarpi externus	Humerus, outer condyloid ridge	Pisiform, upper edge; and inner splint bone, head (two tendons)	Musculo-spiral.
Flexor perforatus	Humerus, inner condyloid ridge	head (two tendons)	Ulnar.
Flexor perforans { ulnar head humeral head radial head	Ulna, olecranon process / Humerus, inner condyloid ridge / Radius shaft	Os corone (bifid tendon)	Ulnar. Median. Median.
		Os pedis, semilunar crest and surface behind it	
Extensor metacarpi magnus	Humerus, outer condyloid ridge, and depression external to coronoid fossa	Large metacarpal bone, upper extremity	Musculo-spiral.
Extensor metacarpi obliquus	Radius, shaft	Inner splint bone, head .	Musculo-spiral.
Extensor pedis	Humerus, depression external to coronoil fossa; external lateral ligament of elbow; and radius, upper extremity	Os pedis, pyramidal process; and to join tendon of extensor suffraginis (two tendons)	Musculo-spiral.
Extensor suffraginis	Lateral ligament of elbow; radius, upper extremity; and line of junction of radius and ulna	Os suffraginis, upper extremity	Musculo-spiral.
Lumbricales (2)	Perforans tendon	Tissue beneath ergot of fetlock	Plantar.
Interossei (2)	Splint bone, head	Suspensory ligament, band sent to extensor pedis tendon	Plantar.

CHAPTER II.

DISSECTION OF THE POSTERIOR LIMB.

In the male subject, the dissection of the perinæum must be completed before the dissector of the hind limb can begin his operations.

THE INNER ASPECT OF THE THIGH.

Position.—The animal should be placed on the middle line of its back, and its hind limbs should be drawn forcibly upwards and outwards by ropes running over pulleys fixed to the ceiling. This is the position most convenient for allowing the dissection of both hind limbs to be pursued at the same time. If only one limb is being dissected, the rope may be unfastened from that limb, and the body allowed to incline to the same side, as in Plate 12.

Surface-marking.—The internal saphena vein ascends on the inner aspect of the thigh; and a few inches below the upper limit of the region, it dips in between the sartorius and gracilis muscles. Pressure at this point in the living animal will produce distension of the vessel, and render its course much more evident. Venesection is sometimes performed on this vessel. Above the point where the before-mentioned vessel disappears from view, the deep inguinal lymphatic glands are situated in the interstice between the sartorius and gracilis muscles. They here cover the femoral artery, and may be very distinctly felt in a case of *lymphangitis*, or "weed."

Directions.—An incision through the skin is to be carried down the middle line of the thigh, and terminated a few inches below the level of the stifle-joint. Here another incision is to be made across the inner aspect of the limb, from its anterior to its posterior border. These incisions, together with those already made in the dissection of the perinæum, will enable the dissector to reflect the skin as an anterior and a posterior flap. The student should then dissect the internal saphena vein with its accompanying artery and nerve, and the cutaneous nerves at the forepart of the region, which are derived from the lumbar nerves. Thereafter the surface of the sartorius and gracilis is to be cleaned, and these muscles are to be examined.

The INTERNAL SAPHENA VEIN (Plate 12). This is a large vessel formed on the inner side of the leg by the junction of an anterior and a posterior root, these being the upward continuations of the inner and outer metatarsal veins. In the thigh it inclines upwards and forwards on the surface of the gracilis, until it disappears between that muscle and the sartorius, to empty itself into the femoral vein.

The SAPHENA ARTERY (Plate 12). This artery lies in front of the vein. It is a long and slender vessel given off by the femoral artery about the middle of the femur. It comes out between the sartorius and gracilis, or it may pierce the edge of one of these muscles. It then descends in front of the saphena vein, and finally divides into two branches, which accompany the roots of that vessel.

The INTERNAL SAPHENOUS NERVE (Plate 12) is a branch of the anterior crural, from which it is given off a little above the brim of the pelvis. At the crural arch (Poupart's ligament) it descends in front of the femoral artery, to which and the sartorius muscle it supplies branches. It then divides into two cutaneous branches, which emerge from between the sartorius and the gracilis, in company with the saphenous artery and vein. The anterior half of the nerve gives off branches for the supply of the thigh in front of the vein, and is continued downwards over the forepart of the inner side of the leg, as far as the hock. The posterior half sends branches backwards for the supply of the posterior part of the thigh, and it then descends behind the anterior half.

CUTANEOUS BRANCHES from the lumbar nerves. These will be found at the forepart of the thigh, the largest (from the 3rd lumbar) being accompanied by the posterior division of the circumflex iliac artery.

The PRECRURAL LYMPHATIC GLANDS. These are superficially placed at the inner side of the front of the thigh, on the track of the above-mentioned branch of the circumflex iliac artery.

FASCIA. At the forepart of the region now exposed, the muscles are overspread by a strong membranous fascia, which is attached superiorly to the tendon of the external oblique muscle of the abdomen, at the line where it is reflected to form *Poupart's ligament*. Round the anterior border of the thigh this fascia is continuous with the strong *fascia lata;* but when traced backwards, it becomes less fibrous, and over the posterior part of the region it is thin and areolar. When it has been examined, the fascia is to be cleaned away from the subjacent muscles.

The DEEP INGUINAL LYMPHATIC GLANDS (Plate 45) are ten or twelve in number, and form a chain connected by areolar tissue, and situated in the upper part of the interstice between the gracilis and sartorius muscles, and over the femoral vessels.

The SARTORIUS (Plate 12). This is a somewhat slender muscle which at present can be dissected only in a part of its course. It is seen descending beneath Poupart's ligament, from its point of origin within the abdominal cavity. It there takes *origin* from the iliac fascia (Plate 45).

In the thigh it lies in front of the gracilis. About the middle of their line of apposition the saphena vessels and nerves emerge, but below that point the muscles are adherent to each other. It is *inserted* into the internal straight ligament of the patella.

Action.—To adduct and flex the hip-joint. To a slight extent it may also rotate the limb inwards at the stifle.

The GRACILIS (Plate 12). This muscle does not possess the slender character from which it is named in human anatomy. It is a large, somewhat four-sided mass, forming the greater part of what is termed the *flat of the thigh.* A linear depression seen on the surface of the muscle when it is cleaned, is often mistaken by students for the line of separation between it and the sartorius. It *arises* from the lower face of the pubis and ischium close to the symphysis, and it is here united to its fellow of the opposite side. Inferiorly it has a broad flat tendon, united in front to that of the sartorius. It is *inserted* with the sartorius into the internal straight ligament of the patella, and into a line on the tibia between its anterior and internal tuberosities. The posterior edge of its tendon is continuous with the deep fascia of the leg. A large branch from the external pudic veins traverses the muscle near its origin, and opens into the femoral vein.

Action.—To adduct the hip, and rotate the limb inwards.

Directions.—The two preceding muscles are to be carefully cut across about their middle, and turned upwards and downwards. On reflecting the proximal half of the gracilis, branches of the obturator nerve and deep femoral artery will be seen penetrating its deep face ; and, in the same way, twigs from the saphena nerve will be found entering the sartorius. The deep inguinal glands are to be removed, and the femoral vessels and anterior crural nerve are to be dissected.

The FEMORAL ARTERY (Plate 13) is the main arterial trunk for the supply of the hind limb. It is the direct continuation of the external iliac, the brim of the pelvis being selected as the arbitrary line of division between the two vessels ; and, in like manner, it is directly continued by the popliteal artery, the vessel changing its name when it passes between the two heads of the gastrocnemius muscle. The lower third of the vessel, however, will not be exposed till the next stage of the dissection. The part of the vessel now seen begins at the pelvic brim, where it is seen issuing from beneath Poupart's ligament. It there rests on the common termination of the iliacus and psoas magnus, having the sartorius in front and the pectineus behind. In the thigh it descends obliquely downwards and backwards, resting first on the common termination of the iliacus and psoas magnus, and then on the vastus internus. It has the sartorius in front ; while posteriorly it is related first to the pectineus, and then to the adductor parvus. In this course it corresponds to the interstice between the gracilis and sartorius muscles, and is covered by the deep inguinal lymphatic glands. It is closely related to the femoral vein, which lies beneath and slightly posterior to it, except at the brim of the pelvis, where the vein is immediately posterior to the artery. In the present stage of the dissection the vessel disappears between the upper and lower insertions of the adductor magnus, where it will subsequently be followed. The following collateral branches of the femoral are here seen :—

1. The PROFUNDA or DEEP FEMORAL ARTERY. This branch is given off under Poupart's ligament at the pelvic brim. At its origin it usually forms a short common trunk with the prepubic artery. It passes downwards and backwards under the pectineus, and will be followed in the next stage of the dissection.

2. MUSCULAR BRANCHES. The largest of these is a vessel of considerable size for the supply of the quadriceps extensor cruris muscle. It comes off at about the same level as the profunda, which it generally exceeds in volume; and passing over the psoas magnus and iliacus, and under the sartorius, it penetrates between the rectus femoris and vastus internus, in company with the anterior crural nerve. Other innominate arteries of smaller size enter the vastus internus, pectineus, gracilis, sartorius, and adductors.

3. The SAPHENA ARTERY already described (page 57).

4. The NUTRIENT ARTERY OF THE FEMUR is given off at the tendon of insertion of the pectineus.

5. An ARTICULAR branch, of slender volume, descends between the vastus internus and adductor magnus to the stifle-joint.

The FEMORAL VEIN ascends in close company with the artery, and receives branches which correspond more or less exactly to those just described. At the brim of the pelvis it lies posterior to the artery, and is continued upwards as the external iliac vein.

The ANTERIOR CRURAL NERVE (Plate 13) is derived from the lumbosacral plexus. It descends between the psoas magnus and parvus ; and passing over the common termination of the iliacus and psoas magnus, where it is covered by the sartorius, it splits into a bundle of branches that together penetrate between the vastus internus and rectus femoris to supply the mass of the quadriceps extensor cruris. While under cover of the sartorius it gives off the internal saphena nerve already described.

Directions.—The pectineus, adductor parvus, adductor magnus, and semimembranosus muscles are now to be cleaned and isolated. These muscles succeed each other from before to behind in the order named. Some little difficulty may be experienced in finding the line of separation between the two adductors, but a reference to Plate 13 will prove of some assistance. Moreover, the fibres of the small adductor are of a paler colour than those of the adductor magnus.

The PECTINEUS (Plate 13). This muscle has a distinctly conical form. It lies posterior to the femoral vessels, and the profunda artery disappears beneath it. It *arises* from the brim and inferior surface of the pubis, and it is there penetrated by the pubio-femoral ligament, from which some of its fibres take origin. Its tapering point is *inserted* into the shaft of the femur in the neighbourhood of the nutrient foramen.

Action.—It adducts the limb, and flexes the hip.

The ADDUCTOR PARVUS (*Adductor brevis* of Percivall) (Plate 13) is situated between the pectineus and the great adductor. It *arises* from the inferior surface of the pubis, and is *inserted* into the posterior surface of the femur about its middle.

Action.—It is an adductor and outward-rotator at the hip-joint.

The ADDUCTOR MAGNUS (*Adductor longus* of Percivall) (Plate 13) *arises* from the inferior surface of the ischium, and from the tendon of origin of the gracilis. It has two *insertions*, between which the femoral artery passes. 1. Its deeper fibres are *inserted* into the posterior surface of the femur, on a quadrilateral area above the smooth groove in which the femoral artery rests. 2. Its more superficial and longer fibres are *inserted* into the forepart of the supracondyloid crest.

Action.—It is an adductor at the hip.

The SEMIMEMBRANOSUS (*Adductor magnus* of Percivall) (Plate 13). This is a muscle of large size. It *arises* from the lower surface of the ischium, including its tuberosity, and by a small slip from the fascia investing the muscles of the tail. It is *inserted* into the inner condyle of the femur, behind the tubercle for the attachment of the internal lateral ligament of the stifle.

Action.—Commonly, it is an adductor and extensor of the hip; but when the femur is fixed, it acts as a lever of the first order, and assists in rearing.

Directions.—The foregoing muscles must now be cut and partially removed as follows :—

The semimembranosus is to be cut transversely, an inch or two above its insertion. The muscle is then to be raised upwards from the semitendinosus, on which it rests; and in doing this, branches of nerves from the great sciatic will be found entering it in front. The central portion of the muscle may then be removed, leaving a few inches at its origin. The other muscles must be served in the same way, leaving only short portions at the origin and insertion, except in the case of the adductor parvus and upper half of the adductor magnus, whose common insertion into the back of the femur is to be entirely removed. Care is to be taken of the femoral artery where it rests on the bone, and in performing the dissection it will be well to refer to Plate 14 as a guide. In reflecting the upper portion of the great adductor, a branch of the obturator nerve will be found entering its deep face, after having passed through the obturator externus muscle. Other branches of the same nerve will be found supplying the small adductor and the pectineus.

The FEMORAL ARTERY (Plate 14). The remaining portion of this vessel is now exposed as it winds round behind the shaft of the femur, leaving its impress on the bone. It is seen passing in between the heads of the gastrocnemius muscle, at which point it takes the name of *popliteal*. In this part of its course it gives off only one vessel of note—the femoro-popliteal.

The FEMORO-POPLITEAL ARTERY. The point of origin of this branch marks the lower limit of the femoral artery. It passes backwards in a horizontal direction, and penetrates the semitendinosus. Near its origin it gives off a considerable branch which ascends behind the femur, supplying the biceps, and anastomosing with the profunda. Other branches descend from it to the gastrocnemius.

POPLITEAL LYMPHATIC GLANDS. A few glands will be found on the track of the femoro-popliteal artery between the semitendinosus and biceps femoris muscles.

The PROFUNDA or DEEP FEMORAL ARTERY. In the preceding stage of the dissection this branch of the femoral was seen at its origin. It passes downwards and backwards, between the adjacent edges of the iliacus and obturator externus, and under cover of the pectineus and adductor parvus. Above the insertion of the quadratus femoris it crosses behind the femur, where its terminal branches descend to supply the biceps. It also furnishes collateral branches to the pectineus, gracilis, and adductors.

VEINS. The foregoing arteries are accompanied by satellite veins of the same names.

The QUADRATUS FEMORIS (Plates 14 and 16). This is a somewhat slender riband-shaped muscle. It *arises* from the lower surface of the ischium in front of the tuberosity, and it becomes *inserted* into an oblique line on the back of the femur, at the level of the third trochanter.

Action.—It is an extensor and outward-rotator at the hip.

The OBTURATOR EXTERNUS (Plate 14). This muscle, which is coarsely fasciculated, covers the obturator foramen, and conceals the obturator nerves and vessels as they emerge from the pelvis. It is traversed by two branches of the obturator nerve, the posterior of which is for the great adductor, while the anterior splits into branches for the supply of the small adductor, pectineus, and gracilis. It ·*arises* from the lower surface of the pubis and ischium, and is *inserted* into the trochanteric fossa.

Action.—It is an extensor and outward-rotator at the hip.

Directions.—The nerves which emerge from the obturator externus should be traced through the substance of that muscle to their origin

from the obturator nerve.   The muscle itself may then be removed to
expose the obturator vessels and nerve.

The OBTURATOR ARTERY (Plates 14 and 46).   This vessel begins at the
pelvic inlet as one of the terminal branches of the internal iliac.   It
leaves the pelvis by the obturator foramen, in company with a vein and
nerve of the same name.   At its point of emergence it is covered by
the obturator externus, and it passes backwards between that muscle
and the bone, and then curves downwards to terminate in the biceps
and semitendinosus.   It gives off the *artery of the corpus cavernosum*.

The OBTURATOR VEIN passes into the pelvis by the obturator foramen,
and aids in forming the internal iliac vein.

The OBTURATOR NERVE is a branch of the lumbo-sacral plexus.
Emerging by the obturator foramen, it divides for the supply of the
obturator externus, adductor parvus, adductor magnus, pectineus, and
gracilis muscles.

Directions.—In this stage of the dissection the great sciatic nerve is
seen in its course downwards through the thigh.   Its examination is
more conveniently undertaken in the dissection of the hip and outer
aspect of the thigh, but attention may also be given to it here.

The GREAT SCIATIC NERVE, which is a branch of the lumbo-sacral
plexus, after passing through the hip (see Plate 16), descends in the
thigh, behind the femur, where it is deeply enclosed between the
biceps and semitendinosus outwardly, and the semimembranosus and
great adductor inwardly.   Under the name of the internal popliteal, it
passes in between the two heads of the gastrocnemius.   The following
branches whose points of origin are not now visible, being situated in
the hip, may be identified by reference to Plate 14:—(1) Branches to the
biceps, semitendinosus, and semimembranosus; (2) the external popliteal; (3) the external saphenous.   The last two will be again seen in
the dissections of the hip, thigh, and leg.

Directions.—The vastus internus, situated at the front of the thigh,
should now be examined.   It is a division of the great muscular mass
termed in man the quadriceps extensor cruris, whose other divisions—the
rectus femoris and vastus externus—will be dissected with the outer
aspect of the thigh.   The dissection in this position of the limb will be
completed by an examination of the common insertion of the iliacus
and psoas magnus.

The VASTUS INTERNUS (and CRUREUS *) (Plates 13 and 14) is a thick
fleshy muscle whose fibres take *origin* from the internal surface and inner
half of the anterior surface of the femur, meeting along the front of the
femur the vastus internus, and with it forming a groove in which the
rectus femoris rests.   Its fibres are *inserted* into the inner ligament of

---

* This is the name given to the fourth division of the quadriceps in human anatomy.   The fibres
that represent it in the horse are in no way separable from the inner vastus.   Under the same name
Percivall describes (inaccurately) the rectus parvus.

the patella, or into that bone along with the other divisions of the quadriceps.

Action.—It is an extensor of the stifle-joint.

PSOAS MAGNUS and ILIACUS (Plate 14). Only the terminal portion of each of these muscles is here seen. They are more fully displayed in the dissection of the sublumbar region, where the psoas magnus *arises* from the last two ribs, and the vertebræ from the 16th dorsal to the 5th lumbar (Plates 44 and 45). The iliacus *arises* from the iliac surface and external angle of the ilium, and from the sacro-iliac ligament. It presents a deep groove for the terminal portion of the psoas magnus. The two muscles pass downwards beneath Poupart's ligament, and have a common *insertion* into the small (*internal*) trochanter of the femur.

Action.—These muscles flex the hip-joint, and rotate it outwards.

THE HIP AND OUTER ASPECT OF THE THIGH.

Position.—The animal should be suspended in imitation of the natural standing posture, by the means mentioned at page 8.

Surface-marking.—A prominent feature of the region is the bony projection formed by the external angle of the ilium (*angle of the haunch*). The tuber ischii may also be felt by pressing deeply at the point of the hip. At the highest part of the croup the internal angle of the ilium (*angle of the croup*) may be felt, and in the middle line the tips of the sacral spines are subcutaneous. In a lean animal a number of grooves are seen marking the divisions of the biceps and the line of opposition of that muscle with the semitendinosus (Plate 15).

Directions.—An incision through the skin is to be carried along the middle line from the root of the tail as far forwards as the lumbar region, where a transverse incision is to be carried outwards and downwards as far as the level of the angle of the haunch. Beginning at the middle line above, the dissector is to reflect the skin from the limb, as far as the middle of the leg. The first few inches of the skin will require to be raised by the use of the scalpel, and then an attempt may be made to tear it downwards off the limb—a method which will show the cutaneous nerves distinctly without further dissection.

CUTANEOUS NERVES. 1. Appearing a few inches from the middle line, are some slender branches derived from the sacral nerves. 2. Two or three branches of considerable size, derived from the lumbar nerves, pass backwards and downwards over the forepart of the gluteal region. 3. A few inches below the point of the hip a cutaneous branch derived from one of the posterior gluteal nerves appears from between the biceps and semitendinosus, and separates into a number of radiating filaments. 4. On a level with the stifle-joint the *peroneal-cutaneous* branch of the external popliteal nerve comes out through the biceps, and is distributed on the outer side of the leg.

Directions.—The dissector should, in the next place, direct his attention to the strong fascia covering the muscles in this region, after which the fascia must be removed, and the muscles cleaned and separated.

GLUTEAL FASCIA and FASCIA LATA. The gluteal fascia forms a bluish-white covering over the muscles of the hip, and by its deep face affords origin to many fibres of the superficial and middle gluteal muscles. It is fixed above to the summits of the sacral spines and to the external angle of the ilium, and between these points it is continuous forwards with the tendon of the latissimus dorsi. It is prolonged downwards over the muscles of the thigh, where it takes the name of the *fascia lata.* This fascia lata receives in front the insertion of the tensor vaginæ femoris muscle, and it should not be removed until that muscle has been examined. It forms a sheath for the muscles of the thigh, and is prolonged downwards over the leg. From its inner face a septum is sent in between the vastus internus and the biceps, to join the tendon of the superficial gluteal muscle, and be inserted into the femur.

The TENSOR VAGINÆ FEMORIS (Plate 15). This muscle is situated at the forepart of the thigh, in front of the superficial gluteal muscle, from which it is somewhat difficult to separate it. It *arises* from the external angle of the ilium, and it is *inserted* into the fascia lata.

Action.—It flexes the hip-joint. It also keeps the fascia lata tense, and mechanically aids in keeping the stifle-joint extended.

Directions.—The gluteal fascia and the fascia lata are now to be removed. It is a matter of some difficulty to remove the former, as its deep face has the muscular fibres taking origin from it, and these are therefore exposed with a rough surface when it is removed.

The SUPERFICIAL GLUTEUS (*Gluteus externus* of Percivall, part of the *gluteus maximus* of human anatomy) (Plate 15). The outline of this muscle is not distinctly recognisable until the gluteal fascia has been removed. It is then seen to have some resemblance to the letter **V**, having in its upper border an indentation that divides it into an anterior and a posterior branch. The anterior branch *arises* from the external angle of the ilium ; the posterior from the gluteal fascia. Both converge to a common tendon, which is *inserted* into the third trochanter of the femur (*trochanter minor externus*). From the posterior branch of the muscle an aponeurotic layer passes backwards beneath the biceps, to be inserted into the sacro-sciatic ligament and the tuber ischii.

Action.—It is an abductor at the hip-joint.

The BICEPS FEMORIS (Plate 15). This is one of the largest muscles in the body. It *arises* from the sacral spines, the fascia enveloping the muscles of the tail, the sacro-sciatic ligament, the tuber ischii, and the gluteal fascia. Inferiorly it has three divisions, one of which is *inserted* into the anterior surface of the patella, a small synovial bursa being interposed between the tendon and the bone, another into the tibial crest, and the third into the fascia of the leg. Besides these, the muscle has an *insertion* into the circular mark behind the third trochanter of the femur, by means of a fibrous band detached from the deep surface of the muscle.

Action.—The anterior half of the muscle, in virtue of its attachment to the patella, is an extensor of the stifle, and an abductor at the hip. The posterior half of the muscle, with its insertions into the tibia and fascia of the leg, is a flexor and an outward-rotator at the stifle. When the stifle-joint is kept extended, the lower end of the muscle becomes its fixed point, and it then extends the pelvis on the femur, and aids in rearing.

The SEMITENDINOSUS (Plate 15). This muscle is placed at the posterior border of the hip and thigh, where it occupies a position between the last-described muscle and the semimembranosus. The muscle is bifid superiorly, where it *arises* by one division from the sacral spines and sacro-sciatic ligament, and by another and shorter branch from the tuber ischii. Inferiorly it has a flat tendon, which is *inserted* into the tibial crest, and whose posterior border blends with the fascia of the leg.

Action.—To flex the stifle and rotate the leg inwards. When the stifle is fixed, it can aid in rearing.

The biceps femoris and semitendinosus muscles represent, apparently, the muscles of the same name in man, *plus* portions of the gluteus maximus. Percivall describes the semitendinosus as representing also the semimembranosus of man.

Directions.—The biceps must be carefully severed at its origin, and pulled downwards. This will expose the aponeurotic layer that passes beneath it from the superficial gluteus. A branch from the posterior gluteal nerves should be found entering the last-named muscle by turning forwards round the middle gluteus. Both branches of the superficial gluteus should then be thrown downwards in order to fully expose the next muscle.

The MIDDLE GLUTEUS (*Gluteus maximus* of Percivall) (Plate 15) is a muscle of great size and strength. It was partly exposed before the removal of the superficial muscle. The fibres of the muscle *arise* from the aponeurosis of the *common mass* of the loins (longissimus dorsi), from the gluteal surface of the ilium, from the two ilio-sacral and the sacro-sciatic ligaments, and from the gluteal fascia. It has three distinct and constant *insertions*: 1. By a tendon, into the *summit* of the great trochanter. 2. By another tendon, which plays over the *convexity* of the same trochanter by means of a synovial bursa, and becomes inserted into the *crest*. 3. By a triangular fleshy slip, into the back of the trochanteric ridge.

Action.—To extend and abduct the hip. In the former of these

F

actions, when the limb is free to move, the femur, and with it the whole limb, is carried backwards; but when the femur is fixed, it raises the trunk, as in rearing.

Directions.—The last-described muscle must be removed in order to expose the deep gluteus and the other structures which it covers. A deep incision should be made through the muscle along the crest of the ilium, and the muscle is to be turned down by severing its fibres at their origin. Care must be taken, in doing this, to avoid cutting the subjacent deep gluteus, whose fibres may be recognised, as soon as they are reached, by their insertion *within* the great trochanter. A reference to Plate 16 may here be useful. The semitendinosus is to be turned down in the same manner, by severing its superior attachments; and this muscle and the biceps may be removed to the extent shown in the Plate. In performing this dissection, the gluteal nerves and vessels and the ischiatic vessels are unavoidably severed, but a look-out should be kept for these, and they should be cut about the points shown in the figure.

The GLUTEAL ARTERY (Plate 16) is a branch given off from the internal iliac within the pelvis. After a very short course it splits into several branches, which, emerging by the great sacro-sciatic opening, are distributed to the gluteal muscles.

The ISCHIATIC ARTERY (Plate 16) is one of the terminal branches of the lateral sacral artery, which, again, is a collateral branch of the internal iliac. It perforates the sacro-sciatic ligament near the edge of the sacrum, and is distributed in the biceps and semitendinosus.

VEINS of the same names accompany the foregoing arteries.

The INTERNAL PUDIC ARTERY. The dissection of this artery belongs to another region, but the vessel is generally visible here in a part of its course. A few inches of it are represented in Plate 16, as showing faintly through the texture of the sacro-sciatic ligament. It is described at page 342.

The GLUTEAL NERVES (Plate 16). These nerves, which are derived from the lumbo-sacral plexus, issue from the great sacro-sciatic opening in company with the gluteal vessels and the great sciatic nerve. They consist of an anterior and a posterior set.

The *Anterior gluteal* nerves are three or four in number. One of them passes downwards and forwards between the middle and internal gluteal muscles, to reach the tensor vaginæ femoris and anterior division of the superficial gluteus. Another branch passes downwards over the deep gluteus, to which it is distributed. One or two other branches supply the middle gluteus.

The *Posterior gluteal* nerves are two in number—an upper and a lower. The *upper* nerve passes backwards on the sacro-sciatic ligament; and after giving branches to the posterior division of the superficial gluteus, and to the posterior fleshy slip of the middle gluteus, it enters the biceps femoris. The *lower* nerve, passing downwards and backwards, divides into an outer and an inner branch ; the former, turning over the tuber ischii on its outer side, becomes cutaneous at the back of the thigh about four or five inches below the tuber ; the latter, after giving twigs to the semitendinosus, joins a branch from the internal pudic nerve to be distributed to the perineal structures.

The GREAT SCIATIC NERVE (Plate 16). This is, at its point of origin, the largest nerve in the body. It is furnished by the lumbosacral plexus, and appears at the great sacro-sciatic opening as a broad riband. In its downward course in the hip it is covered by the middle gluteus, and rests in succession on the sacro-sciatic ligament, the gluteus internus, the gemelli and common tendon of the obturator internus and pyriformis, and the quadratus femoris. In the thigh it is included between the biceps and semitendinosus outwardly, and the semimembranosus and great adductor inwardly. The trunk of the nerve is continued as the internal popliteal nerve between the two heads of the gastrocnemius, where it will be followed in the dissection of the leg. It gives off in succession the following branches :—1. A nerve for the supply of the obturator internus, pyriformis, gemelli, and quadratus femoris. This slender branch is given off about midway between the great and small sciatic openings, and it descends at the posterior border of the parent trunk, or between that and the ligament. The nerves to the quadratus and gemelli may arise from the sciatic independently, and the branch to the first of these muscles passes under the gemelli and the common tendon of the pyriformis and obturator internus. 2. The *external popliteal nerve* is a large branch that separates from the great sciatic about the level of the small sacro-sciatic opening ; and, descending in front of the parent nerve, it passes between the biceps and the outer head of the gastrocnemius, where it will be followed at a later stage. The *peroneal cutaneous* branch of this nerve has already been seen perforating the lower part of the biceps, at the level of the stifle. 3. A branch that divides to supply the semimembranosus and lower portions of the biceps and semitendinosus (Plate 14). 4. The *external saphenous nerve*, which will be followed in the dissection of the leg.

Directions.—The great sciatic nerve should now be cut at the upper border of the gluteus internus, and turned downwards with its branches. The gluteus internus, and the common tendon of the pyriformis and obturator internus, together with the gemelli, should be carefully cleaned and defined ; and to facilitate this, the limb should be rotated inwards as far as possible, by pulling the point of the hock outwards. This will put these muscles on the stretch.

The DEEP GLUTEUS, or gluteus internus (Plate 16, and fig. 2), is placed above the hip-joint, in immediate contact with the capsular ligament.

It is a comparatively small muscle, with coarse fasciculi having a slightly spiral direction. It *arises* from the rough lines on the gluteal surface of the shaft of the ilium just above the cotyloid cavity, and from the supra-cotyloid ridge (superior ischiatic spine). It is *inserted* to the inner side of the convexity of the great trochanter.

Action.—It is an abductor and inward-rotator at the hip-joint.

The OBTURATOR INTERNUS and the PYRIFORMIS (Plate 16, and fig. 2) are two muscles arising within the pelvis, the former taking *origin* from the bone around the obturator foramen, and the latter from the pelvic surface of the ilium. They have a common tendon, which emerges from the pelvis by the lesser sacro-sciatic opening, where it plays over a smooth portion on the external border of the ischium. The tendon is *inserted* into the trochanteric fossa.

Action.—To produce outward rotation at the hip.

The GEMELLI. In Plate 16 a bundle of muscular fibres is seen at each edge of the above-mentioned common tendon. If this common

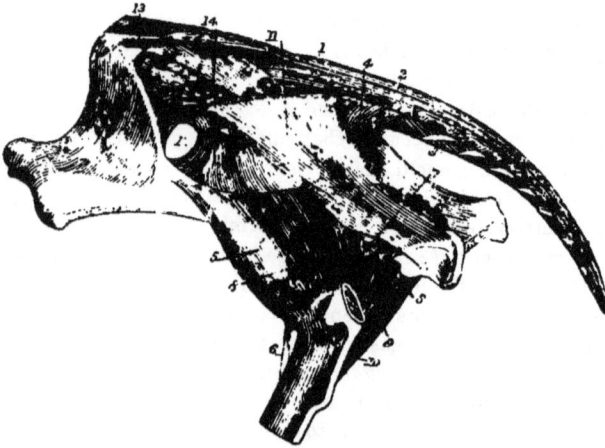

FIG. 2.

MUSCLES OF THE TAIL, DEEP MUSCLES OF THE HIP, AND PELVIC LIGAMENTS (*Chauveau*).

1. Erector coccygis; 2. Curvator coccygis; 3. Depressor coccygis; 4. Compressor coccygis; 5. Deep glutens; 6. Rectus parvus; 7. Common tendon of obturator internus and pyriformis; 8. Gemelli; 9. Accessory fasciculus of the same; 10. Quadratus femoris; 11. Sacro-sciatic ligament; 12. Great sacro-sciatic foramen; 13. Superior ilio-sacral ligament; 14. Inferior ilio-sacral ligament.

tendon be cut where it appears at the lesser sciatic opening, and raised outwards, what previously seemed two distinct muscular bundles will now be seen to be the edges of a single flattened muscle, which *arises* from the ischium below and at the edges of the smooth surface for the passage of the common tendon, and becomes *inserted* in common with that tendon.*

Action.—The same as the two preceding muscles.

Directions.—The tensor vaginæ femoris and the gluteus internus should now be cut away, care being taken not to injure the capsular ligament, on which the latter muscle rests. The rectus femoris, vastus externus, and rectus parvus are then to be dissected. The last-mentioned muscle will be found by dissecting deeply into the upper part of the interstice between the other two muscles, and at the same point the iliaco-femoral artery will be found.

The ILIACO-FEMORAL ARTERY is one of the terminal branches of the internal iliac (Plate 48). It comes out between the iliacus and the shaft of the ilium, and penetrates between the rectus femoris and the vastus externus.

The RECTUS FEMORIS (Plate 18) *arises* by two heads—one from each of the depressions on the shaft of the ilium, above and in front of the cotyloid cavity. The central portion of the muscle is thick and fleshy, and rests in a groove formed by the two vasti, with which it is confounded at its lower extremity. It is *inserted* into the anterior face of the patella.

Action.—To flex the hip-joint and extend the stifle.

The VASTUS EXTERNUS (Plate 18) *arises* from the outer surface of the femur, and from the outer half of the anterior surface of the same bone. Its fibres become *inserted* along with the rectus femoris into the patella.

Action.—To extend the stifle.

The RECTUS PARVUS (fig. 2) is very slender when compared with the muscles between which it is placed, being about the thickness of a human finger. It *arises* from the ilium, external to the pit from which the outer head of the rectus femoris takes origin. Passing in front of the capsular ligament of the hip-joint, to which it adheres, it insinuates itself between the two vasti muscles, and is *inserted* into the anterior surface of the femur.

Action.—The muscle is of too slender a size to exert any appreciable flexor action on the hip-joint, and probably its function is to raise the capsular ligament during flexion of the joint. (See footnote, page 62.)

Directions.—The dissector is now in a position to detach the limb from the trunk, and this should be done by cutting through the bone and soft structures, below the level of the internal trochanter. It is necessary to make the section at this point, in order to leave the hip-joint and the common insertion of the iliacus and psoas magnus intact for examination by the dissector of the abdomen and pelvis. The limb

---

* In man this muscle consists of two separate slips, and from this disposition it is named. I do not hesitate to give it the same designation here, although I have never found it double as it is usually described.

having been removed, it should be placed on a table, and the cut muscles connected with it may be completely cut away after they have been identified. In doing this, a better opportunity will be afforded to observe accurately the insertion of each muscle. In removing the lower portion of the biceps, particular care must be taken not to cut the external popliteal and external saphenous nerves, which are included between it and the outer head of the gastrocnemius (Plate 18).

THE LEG.

Surface-marking.—The bones of the leg are clothed by muscles except at the inner side of the limb, where the tibia is subcutaneous. This unprotected area of bone corresponds to the *shin* in man. The superficial muscles of the region (see Plates 17 and 18) form prominences more or less distinct, especially in the neighbourhood of the hock, where the various tendons stand out distinctly during the movements of the living animal.

Position.—In the further dissection of the limb, it may be placed on a clean table, and laid on either side as may be convenient; or a cord may be passed round the femur, and the limb suspended at such a height as just to permit the hoof to come into contact with the table. This latter method has the advantage of keeping the part clean; and while dissection is being carried on, the leg may be steadied in any position by an assistant.

Directions.—An incision through the skin is to be carried down the middle line of the limb on its inner side, and terminated a few inches below the hock, where a circular incision may be carried round the limb. The whole of the skin above the circular incision is then to be removed, and the cutaneous nerves and vessels of the region are to be examined.

The INTERNAL SAPHENA VEIN. On the inner side of the leg, above the hock, two venous branches will be seen to converge and unite to form the internal saphena vein, which is continued up the leg to the thigh, where it has already been dissected. The vessels by whose union the main vein is formed, are the upward continuations of the internal and external metatarsal veins. Slender branches of the saphena artery accompany these veins.

The EXTERNAL SAPHENA VEIN. This vessel begins at the hock, where it communicates with the internal saphena vein, and with the posterior tibial vein. It ascends at the outer side of the gastrocnemius tendon, and, passing between that muscle and the biceps, it empties itself into the femoro-popliteal vein.

CUTANEOUS NERVES. 1. The ramifications of the *internal saphenous* nerve cover the inside of the leg, and descend over the inside of the hock. 2. The *external saphenous* nerve (Plate 18) is a branch of the great sciatic. It descends over the outer head of the gastrocnemius, where it is covered by the biceps, and is reinforced by a branch from the external popliteal (or from the peroneal cutaneous division of that nerve). It then continues to descend, lying in company with the vein of the same name, in front of the outer edge of the gastrocnemius tendon ; and passing over the hock, it is distributed to the skin on the outer side of the metatarsus. 3. The cutaneous termination of the *musculo-cutaneous* division of the external popliteal nerve (Plate 18) pierces the deep fascia on the outer side of the limb at the lower third of the leg ; and, passing over the hock, it is distributed to the skin on the front of the metatarsus.

DEEP FASCIA OF THE LEG. This forms a close-fitting, fibrous envelope to the muscles of the region. Its inner face furnishes septa that pass in between the muscles ; and over the inner surface and crest of the tibia, it is adherent to the bone. Above it is continuous with the fascia lata and tendons of the gracilis and semitendinosus on the inside, and with the tendon of the biceps on the outside. As it passes over the hock it becomes thinner, and is continuous with the fascia of the meta-tarsal region.

Directions—The muscles on the back of the leg may now be dissected, the vessels and nerves shown in Plates 17 and 18 being at the same time carefully preserved. By a reference to Plate 18, the student should note the position of the small *soleus* muscle, so as to avoid its removal with the fascia.

The GASTROCNEMIUS (Plates 17 and 18). At its origin this muscle consists of two distinct fleshy heads, which terminate in a single inferior tendon. The outer head *arises* from the outer lip of the supracondyloid fossa of the femur, the inner head from the supracondyloid crest. The cord-like tendon is joined by that of the soleus, and is *inserted* into the back part of the summit of the os calcis. When the hock is strongly flexed, the tendon for an inch or two above its insertion rests on the forepart of the summit, and a small synovial bursa is here interposed between the tendon and the bone. The tendon of the perforatus is at first beneath that of the gastrocnemius ; but, passing to the inner side, it places itself superficial to the latter, which it completely covers at the summit of the os calcis. In thus changing positions, the two tendons form a half twist, and indent each other like the strands of a rope. This tendon of the gastrocnemius corresponds to the firm tendon extending upwards from the human heel, and known as the *tendo Achillis.*

Action.—To extend the hock-joint.

The SOLEUS (Plate 18). In British veterinary text-books this muscle is erroneously termed *plantaris.* It is a small muscle of delicate texture, and it is often partially or entirely removed in cleaning the

72 THE ANATOMY OF THE HORSE.

gastrocnemius. It *arises* from the head of the fibula, and its tendon joins that of the preceding muscle, which it assists in extending the hock.

Directions.—The inner head of the gastrocnemius is to be severed at its origin, and turned downwards in the manner shown in Plate 17.

The SUPERFICIAL FLEXOR of the digit (flexor perforatus) (Plates 17 and 18) is remarkable in that, throughout nearly the whole of its extent, it exists as a strong tendinous cord with a sparing admixture of muscular fibres at its upper fifth only. It *arises* from the bottom of the supra-condyloid fossa; and, winding round the gastrocnemius tendon in the manner already described, it gains the summit of the os calcis, over the extreme posterior portion of which it plays by means of a synovial bursa. At the os calcis it detaches on each side a slip to be inserted into the bone. It is continued downwards in the meta-tarsal and digital regions in the same manner as the perforatus of the fore limb, becoming finally *inserted* by a bifid termination into the second phalanx. In front of the tendons of the superficial flexor and gastrocnemius there will be noticed a strong fibrous band, which is united to these muscles above, and inserted into the os calcis below, while laterally it is continuous with the deep fascia of the leg.

Action.—It flexes successively the pastern and fetlock joints; and, by its insertion into the os calcis, it is also an extensor of the hock-joint. It also plays an important part in mechanically maintaining the hock in a state of extension so long as the hip and stifle joints are kept extended by muscular contraction.

The flexor perforatus of the horse is represented in man by two distinct muscles—the *plantaris* and the *flexor brevis digitorum.*

Directions.—The deep layer of muscles at the back of the leg consists of the popliteus, the flexor perforans, and the flexor accessorius; and these should now be examined as far as possible without disturbance of the vessels and nerves. The superficial muscles must therefore, in the meantime, be allowed to remain in position.

The POPLITEUS (Plate 17). This muscle is placed immediately behind the stifle-joint, whose posterior ligament it covers. It *arises* by a tendon from the lower and most anterior of the two pits situated on the outer side of the external condyle of the femur. (The other pit is for the attachment of the external lateral ligament of the stifle, the ligament concealing the origin of the tendon.) The tendon is partly invested by the synovial membrane of the joint, and plays round the external semi-lunar cartilage, and over the articular surface of the tibia. The fibres of the muscle have an oblique direction downwards and inwards, and are *inserted* into the comparatively smooth triangular area at the upper part of the posterior surface of the tibia, and into the inner edge of the bone at the same level. The terminal portion of the popliteal artery is concealed by the muscle.

Action.—It flexes the stifle, and to a slight extent rotates it inwards. The DEEP FLEXOR of the digit (flexor perforans) (Plates 17 and 18). This muscle is indistinctly divided into an outer and an inner division, the former being the larger of the two. It *arises* from the ridged area on the posterior surface of the tibia, from the external tuberosity at the upper end of the same bone, from the fibula, and from the interosseous membrane uniting the two bones. At the lower third of the tibia the muscular divisions are succeeded by tendons, which soon unite ; and the single tendon thus formed glides through the tarsal sheath at the inner side of the os calcis, and then descends at the back of the metatarsus and digit, to be *inserted* into the solar surface of the os pedis, in a manner exactly similar to the flexor perforans of the fore limb. Like that muscle, it receives, at the upper part of the metatarsus, a reinforcing or check band—the *subtarsal ligament*, which is the downward continuation of the *posterior tarso-metatarsal* ligament of the hock. This band is not so strong as the subcarpal ligament of the fore extremity.

Action.—It flexes successively from below upwards the interphalangeal joints and the fetlock, and finally extends the hock.

The TARSAL SHEATH, through which the tendon passes at the inner side of the back of the hock, is formed outwardly by the grooved surface of the os calcis, in front by the posterior ligament of the tibio-tarsal articulation and by the posterior tarso-metatarsal ligament, and it is completed inwardly by a fibrous arch that converts the groove into a complete canal. An extensive synovial membrane here invests the tendon and lines the passage, extending upwards for a few inches at the lower extremity of the tibia, and downwards below the middle of the metatarsus. A dropsical condition of this synovial sac gives rise to the condition termed "thorough-pin."

The FLEXOR ACCESSORIUS (Plate 17) is a somewhat slender muscle extending obliquely downwards at the back of the leg, between the popliteus and the perforans. It *arises* from the back of the external tuberosity at the head of the tibia. Its tendon, which begins at the lower third of the leg, descends first in a groove on the deep flexor, and then through a synovial passage at the inner side of the tarsus, and finally blends with the tendon of the deep flexor at the back of the metatarsus.

Action.—To assist the deep flexor.

Directions.—The vessels and nerves of the region must now be noticed, and it will be convenient to begin with the latter.

The INTERNAL POPLITEAL NERVE (Plates 17 and 18) is the continuation of the *great sciatic*. It passes in between the two heads of the gastrocnemius muscle, follows for a short distance the posterior border of the perforatus, and at the level of the lower border of the popliteus it is continued under the name of the posterior tibial nerve. The nerve furnishes branches to all the muscles at the back of the leg, viz., both heads of the gastrocnemius, the soleus, the perforatus, the popliteus, the perforans, and the flexor accessorius. The branch to the soleus gains its muscle by passing between the popliteus and the outer head of the gastrocnemius.

The POSTERIOR TIBIAL NERVE (Plate 17) continues the internal popliteal. It is at first deeply placed beneath the inner head of the gastrocnemius, where it crosses the perforatus. Becoming more superficial by emerging from beneath the first-named muscle, it descends on the inner side of the leg, in front of the *tendo Achillis*, being covered by the deep fascia of the leg. At the tarsus it bifurcates to form the external and internal plantar nerves. These accompany the perforans tendon through the tarsal sheath, and are continued through the metatarsal and digital regions like the corresponding nerves of the fore limb. The only collateral branches of the posterior tibial nerve are slender *cutaneous filaments*, one of which is shown in Plate 17, descending over the inner side of the hock.

The EXTERNAL POPLITEAL NERVE and the EXTERNAL SAPHENOUS NERVE cross the external head of the gastrocnemius on its outer side (Plate 18). The latter nerve has already been described, and the former should be preserved to be followed in the dissection of the front of the leg.

Directions.—The outer head of the gastrocnemius and the perforatus should now be detached close to their origin, and turned downwards. This will expose the whole of the popliteus, which must be dissected carefully from the posterior ligament of the stifle and from the tibia, in order to follow the popliteal artery.

The POPLITEAL ARTERY (Plate 17) is the direct continuation of the femoral. In veterinary anatomy the arbitrary line of distinction is usually drawn at the point where the vessel passes in between the heads of the gastrocnemius. It passes over the posterior ligament of the stifle, where it is covered by the popliteus; and at the tibio-fibular arch it bifurcates to form the anterior and posterior tibial arteries. It gives off—(1) *articular branches* to the stifle, and (2) *muscular branches* to the superficial muscles at the back of the leg.

The POSTERIOR TIBIAL ARTERY (Plate 17) is much the smaller of the two terminal branches of the popliteal. In the first part of its course it is deeply placed beneath the popliteus and the deep and accessory flexors. As it descends, it becomes more superficial, and appears at the posterior border of the flexor accessorius, whose tendon it follows in the same position. A little above the hock it forms an S-shaped curve that brings it into company with the terminal part of the posterior tibial nerve; and passing with that nerve into the tarsal sheath, it divides at the back of the hock into the two plantar arteries. The collateral branches of the posterior tibial are—(1) *muscular branches* to the deep muscles at the back of the leg; (2) the *nutrient artery* of the tibia; (3) *a retrograde branch* which, emanating from the second curve of the sigmoid flexure, ascends in front of the *tendo Achillis;* (4) *articular branches* to the tarsus.

VEINS. The foregoing arteries run in company with satellite veins bearing the same names.

Directions.—The front of the leg must now be dissected; and as the first step, the muscles of the region should be cleaned and isolated. These are—the extensor pedis, the flexor metatarsi, and the peroneus. The first of these is superposed to the second on the front of the leg, while to the outer side of both is the smaller peroneus. In dissecting the tendons of these muscles in the region of the hock, care should be taken of three transverse fibrous bands that retain the tendons in position (Plates 18 and 19). The first of these bands is fixed by its extremities to the lower end of the tibia, and beneath it pass the tendons of the extensor pedis and flexor metatarsi. The second is fixed outwardly to os calcis; and, passing over the extensor pedis tendon, it is attached to the flexor metatarsi. The third retains the tendons of the extensor pedis and peroneus in position at the upper end of the large metatarsal bone, to which its extremities are attached.

The EXTENSOR PEDIS (Plate 18). This muscle *arises*, in common with the tendinous portion of the flexor metatarsi (fig. 3, page 76), from the pit between the trochlea and external condyle of the femur. It has a thick, fusiform muscular belly, which at the lower third of the leg is succeeded by a strong tendon. This passes over the front of the hock, and under the three annular bands just described. It then descends over the front of the metatarsus, where it receives the insertion of the short extensor of the digit, and is joined by the tendon of the peroneus. In the dissection of the digit it will be pursued to its *insertion* into the pyramidal eminence of the os pedis.

Action.—It extends in succession from below upwards the interphalangeal joints and the fetlock, and finally flexes the hock.

The PERONEUS (Plate 18). This is a much smaller muscle than the preceding, to whose outer side it lies. Its muscular fibres have a penniform arrangement, and *arise* from the external lateral ligament of the stifle, from the fibula, and from the aponeurotic septum between it and the deep flexor of the phalanges. Its tendon passes through the groove on the external tuberosity (external malleolus) at the lower end of the tibia, and then over the outer side of the hock, where it plays in a synovial canal formed in the external lateral ligament. Below the hock it is directed obliquely forward, and joins the tendon of the extensor pedis about the middle of the metatarsus.

Action.—The same as the preceding muscle.

Directions.—Cut the extensor pedis about the middle of the leg, and reflect it upwards and downwards to expose the next muscle.

The FLEXOR METATARSI (fig. 3). This muscle consists of two parallel portions—a superficial and a deep. The *superficial* division exists in the form of a tendinous cord with little or no muscular tissue, and *arises*, in common with the extensor pedis, from the pit between the trochlea and external condyle of the femur. This tendon of origin passes through the notch between the anterior and external tuberosities at the upper end of the tibia, and is there enveloped by the synovial membrane of the femoro-tibial joint. In the leg the tendinous cord rests on the deep division of the muscle, and passes under the annular band at the lower extremity of the tibia, in company with the tendon of the extensor pedis. At the front of the hock it is perforated by the tendon of the deep division, and then bifurcates, one branching continuing downwards to be *inserted* into the upper extremity of the large metatarsal bone, the other deviating outwards to be *inserted* into the cuboid.

The *deep* division of the flexor metatarsi rests on the tibia, and its muscular fibres *arise* from the upper part of the outer surface of that bone. At the lower end of the tibia it is succeeded by a tendon which perforates that of the superficial division of the muscle, and divides, one branch passing to be *inserted* into the head of the large metatarsal bone, along with the large division of the superficial cord, while the other branch is carried inwards to be *inserted* into the cuneiform parvum.

Action.—To flex the hock. In this action the superficial tendinous cord plays merely a mechanical part, flexing the hock when the stifle is flexed.

The EXTERNAL POPLITEAL NERVE (Plate 18). This nerve has already been seen in the hip and thigh. It is a branch given off by the great sciatic; and, descending in front of the parent nerve, it passes between the biceps and the outer head of the gastrocnemius, where, a little behind the external lateral ligament of the stifle, it divides into the musculo-cutaneous and anterior tibial nerves.

The MUSCULO-CUTANEOUS NERVE descends along the line of contact of the extensor pedis and peroneus, supplying filaments to the latter muscle. At the lower third of the leg, as has already been seen, the

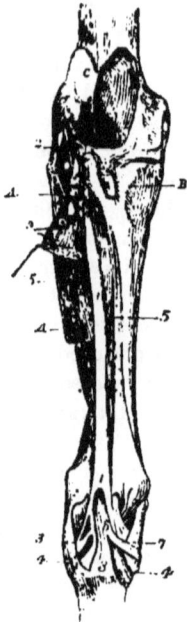

FIG. 3.

FLEXOR METATARSI MUSCLE (*Chauveau*).

1. Superficial division of the muscle; 2. Its origin from the femur; 3. Its cuboid branch; 4. Its metatarsal branch; 5. Deep division of the muscle; 6. Its tendon passing through that of the superficial division; 7. Cuneiform branch of this tendon; 8. Metatarsal branch of the same; 9. Extensor pedis; A. Peroneus; B. Insertion of middle straight patellar ligament; C. Femoral trochlea.

cutaneous division of the nerve pierces the deep fascia, and passes over the hock to supply the skin on the outer side of the metatarsus.

The ANTERIOR TIBIAL NERVE separates from the preceding at an acute angle, and a few inches below the stifle it passes under cover of the extensor pedis. It supplies twigs to the last-named muscle, the flexor metatarsi, and the short extensor ; and descends at the outer side of the tibial vessels, afterwards accompanying the large metatarsal artery to terminate in the skin on the outer side of the digit.

The ANTERIOR TIBIAL ARTERY (Plate 19). This, it will be recollected, is one of the terminal branches of the popliteal artery. Originating behind the upper extremity of the tibia, it is here seen coming forwards through the tibio-fibular arch. It descends on the tibia, under cover of the flexor metatarsi, and accompanied by the nerve and vein of the same name. Gaining the front of the hock, it rests on the anterior tibio-tarsal ligament, covered by the flexor metatarsi and extensor pedis at their line of contact. Here it deviates outwards under the tendon of the last-mentioned muscle, and divides into two vessels of unequal size. The larger of these, which continues the direction of the parent vessel, is the large metatarsal artery ; the other is the perforating metatarsal artery ; and both will be dissected with the metatarsus. The anterior tibial artery gives off numerous un-named *muscular branches* to the extensor pedis, flexor metatarsi, and peroneus ; and *articular branches* to the hock.

The ANTERIOR TIBIAL VEIN, which may be double, keeps close company with the artery. It is formed at the front of the hock by the fusion of several rootlets. The largest of these is the upward continuation of the deep metatarsal vein, which comes forwards through the vascular canal between the tarsal bones. After passing backwards through the tibio-fibular arch, the anterior joins the posterior tibial vein to form the popliteal.

THE METATARSUS AND DIGIT.

The distal portion of the horse's hind limb, beyond the lower extremity of the tibia, is technically termed the *pes*, as it corresponds to the foot of man. The tarsus, or hock, represents the human ankle ; the part between the tarsus and fetlock corresponds to the body of the human foot, and is termed the metatarsus ; while the rest of the limb, beyond the fetlock, is the digit, and is the homologue of man's third toe.

Surface-marking.—Extending down the middle line in front is the tendon of the extensor pedis, which, a little below the tarsus, is joined obliquely by the tendon of the peroneus. Behind the metatarsus, and resting on the bone, is the suspensory ligament ; and behind that again are the deep and superficial flexors of the foot. The edges of these structures can be distinctly seen or felt in the living animal, and in the dead subject they may be identified by a reference to Plate 19.

At the upper part of the inner face of the metatarsus is a flattened
horny callosity, or *chestnut;* and another horny excrescence, in the
form of a spur, or *ergot,* is concealed in the tuft of hair behind the
fetlock.   By manipulation in the neighbourhood of the heels, the lateral
cartilages may be felt.

Directions.—Remove the entire remaining portion of skin from the
limb; and if it is intended to study on the same preparation the parts
contained within the hoof, this must, before the removal of the skin, be
detached by force in the manner described at page 35.   The various
structures are now to be defined by dissection, in the order of the
following description; and while the vessels and nerves are being
cleaned, care must be taken of the small lumbricales muscles, which lie
on the tendon of the deep flexor, above the fetlock.

CUTANEOUS NERVES.—Descending over the inner side of the hock and
metatarsus are twigs of the internal saphenous and posterior tibial
nerves, and on the outer side of the same regions are branches of the
external saphenous and musculo-cutaneous nerves.

The LARGE METATARSAL ARTERY (*Dorsalis pedis* of man) (Plate 19)
is the larger branch resulting from the division of the anterior tibial
artery at the front of the tarsus.   It inclines outwards and downwards under the extensor brevis and the peroneus, and places itself in
the groove formed on the outer side of the metatarsus by the junction of
the large and outer small metatarsal bones.   Along this groove it
descends in company with the slender continuation of the anterior
tibial nerve, until, a little above the button of the smaller bone, it
passes to the back of the metatarsus by penetrating between the two
bones.   Finally, it bifurcates above the fetlock, between the two
divisions of the suspensory ligament, to form the digital arteries.   It
gives off numerous un-named twigs to the skin, tendons, etc.

Descending in the metatarsal region, there are other four arteries
besides the vessel just described.   They will be found, one at each side
of the flexor tendons, in company with the vein and nerve, and another
at each edge of the suspensory ligament, within the splint bone of the
same side.   All of these are branches of an arterial arch formed across
the origin of the suspensory ligament from the back of the tarsus.   The
arch corresponds to the subcarpal arch of the anterior limb, and is
formed as follows :—

The PERFORATING METATARSAL ARTERY, the smaller branch resulting
from the division of the anterior tibial artery, passes from the front to
the back of the tarsus by the canal between the cuboid, scaphoid, and
cuneiform bones.   Here it unites with the outer and inner plantar
divisions of the posterior tibial, which descend in the tarsal sheath, one
on each side of the perforans tendon.   Of the four vessels that spring
from the arch thus formed, the two that descend with the plantar

nerves at the side of the flexor tendon are un-named and slender (Plate 19). The other two are termed the plantar interosseous metatarsal arteries. This may be regarded as the most typical arrangement of the arteries here, but it is not constant. Sometimes the inner plantar artery is directly continued as the satellite vessel of the internal plantar nerve in the metatarsus, the outer plantar artery alone uniting with the perforating metatarsal artery.

The EXTERNAL PLANTAR INTEROSSEOUS ARTERY is very slender. It descends, as beforesaid, between the outer splint bone and the edge of the suspensory ligament; and above the fetlock it anastomoses with a recurrent twig from the large metatarsal artery.

The INTERNAL PLANTAR INTEROSSEOUS ARTERY, a vessel of considerable size, descends between the inner splint bone and the edge of the suspensory ligament. Above the lower extremity of that bone it inclines towards the middle of the limb to join the large metatarsal artery. It supplies the nutrient artery of the large metatarsal bone.

The DIGITAL ARTERIES (Plate 19). These arteries separate at an acute angle, in passing backwards between the branches of bifurcation of the suspensory ligament. For the remainder of their course they are identical with the homonymous vessels of the fore limb. For their description, turn to page 28.

The DIGITAL VEINS (Plate 19). These are the satellites of the digital arteries, in front of which they ascend. They drain away the blood from the venous plexuses within the hoof; and, uniting with one another above the fetlock, they form an arch between the deep flexor and the suspensory ligament. From this arch spring the metatarsal veins.

The METATARSAL VEINS are three in number:—

1. The *Internal Metatarsal Vein* ascends in front of the inner edge of the deep flexor tendon, in company with the internal plantar nerve and a slender artery. The vein is the most anterior of the three structures, and the slender artery is between the vein and the nerve. At the upper third of the metatarsus the vein deviates forwards, crossing the inner splint bone and the large metatarsal obliquely, to gain the inner side of the hock, above which it is continued as the anterior root of the internal saphena vein. The course of the vein over the hock is generally apparent in the living animal, and when very prominent it constitutes the so-called "blood-spavin."

2. The *External Metatarsal Vein* (Plate 19) ascends on the inner edge of the deep flexor, having the same relationship to nerve and artery as the internal vein. After communicating with the deep vein, it is continued through the tarsal sheath to become the posterior root of the internal saphena.

3. The *Deep Metatarsal Vein* ascends between the suspensory ligament and the large metatarsal bone; and passing from the back to the

front of the hock, by the vascular canal for the perforating metatarsal artery, it is continued as the anterior tibial vein.

The PLANTAR NERVES. These nerves result from the bifurcation of the posterior tibial nerve when it gains the back of the tarsus. They accompany the perforans tendon in the tarsal sheath; and diverging from one another, they descend in the metatarsal region, one at each side of the deep flexor tendon. Each is accompanied in the metatarsus by the metatarsal vein of that side, and by a slender artery from the vascular arch at the back of the tarsus. A little below the middle of the metatarsus the inner nerve detaches a considerable branch that winds obliquely downwards and outwards behind the flexor tendons to join the outer nerve above the level of the button of the splint bone. At the fetlock each nerve, coming into relation with the digital vessels, resolves itself into three branches for the supply of the digit. These are identical in their arrangement with the like branches of the plantar nerves in the fore limb, for the description of which, turn to page 30.

The student must now pursue the dissection of the following muscles which have already been dissected in the leg, viz., the extensor pedis and peroneus on the front of the limb, and the superficial and deep flexors behind. In addition to these, there are the short extensor of the foot, the lumbricales, and the interossei, which entirely belong to this region; and since they are of small size, and might easily be overlooked or injured, their dissection must be first undertaken.

The LUMBRICALES (Plate 19) and INTEROSSEI MUSCLES. These exactly resemble the muscles of the same name in the anterior member. Turn, therefore, to the description of the latter given at page 31, substituting the word *foot* for *hand*, *toes* for *fingers*, and *metatarsal* for *metacarpal*.

The SHORT EXTENSOR of the foot (*extensor brevis digitorum* of man) (Plate 19). Look for this small muscle at the front of the tarsus, in the angle of union of the extensor pedis and peroneus tendons. It *arises* from the os calcis and astragalus, and is *inserted* into the united tendon of the above-mentioned muscles, to whose action it is auxiliary.

The EXTENSOR PEDIS tendon (Plate 19) descends along the middle line of the limb in front, to be *inserted* into the pyramidal eminence of the os pedis. Above the middle of the metatarsus it receives on its outer side the tendon of the peroneus, and at the same point it is joined by the short extensor. A small synovial bursa is interposed between the tendon and the anterior ligament of the fetlock, but at the front of the interphalangeal joints the ligament supports directly the articular synovial membranes. At the middle of the first phalanx the tendon is joined on each side by a strong band that descends from the suspensory ligament.

Action.—It extends in succession from below upwards the interphalangeal joints and the fetlock, and finally it flexes the hock.

The PERONEUS (Plate 19). The tendon of this muscle emerges from the thecal canal in the external lateral ligament of the tarsus, and joins the tendon of the last-described muscle about the middle of the metatarsus.

Action.—The same as the preceding muscle.

The SUPERFICIAL FLEXOR (flexor pedis perforatus) (Plate 19). The tendon of this muscle, after playing over the os calcis, descends on the middle line of the limb to the back of the fetlock, where it forms a remarkable ring for the passage of the tendon of the deep flexor. Beyond this point the tendon bifurcates, and each half is *inserted* into the upper extremity of the second phalanx, on its lateral aspect. In connection with the tendon of this and the next muscle there is developed an extensive synovial apparatus, termed the metatarso-phalangeal sheath, which exactly resembles the metacarpo-phalangeal sheath of the fore limb, described at page 34.

Action.—It flexes successively the pastern and fetlock joints; and, by its insertion into the os calcis, it is also an extensor of the hock-joint. It also mechanically maintains the hock in a state of extension as long as the hip and stifle joints are kept extended by their proper muscles.

The DEEP FLEXOR (flexor perforans) (Plate 19). The tendon of this muscle, after its passage through the tarsal sheath, descends between the suspensory ligament and the superficial flexor. At the fetlock it passes through the ring of the last-named muscle, descends behind the digit, plays over the navicular bone, and finally becomes *inserted* into the solar surface of the os pedis (see page 42). At the upper part of the metatarsus it receives the check band, or subtarsal ligament, which is analagous to the subcarpal ligament of the fore limb, but not so strong. Like the analagous structure in the fore limb, it is involved in sprain of the back tendons. A little lower the deep flexor is joined on its outer side by the tendon of the flexor accessorius.

Action.—It flexes successively from below upwards the interphalangeal joints and the fetlock.

Directions.—For the description of the foot, which is identical in the fore and hind limbs, turn to page 35. If the student has already dissected the foot in a fore limb, he may proceed at once to the articulations.

THE STIFLE-JOINT (PLATES 17 AND 18).

This corresponds to the knee-joint of man. It comprises—(1) the articulation between the patella and the femoral trochlea; and (2) the articulation between the condyles of the femur and the proximal end of the tibia.

Directions.—The various structures in connection with the joint are to be examined in the order of the following description; and in order to expose them, the muscles, fat, etc., are to be removed from around the joint, care being taken, in the first stage of the dissection, to preserve the thin femoro-patellar capsule intact.

G

FEMORO-PATELLAR ARTICULATION.

MOVEMENTS.—This joint is commonly classified as an *arthrodia*. The movements (see page 43) of the patella on the trochlea, however, are not those of simple gliding, but of *gliding* with *coaptation*. In the latter movement, while the patella moves as a whole upwards or downwards, successive areas of its articular surface come into contact with the trochlea. These movements take place at the same time as the movements in the femoro-tibial articulation. In complete extension of that joint the patella lies at the upper part of the trochlea, and the three straight patellar ligaments are tense. When flexion takes place, these ligaments become relaxed, and the patella descends over the trochlea till it rests at its lower part.

The ligaments of the joint are—one capsular, two lateral, and three straight.

The CAPSULAR LIGAMENT is loose and membranous, and it supports the synovial membrane. It is attached, on the one hand, to the margin of the patellar articular surface, and, on the other, at the periphery of the trochlea.

The LATERAL LIGAMENTS are two thin, riband-shaped bands, stretching, one on each side of the joint, from the femur to the patella. They serve to strengthen the capsular ligament, from which they are not distinct.

The STRAIGHT PATELLAR LIGAMENTS. These correspond to the single *ligamentum patellæ* of the human knee. They are three in number, and are distinguished as *external*, *middle*, and *internal*. All three ligaments are attached superiorly to the anterior surface of the patella, the inner one having a fibro-cartilaginous thickening which extends the articular surface of the patella, and glides on the inner ridge of the femoral trochlea. The middle ligament lies on a deeper plane than the other two, and rests inferiorly in the vertical groove on the anterior tuberosity of the tibia. It is inserted into the lower part of this groove, and a small synovial bursa is developed between the ligament and the bone above the point of insertion. The external and internal ligaments are inserted into the same tuberosity, one on each side of the attachment of the middle ligament. These three ligaments may be regarded as the terminal tendon of the quadriceps extensor cruris, whose action they transmit to the bones of the leg.

SYNOVIAL MEMBRANE. This will be exposed by incising the capsular ligament. It lines the inner surface of that ligament, and extends upwards beyond the trochlea, forming a protrusion under the quadriceps extensor cruris. Inferiorly it is in contact with the synovial membranes of the femoro-tibial joint, and sometimes it communicates with them.

It is a point worthy of notice in connection with the interior of the joint, that the inner ridge of the femoral trochlea is much higher than the outer ; and when the patella is' dislocated, it is carried outwards over the external ridge.

THE FEMORO-TIBIAL ARTICULATION (FIG. 4).

MOVEMENTS.—This is a ginglymus, or hinge joint, in which the movements are principally *flexion* and *extension*.   In *extension* the bones of the leg are carried forwards, but cannot be brought into a straight line

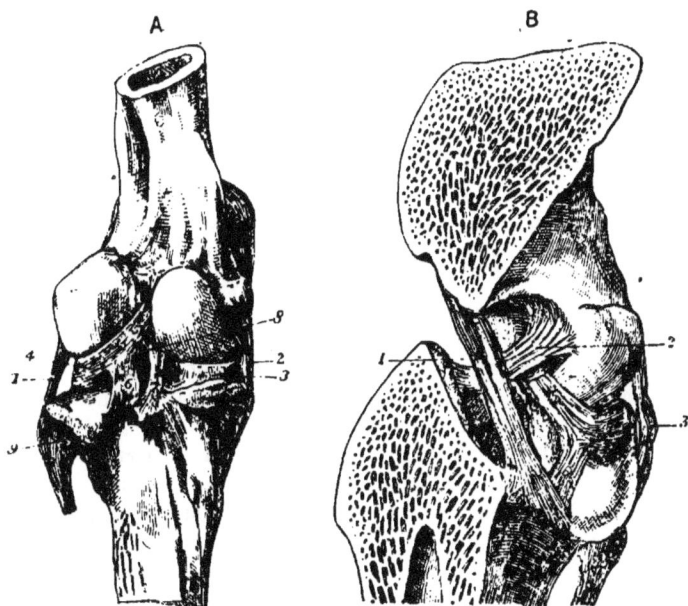

FIG. 4.

A. FEMORO-TIBIAL LIGAMENTS, BACK VIEW.

1. External lateral ligament ; 2. Internal lateral ligament ; 3. Inner semilunar fibro-cartilage ; 4. Outer semilunar fibro-cartilage, with 5, and 6, the femoral and tibial attachments (coronary ligaments) of its posterior extremity ; 7. Posterior crucial ligament ; 8. Anterior crucial ligament ; 9. Head of fibula.

B. ANTERO-POSTERIOR VERTICAL SECTION OF THE FEMORO-TIBIAL ARTICULATION TO SHOW THE CRUCIAL LIGAMENTS.

1. The posterior crucial ligament ; 2. Anterior crucial ligament ; 3. External lateral ligament ; 4, 5, and 6, as in A.

with the femur, the movement being arrested by tension of the lateral ligaments and of the anterior crucial ligament.   The contrary movement, *flexion*, is finally arrested by tension of the posterior crucial ligament. A slight degree of *lateral movement* and *rotation* can be produced when the joint is flexed.

LATERAL LIGAMENTS.—These are two strong fibrous cords, placed one on each side of the joint.   The *external* is fixed above to the higher of the two pits on the external condyle of the femur, where it covers the origin of the popliteus from the lower pit. It descends over the external tuberosity of the tibia, a synovial bursa being interposed, and is inserted into the head of the fibula. The *internal* is longer, but more slender, than the preceding. It is fixed above to a small tubercle on the inner condyle, plays over the inner edge of the tibial articular surface, and is inserted into the internal tuberosity of the tibia.

The POSTERIOR LIGAMENT is of a flattened, membranous character, and consists of a superficial and a deep layer, which are separable from each other superiorly, but blended below. Superiorly the ligament is attached to the posterior surface of the femur above the condyles; below it is inserted into the corresponding surface of the tibia, just below the margin of the articular surface; while laterally its margins blend with the lateral ligaments. The superficial surface of the ligament is related to the popliteal vessels, and to the gastrocnemius, flexor perforatus, and popliteus muscles. Its deep face serves to support the synovial membranes of the joint, and is partly adherent to the semilunar cartilages and posterior crucial ligament. The ligament presents apertures for the transmission of vessels to the interior of the joint.

SYNOVIAL MEMBRANES. These are two in number, one for each condyle of the femur and corresponding part of the articular surface of the tibia. They are separated from each other by the crucial ligaments in the interior of the joint; while behind, and at the sides, they line the posterior and lateral ligaments. In front they are in contact with the synovial capsule of the femoro-patellar articulation, and are supported by a pad of fat, which separates them from the straight ligaments of the patella. A communication frequently exists in front between these synovial capsules and that for the gliding of the patella. These synovial membranes invest the semilunar cartilages; and the external one covers, in addition, the tendon of origin of the popliteus, and the common tendon of origin of the flexor metatarsi and extensor of the digit.

Directions.—The posterior ligament should now be cut away; and the patella being thrown down, the synovial membrane and fat should be removed from the front of the joint. The joint should then be strongly flexed, in order to expose, as far as possible, the crucial ligaments in the intercondyloid groove. The rims of the semilunar cartilages and their coronary ligaments will at the same time be exposed.

The CRUCIAL LIGAMENTS are two strong fibrous cords stretching between the femur and the tibia, and lodged in the intercondyloid groove. They cross one another somewhat like the limbs of the letter X, and hence their name. They are distinguished as anterior and posterior. The *anterior*, the most external of the two, is attached superiorly to the intercondyloid groove, and to the external condyle of the femur where it bounds that groove. Its fibres have a slightly spiral arrangement, and extend downwards and forwards to be inserted into the summit of the tibial spine. The *posterior* ligament is longer than the anterior, and is fixed superiorly to the intercondyloid groove and inner condyle. It extends downwards and backwards to be fixed to a special tubercle on the back of the tibia below the rim of its articular surface. These two ligaments bind the femoral and tibial articular surfaces closely together, and at the same time restrict the movements of the joint, the anterior ligament being put upon the stretch during extension, and finally arresting that movement, while the posterior ligament plays the same part with regard to flexion.

The INTER-ARTICULAR or SEMILUNAR FIBRO-CARTILAGES. These are two crescentic or sickle-shaped pieces of fibro-cartilage, interposed between the condyles of the femur and the articular surface of the tibia. The convex margin of each is turned outwards, and is much thicker than the concave edge, which embraces the tibial spine, and is so thin as to be translucent. The lower surface of each is flattened to rest on the tibia, but the upper surface is hollowed to embrace the femoral condyle. They are fixed in position as follows :—The anterior extremity of the *inner* fibro-cartilage is fixed into an excavation in front of the tibial spine, while its posterior end is similarly fixed behind the spine. The *outer* cartilage is fixed by its anterior extremity in front of the spine, while its posterior extremity is bifid, having an upper slip inserted into a depression at the posterior part of the intercondyloid groove, and a lower into the rim of the tibial articular surface, partly under cover of the posterior interosseous ligament. These slips of insertion at the extremities of the fibro-cartilages are sometimes termed the *coronary ligaments*, three of which belong to the outer, and two to the inner, fibro-cartilage. Although these insertions serve to prevent the total displacement of the fibro-cartilages, some degree of movement is, nevertheless, permitted to the latter; for it will be noticed, that during flexion they are, as it were, squeezed towards the front of the joint, while in extension they are carried backwards.

Directions.—If the internal lateral ligament be now cut, and the internal condyle removed with the saw, a better view will be obtained of the crucial ligaments; after which, complete separation of the femur and tibia should be effected by cutting the remaining lateral ligament, the crucial ligaments, and the slip of insertion of the external fibro-cartilage at the back of the intercondyloid groove. This will expose thoroughly the semilunar fibro-cartilages.

TIBIO-FIBULAR ARTICULATION. In the horse the amount of movement permitted between the bones of the leg is very restricted, and not appreciable on the general movements of the limb. Where the head of the fibula is opposed to the rough diarthrodial facet on the external tuberosity at the upper end of the tibia, short and strong *peripheral fibres* pass between the two bones, and bind them closely together. An *interosseous membrane* extends across the tibio-fibular arch, and is perforated by the anterior tibial vessels. Just above the aperture for the transmission of these vessels the fibres of the ligament are disposed in opposite directions, like the limbs of the letter **X**.

Where the osseous substance of the fibula ceases, a fibrous cord begins, and this is carried downwards to the region of the external tuberosity at the lower end of the tibia, where it mixes its fibres with the external lateral ligament of the tibio-tarsal joint.

THE TARSUS (FIG. 5).

Several articulations are formed in the tarsus, or hock; and these are of very unequal importance as regards the amount of movement permitted. The most important of them is that corresponding to the ankle-joint of man, which is formed between the astragalus and the lower extremity of the tibia; and attention should first be given to the movements that take place here. This is one of the most typical ginglymoid joints in the body, the movements being limited to *flexion* and *extension.* It will be observed that in flexion the distal part of the limb does not move in the plane of the leg, but deviates a little outwards, and that in extension the movement is arrested by tension of the lateral ligaments before the distal portion of the limb is brought into the same straight line as the leg.

In the other articulations found in connection with the tarsus the movements are of a very restricted character, and are not concerned in the general movements of the limb. They, however, serve a no less important purpose in the joint, distributing and equalising pressure, and obviating the bad effects which concussion would have been likely to produce in the tarsus, had it been one rigid structure.

Directions.—The ligaments of the tarsus are both numerous and complicated, and the best order of their dissection is that in which they are hereafter described. Since one set of ligaments must be removed in order to expose the following set, the dissector should not proceed with undue rapidity.

TIBIO-TARSAL LIGAMENTS.—These are four in number, viz., two lateral, an anterior, and a posterior.

The *External Lateral Ligament* consists of a superficial and a deep fasciculus, which cross one another like the legs of the letter **X**. The superficial division, which is the larger of the two, is fixed superiorly to the posterior part of the external tuberosity at the lower end of the tibia, while inferiorly its fibres are inserted into the astragalus, os calcis, cuboid, large metatarsal bone, and external small metatarsal. It is perforated by the thecal canal for the passage of the peroneus tendon. The deep division of the ligament extends downwards and backwards from its point of attachment to the forepart of the external tuberosity of the tibia, and it becomes inserted by distinct slips into the astragalus and os calcis. In order to expose it thoroughly, the superficial division should be cut at its point of attachment to the tibial tuberosity, and dissected downwards, the difference of direction serving to distinguish the fibres of the two divisions.

The *Internal Lateral Ligament* is, like the preceding, a composite ligament, and consists of three divisions, which may be distinguished as

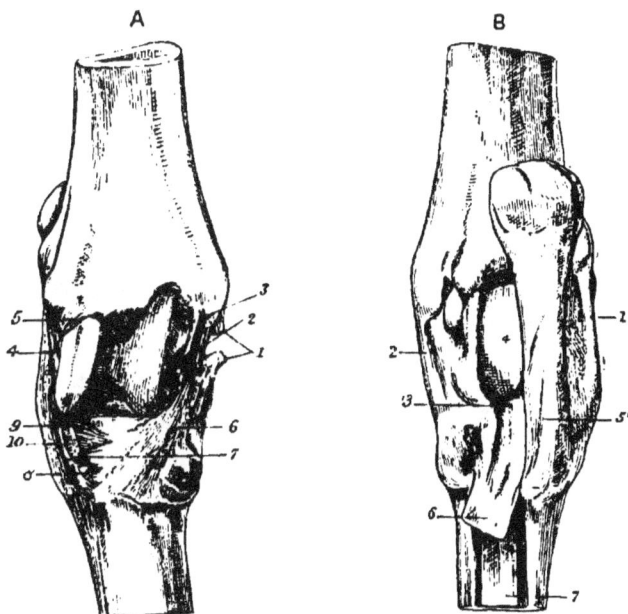

FIG 5.

A.—LIGAMENTS OF THE TARSUS, FRONT VIEW.

1. Superficial fasciculus of the internal lateral ligament (cut); 2. Middle fasciculus of the same (two slips); 3. Deep fasciculus of the same; 4. Superficial fasciculus of the external lateral ligament; 5. Deep fasciculus of the same; 6. Astragalo-metatarsal ligament; 7. Canal for the perforating metatarsal artery; 8. Anterior cuboido-cunean ligament; 9. Anterior cuboido-scaphoid ligament; 10. Cuboid insertion of the flexor metatarsi.

B.—LIGAMENTS OF THE TARSUS, BACK VIEW.

1. External lateral ligament; 2. Internal lateral ligament; 3. Tarso-metatarsal ligament; 4. Fibro-cartilaginous thickening of the posterior ligament. 5. Calcaneo-metatarsal ligament; 6. Subtarsal ligament, or check-band to perforans tendon; 7. Suspensory ligament.

superficial, middle, and deep. The superficial division, the largest of the three, is fixed, on the one hand, to the internal tuberosity at the lower end of the tibia, and, on the other, to the astragalus, scaphoid, large and small cuneiforms, and large and internal small metatarsal bones. The middle division is of intermediate size; and in order to expose it, the superficial division must be cut, and dissected downwards. Above it is attached to the internal tuberosity of the tibia; and, passing downwards and backwards, it is inserted by distinct slips into the astragalus and os calcis. The deep division is very slender, and stretches between the tibia and the astragalus, under cover of the middle fasciculus, which must be removed in order to expose it.

The *Anterior Ligament* is membranous and four-sided. It is fixed above to the tibia; and below to the astragalus, scaphoid, cuneiform magnum, and astragalo-metatarsal ligament; while on each side it blends with the lateral ligament. The posterior surface of the ligament is lined by the synovial membrane of the joint. The anterior surface is related to the anterior tibial vessels, and to the flexor metatarsi and extensor pedis tendons. Towards its inner side the ligament is unsupported; and hence, when the synovial membrane becomes dropsical, the distension shows at that point, constituting a "bog-spavin."

The *Posterior Ligament* is of a similar form to the preceding. It is fixed above to the tibia, below to the astragalus and os calcis, and at the sides to the lateral ligaments. Its anterior surface supports the synovial membrane of the joint; while the posterior is lined by the synovial membrane of the tarsal sheath, and presents a fibro-cartila-ginous thickening where the perforans tendon plays over it. This tendon affords support to the posterior ligament, which therefore does not bulge so readily as the anterior ligament; but in a case of extreme distension of the synovial membrane, the swelling shows itself at the back of the joint.

The *Synovial Membrane* is supported by the anterior, posterior, and lateral ligaments; and it communicates with the synovial membrane that lubricates the articulations between the os calcis and the astragalus on the one hand, and the cuboid and scaphoid on the other. It also sometimes supplies the two upper facets between the os calcis and astragalus.

Directions.—The anterior and posterior ligaments should be incised in order to expose the synovial membrane; and, thereafter, these and the lateral ligaments should be cut away. This will effect the separation of the tibia; and the next group of ligaments may then be examined.

The following ligaments can hardly be classified as belonging specially to any one articulation or set of articulations. For the most part they bind together the series of tarsal bones, and also serve to bind these to the metatarsal bones.

The *Astragalo-metatarsal Ligament.*—This is a flat, radiating ligament, situated on the inner side of the tarsus. Its fibres are attached above to the tubercle on the inner side of the astragalus ; and, widening as it descends, it becomes inserted into the scaphoid, cuneiform magnum, and large metatarsal bone.

The *Calcaneo-metatarsal* or *Calcaneo-cuboid Ligament.*—This is a strong, cord-like ligament, situated at the outer side of the back of the hock, and attached to the posterior border of the os calcis, the cuboid, and the head of the external small metatarsal bone.

The *Tarso-metatarsal Ligament* will be seen covering the tarsal bones at the back of the hock. It forms a thick mass of fibrous tissue intimately adherent to these bones and to the heads of the metatarsal bones. Its inner border is blended with the lateral ligament of the tibio-tarsal joint; and its outer with the calcaneo-metatarsal ligament. Below it is continued as the *subtarsal ligament*, which joins the perforans tendon. The anterior face of the ligament, where not adherent to the bones, is lined by synovial membrane ; and its posterior face is similarly lined by the synovial membrane of the tarsal sheath.

Directions.—At the front of the hock the point of a scalpel should be introduced into the articulation between the astragalus and the scaphoid ; and by cutting round the hock through the three ligaments just described, an attempt should be made to separate the astragalus and os calcis, as a single piece, from the rest of the tarsal bones. Before this can be effected, however, there must be cut an *interosseous ligament*, which is composed of short and strong fibres passing between the os calcis and astragalus on the one hand, and the cuboid and scaphoid on the other. At the same time the *synovial membrane* belonging to the articulations between these two sets of bones will be opened into. This capsule communicates in front with that of the tibio-tarsal joint, and "is prolonged superiorly between the calcis and astragalus, to lubricate two of the facets by which these bones come into contact. In addition, it descends between the cuboid and scaphoid bones, to form a prolongation for the anterior cuboido-scaphoid arthrodia."—*Chauveau.*

Ligaments uniting the Os Calcis and Astragalus.—There are four of these—a *superior*, two *lateral*, and an *interosseous.* The first of these is composed of fibres passing between the two bones above their surfaces of contact ; the lateral ligaments pass between them on each side ; while the interosseous ligament cannot be seen in its entirety, as it passes between the rough impressions on the surfaces of apposition of the bones, and must be cut before these can be separated.

Directions.—Attention should next be turned to the following ligaments, which bind together the other four tarsal bones.

The *Anterior Cuboido-scaphoid Ligament* is of small size, and passes between the two bones from which it is named, above the entrance to the canal by which the perforating metatarsal artery passes through the hock.   The same bones are joined by an *interosseous ligament*, which forms the roof of that canal.

The *Anterior Cuboido-cunean Ligament* connects the cuboid and cuneiform magnum bones below the entrance to the above-mentioned vascular canal; and an *interosseous cuboido-cunean ligament* forms the floor of the canal.

The *Scaphoido-cunean Interosseous Ligament* joins the scaphoid and two cuneiform bones.

The *Intercunean Ligament* passes between the two cuneiforms.

These and the other interosseous ligaments are concealed in the interstices between the different bones which they bind together, and cannot be fully seen until the bones are separated.

Synovial Membranes.—"There is a proper synovial membrane for the facets by which the scaphoid and cuneiform magnum bones correspond; this synovial membrane belongs also to the two cuboido-scaphoid, and posterior cuboido-cunean arthrodiæ.   The anterior cuboido-scaphoid diarthrosis receives a prolongation from the synovial membrane between the os calcis and astragalus on the one hand, and the cuboid and scaphoid on the other.   The play of the anterior cuboido-cunean, and inter-cunean facets is facilitated by two prolongations of the tarso-metatarsal synovial membrane."—*Chauveau.*

THE TARSO-METATARSAL ARTICULATION.

An *Interosseous Ligament* binds the heads of the metatarsal bones to the tarsal bones with which these articulate, and the union is further secured by many of the ligaments, already dissected, which, though they belong to the hock, have points of insertion into the heads of the metatarsal bones.

Synovial Membrane.—This not only supplies the tarso-metatarsal joint, but also ascends between the two cuneiforms, and into the anterior facet between the cuboid and cuneiform magnum.   It also descends into the articulations between the large and small metatarsal bones.

Directions.—For a description of the remaining joints of the hind limb (except the hip), turn to the description of the corresponding articulations of the fore limb (page 50).   The hip-joint is described with the pelvis, at page 338.

TABULAR VIEW OF THE MUSCLES IN THEIR ACTION ON THE JOINTS
OF THE HIND LIMB.

HIP.

Flexors . . . .
- Sartorius.
- Pectineus.
- Psoas magnus.
- Iliacus.
- Tensor vaginæ femoris.
- Rectus femoris.
- Rectus parvus (?)

Extensors . . .
- Semimembranosus.
- Quadratus femoris.
- Middle gluteus.
- Obturator externus.

Abductors . . .
- Superficial gluteus.
- Biceps femoris (anterior half).
- Middle gluteus.
- Deep gluteus.

Adductors . . .
- Sartorius.
- Gracilis.
- Adductor magnus.
- Semimembranosus.
- Pectineus.
- Adductor parvus.

Rotators inwards
- Sartorius.
- Gracilis.

Rotators outwards
- Deep gluteus.
- Adductor parvus.
- Quadratus femoris.
- Psoas magnus.
- Iliacus.
- Obturator externus.
- Obturator internus.
- Pyriformis.
- Gemelli.

STIFLE.

Flexors . . . .
- Biceps femoris (posterior half).
- Semitendinosus.
- Popliteus.

Extensors . . .
- Vastus internus.
- Vastus externus.
- Rectus femoris.
- Biceps femoris (anterior half).

Rotator outwards — Biceps femoris (posterior half).

Rotators inwards
- Semitendinosus.
- Popliteus.

HOCK.

Flexors . . .
- Extensor pedis.
- Peroneus.
- Flexor metatarsi.

Extensors . . .
- Gastrocnemius.
- Soleus.
- Flexor perforatus.
- Flexor perforans.
- Flexor accessorius.

FETLOCK.

Flexors . . . .
- Flexor perforatus.
- Flexor perforans.
- Flexor accessorius.

Extensors . . .
- Extensor pedis.
- Peroneus.
- Extensor brevis.

PASTERN.

Flexors . . . .
- Flexor perforatus.
- Flexor perforans.
- Flexor accessorius.

Extensors . . .
- Extensor pedis.
- Peroneus.
- Extensor brevis.

COFFIN-JOINT.

Flexors . . . .
- Flexor perforans.
- Flexor accessorius.

Extensors . . .
- Extensor pedis.
- Peroneus.
- Extensor br evis.

Name of Muscle.	Origin.	Insertion.	Source of Nerve.
Sartorius	Iliac fascia	Inner straight ligament of patella	Internal Saphenous.
Gracilis	Pubis and ischium, lower face	Inner straight ligament of patella; and tibia, line between anterior and internal tuberosities	
Pectineus	Pubis, brim and inferior surface; and pubio-femoral ligament	Femur, shaft near nutrient foramen	Obturator.
Adductor parvus	Pubis, inferior surface	Femur, posterior surface of shaft	
Adductor magnus	Ischium, inferior surface; and tendon of origin of gracilis	Femur, posterior surface of shaft and supra-condyloid crest (two insertions)	
Obturator externus	Pubis and ischium, lower surface	Femur, trochanteric fossa	
Semimembranosus	Ischium, inferior surface and tuber; and fascia of coccygeal muscles	Femur, inner condyle	Great sciatic.
Quadratus femoris	Ischium, lower surface	Femur, posterior surface of shaft	
Psoas magnus	Ribs, last two; and vertebræ, 16th dorsal to 5th lumbar	Femur, small (internal) trochanter	Lumbar nerves.
Iliacus	Ilium, iliac surface and external angle; and sacro-iliac ligament		
Vastus externus	Femur, outer and anterior surfaces of shaft	Patella	Anterior crural.
Vastus internus	Femur, inner and anterior surfaces of shaft	Patella and its inner straight ligament	
Rectus femoris	Ilium, above acetabulum (two heads)	Patella	
Rectus parvus	Ilium, above acetabulum	Femur, anterior surface of shaft	
Tensor vaginae femoris	Ilium, external angle	Fascia lata	
Superficial gluteus	Ilium, external angle; and gluteal fascia	Femur, third trochanter	Gluteal nerves.
Middle gluteus	Ilium, gluteal surface; ilio-sacral and sacro-sciatic ligaments; gluteal fascia and fascia of longissimus dorsi	Femur, summit, crest, and ridge of great trochanter	
Deep gluteus	Os innominatum, shaft of ilium and supra-cotyloid ridge	Femur, inner side of convexity of great trochanter	

Muscle	Origin	Insertion	Nerve
Biceps femoris	Sacral spines; tuber ischii; sacro-sciatic ligament; gluteal and coccygeal fascia	Femur, posterior surface of third trochanter; patella; tibia, crest; and fascia of leg	Gluteal nerves and great sciatic
Semitendinosus	Sacral spines; tuber ischii; and sacro-sciatic ligament	Tibia, crest	
Obturator internus	Pubis and ischium, pelvic surfaces		
Pyriformis	Ilium, pelvic surface		
Gemelli	Ischium, outer edge	Femur, trochanteric fossa	Great sciatic
Gastrocnemius	Femur, outer lip of supracondyloid fossa, and supracondyloid crest (two heads)	Os calcis, summit	
Soleus	Fibula, head	Tendon of gastrocnemius	
Flexor perforatus	Femur, bottom of supracondyloid fossa	Os calcis (slip to either side); and os coronæ (bifid tendon)	
Flexor perforans	Tibia, external tuberosity and posterior surface of shaft; fibula; and interosseous membrane	Os pedis, semilunar crest and surface behind it	
Flexor accessorius	Tibia, external tuberosity	Tendon of perforans	
Popliteus	Femur, outer condyle	Tibia, posterior surface and inner edge of shaft	Internal popliteal
Peroneus	External lateral ligament of stifle; fibula; and intermuscular septum	Tendon of extensor pedis	Musculo-cutaneous
Extensor pedis	Femur, between trochlea and outer condyle	Os pedis, pyramidal eminence	
Flexor metatarsi { Superficial division	Femur, in common with extensor pedis	Large metatarsal, upper extremity; and cuboid (bifid tendon)	
Flexor metatarsi { Deep division	Tibia, outer surface	Large metatarsal, upper extremity; and cuneiform parvum (bifid tendon)	Anterior tibial
Extensor brevis	Perforans tendon	Tendon of extensor pedis	
Lumbricales (2)		Tissue beneath ergot of fetlock	
Interossei (2)	Splint bone, head	Suspensory ligament, band sent to extensor tendon	Plantar

CHAPTER III.

DISSECTION OF THE BACK AND THORAX.

THE dissection of the thorax should be begun at the same time as that of the outer scapular region (see page 8).

(see page 8).

THE CHEST-WALL AND BACK.

Directions.—The portion of skin remaining on the chest-wall and loins should be removed, the operation being commenced at an incision carried along the middle line, from the withers to the croup. The cutaneous nerves must then be sought, after which the other structures are to be taken up in the order of their description.

CUTANEOUS NERVES. In the back these are derived from the dorsal nerves. One set of branches appears close to the spinous processes; and another a few inches outwards, along the course of the longissimus dorsi muscle. Both of these are derived from the superior primary branches of the dorsal nerves. Over the sides of the chest the cutaneous nerves are derived from the perforating branches of the intercostal nerves, which are dissected with the abdominal muscles (see page 288).

In the loins the cutaneous nerves are derived from the superior primary branches of the lumbar nerves, and the most posterior of them are continued backwards to the skin over the gluteal region.

The PANNICULUS CARNOSUS (Plate 38). This is an extensive muscle adherent to the deep surface of the skin over a large part of the abdomen, thorax, and shoulder. It is fully described at page 287, which see.

Directions.—The panniculus should now be entirely removed, beginning at its upper border.

The LATISSIMUS DORSI. This muscle is partly described at pages 9 and 14, in connection with the dissection of the fore limb. It *arises* by a broad aponeurotic tendon from the series of vertebral spines, beginning about the 4th dorsal, and extending back to the last lumbar. This tendon is not well defined at its inferior border, where it is adherent to the ribs, and blends with the oblique muscles of the abdomen. Posteriorly the tendon becomes continuous with the gluteal fascia. The tendon is succeeded by a thick muscular portion, which contracts and passes to the inner side of the fore limb, where it becomes *inserted* into the internal tubercle of the humerus.

Action.—It is a flexor and an inward-rotator of the shoulder-joint.

The SERRATUS MAGNUS (Plate 4). This muscle will be seen here, as left by the dissector of the fore limb. The student should notice its mode of origin from the ribs (see page 7), and then carefully remove it.

Directions.—The latissimus dorsi must now be removed, beginning below, where its muscular portion was cut by the dissector of the fore limb. This operation must be conducted with care, in order to leave intact the anterior and posterior serratus muscles, whose thin tendons are adherent to that of the latissimus. Indeed, over the last ribs, in an old subject, it will be found impossible to separate the latissimus from the underlying serratus, and the former may there be cut off.

The SERRATUS POSTICUS (Plate 20). (This and the succeeding muscle are described together by Percivall, under the name *superficialis costarum.* It corresponds to the *serratus posticus inferior* of man.) This muscle is provided with an aponeurotic tendon, by which it *arises* from the summits of the vertebral spines from the 11th dorsal to the 2nd lumbar. The inferior border of the tendon has a muscular fringe with eight or nine distinct slips, which are *inserted* into the posterior borders and outer surfaces of the same number of ribs at the end of the series.

Action.—It is a muscle of expiration.

The SERRATUS ANTICUS (*Serratus posticus superior* of man) (Plate 20). This muscle is partly covered by the preceding, whose three anterior slips should therefore be carefully removed, as has been done in Plate 20. It repeats the form of the posticus, having a thin, translucent aponeurotic tendon, which, in front, is confounded with the splenius. By the upper border of this tendon it *arises* from the summits of the dorsal spines from the 2nd or 3rd to the 13th. The inferior border of the tendon is succeeded by the fleshy portion of the muscle, and this is *inserted* into the anterior borders and outer surfaces of the ribs from the 5th to the 13th inclusive.

Action.—To assist in inspiration.

Directions.—The two muscles just described must be removed in order to expose the next layer; and this is to be done by incising the aponeurotic portion of each horizontally, an inch or two above its point of junction with the muscular portion. The portions above the incision can then without difficulty be stripped upwards from the surface of the longissimus dorsi. The lower portions must next be dissected downwards in order to expose the transversalis costarum. In doing this, it will be found that a fibrous septum passes from the aponeurosis of the serratus anticus near its lower border, and, penetrating between the two muscles now exposed, becomes attached to the ribs. This must be cut, and the muscular slips of the serrati must be carefully raised from the transversalis costarum.

VESSELS AND NERVES. A set of *nerves* will be found at the inner

edge of the longissimus dorsi, and another perforating its substance.
Both sets are derived from the superior primary branches of the dorsal
or lumbar nerves.

The *arteries* and *veins* are branches of the dorso-spinal divisions of the
intercostal or lumbar vessels.

The TRANSVERSALIS COSTARUM (Plate 21). (This muscle corresponds
to the *ilio-costalis* and *musculus accessorius* of man.)   This is a composite
muscle extending across the entire series of ribs, being five or six inches
removed from the spine posteriorly, but close to it in front.   Its fibres
are directed forwards and slightly downwards, and it possesses two
series of tendons.   One set, forming slips of origin, is concealed at the
upper edge of the muscle; the other, serving as slips of insertion, is
visible at its lower edge.   By the upper set of tendons it *arises* from the
transverse processes of the first two lumbar vertebræ, and from the
anterior borders of the ribs.   By the lower set of tendons it is *inserted*
into the hinder edges of the ribs anterior to the 14th, and to the transverse process of the last cervical vertebra.

Action.—To pull the ribs backwards, and thus assist in expiration.
Both muscles acting together may also assist in extending the spine;
or acting singly, they may incline it laterally.

The LONGISSIMUS DORSI (Plate 21).   This is the longest and
strongest muscle in the body, and it is also the most complex.   It
extends along the spine, from the sacrum to the neck.   In the loins it
forms a great muscular and tendinous mass (the *common mass* of man);
and anteriorly it is bifurcate, the trachelo-mastoid and complexus
muscles getting origin between its two branches.   Its fibres *arise* from •
the sacral surface of the ilium between the crest and the sacro-iliac joint,
and from a strong, glistening fascia covering the surface of the muscle,
this fascia being fixed to the lumbar and dorsal spines, or to the supraspinous ligament.   Its fibres are *inserted* into the lumbar transverse and
articular processes, the dorsal transverse processes, and the ribs as far
outwards as the edge of the transversalis costarum.   About the 5th rib it
divides; and the lower branch, continuing the outer series of attachments,
is *inserted* into the ribs, and the transverse processes of the first four
dorsal and last four cervical vertebræ; while the upper division, getting
many new fibres from the first four dorsal spines, becomes *inserted* into
the spinous processes of the four cervical vertebræ in front of the last.

Action.—Acting with the opposite muscle, it is the great extensor of
the dorso-lumbar portion of the spine, being, in this respect, the chief
antagonist of the sublumbar and abdominal muscles.   By its costal
attachments it may also assist in expiration.   By its cervical attachments it raises the neck.   Acting singly, it inclines the spine to the
side of the acting muscle.

The RETRACTOR COSTÆ (Plate 45).   This is a small triangular muscle which lies under cover of the last slip of the serratus posticus. It is thin and aponeurotic at its upper edge, where it *arises* from the first two or three lumbar transverse processes. The remainder of the muscle is fleshy, and it is *inserted* by its anterior edge into the posterior border of the last rib. Its lower edge is parallel to the highest fibres of the internal oblique muscle of the abdomen, and it is generally described as a part of that muscle.

Action.—To assist in expiration.

Directions.—Two sets of muscles lie under cover of the longissimus dorsi, viz., the semispinalis of the back and loins, and the levatores costarum. A segment of the longissimus, from the 13th to the 17th rib, should be excised after the fashion of Plate 21; or if it be desired to expose the whole of each series, the longissimus dorsi must be entirely removed.

The LEVATORES COSTARUM (Plate 21). These form a series of small muscles, each occupying the extreme upper part of an intercostal space, and at that point taking the place of the external intercostal muscle. Each *arises* from the transverse process of a dorsal vertebra; and passing downwards and backwards, it expands, and becomes *inserted* into the outer surface of the rib posterior to the vertebra from which it takes origin. In the first two or three spaces the muscles are rudimentary or absent.

Action.—To assist in inspiration.

The SEMISPINALIS of the back and loins (Plate 21). This is a composite muscle, covering the sides of the vertebral spines from the sacrum to the neck, and consisting of numerous fasciculi directed obliquely upwards and forwards. Anteriorly these fasciculi are in series with the semi-spinalis colli, and posteriorly with the curvator coccygis. The fasciculi take *origin* from the lateral lip of the sacrum, from the articular tubercles of the lumbar vertebræ, and from the transverse processes of the dorsal vertebræ. They become *inserted* into the vertebral spines, each fasciculus being inserted into the 3rd or 4th vertebra anterior to the one from which it takes origin. In the forepart of the dorsal region (Fig. 18, page 156) the insertion is into the sides of the spines, but elsewhere it is into or near the summits of the processes.

Action.—It is an extensor or a lateral flexor of the spine, according as the right and left muscles act together or singly.

Directions.—Clean the outer surfaces of a few of the external intercostal muscles about the middle of the series, and at the side of the sternum define the lateralis sterni muscle.

The LATERALIS STERNI. This is a thin, flat muscle, a few inches broad. *Arising* from the outer surface of the 1st rib above its cartilage, it passes obliquely downwards and backwards over the 2nd chondro-costal joint, and over the 3rd and 4th costal cartilages, and

H

is *inserted* into the side of the sternum. Frequently some of its fibres terminate on the 3rd and 4th costal cartilages, or on the aponeurosis over the internal intercostal between these cartilages.

Action.—Acting from its attachment to the 1st rib as its fixed point, the muscle will exert a feeble inspiratory action.

The EXTERNAL INTERCOSTAL MUSCLES (Plate 21). Each muscle of this set occupies an intercostal space, extending from near the spine as far as the lower extremities of the ribs. The muscular fibres of each are fixed by their extremities to the margins of the ribs that bound the intercostal space. They pass obliquely downwards and backwards; and may be considered as having their point of *origin* from the anterior rib, and their *insertion* into the posterior rib.

Directions.—In one or two of the spaces the external intercostal should be removed (see Plate 21) in order to expose the internal muscle, which will readily be distinguished by the different direction of its fibres.

The INTERNAL INTERCOSTAL MUSCLES (Plate 21). These equal in number the external set, one being lodged in each intercostal space. They differ from the external set in that they are prolonged beyond the lower extremities of the ribs to fill the interspaces of the costal cartilages, while in the extreme upper part of the intercostal spaces they are absent or much reduced in thickness. They differ, moreover, in the direction of their fibres, which is oblique downwards and forwards; and each may be viewed as having its *origin* from the posterior rib and cartilage, and its *insertion* into the anterior rib and cartilage, of the space that it occupies. The inner surface of each is lined by pleura, but at present no attempt need be made to expose this.

Action of the intercostal muscles.—The external set and the inter-cartilaginous portions of the internal set are muscles of inspiration. The interosseous portions of the internal set are muscles of expiration.

Directions.—In a few of the intercostal spaces the vessels and nerves should be exposed. They will be found at the hinder edge of the rib, and should be followed upwards and downwards.

INTERCOSTAL ARTERIES. There are seventeen intercostal arteries on each side, one for each space. The first is derived from the superior cervical artery; the second, third, and fourth from the dorsal artery or its subcostal branch; and the remaining thirteen from the posterior aorta. Their points of origin will be seen in the dissection of the cavity of the thorax. Each vessel on gaining the upper extremity of the inter-costal space gives off a large dorso-spinal branch, and then descends behind the rib, with the vein and nerve. The *dorso-spinal* artery sends a branch into the spinal canal by the vertebral foramen, and is then expended in the muscles occupying the costo-vertebral groove at the side of the dorsal spines. In the intercostal space the intercostal artery is accompanied by a vein and nerve, the vein being in front, and the nerve posterior. In the upper third of the space the vessels descend between the outer and inner muscles, and rest in the groove at the posterior edge of the rib. For the rest of their course they are under cover of the hinder edge of the rib, and, generally, between the inner muscle and the pleura ; but, here and there, slips of the inner muscle may pass between the vessels and the pleura. At the lower extremities of the intercostal spaces the arteries behave as follows :—The first six (or seven) anastomose with ascending branches from the internal thoracic artery ; the remainder as far as the thirteenth anastomose with similar branches from the asternal artery ; and the last four run into the abdominal wall and are expended in its muscles, anastomosing with the abdominal and circumflex iliac arteries. In their descent the intercostal arteries give off costal, pleural, muscular, and cutaneous branches.

The INTERCOSTAL VEINS accompany the arteries. On the left side the first joins the superior cervical vein, the next ten or eleven join the left dorsal vein, and the last five or six the great vena azygos. On the right side the first joins the superior cervical vein, the next three join the dorsal vein, and the remaining thirteen the great vena azygos.

The DORSAL NERVES. There are eighteen dorsal nerves, one emerging by the intervertebral foramen behind each dorsal vertebra. Each divides in the foramen to form a superior and an inferior primary branch. The superior primary branch supplies the muscles in the costo-vertebral groove, and the superjacent skin. The inferior primary branch of the 1st nerve, after detaching a very slender intercostal twig, joins the brachial plexus. The 2nd nerve gives a slender branch to the brachial plexus, and is continued as the intercostal nerve of the second space. The inferior primary branches of the succeeding nerves, except the last, are directly continued as *intercostal* nerves. The inferior primary branch of the last (18th) dorsal nerve descends behind the last rib (see pages 292 and 324).

The *Intercostal Nerves.*—These accompany the intercostal vessels, and terminate thus :—The 1st intercostal nerve is very slender and does not reach the bottom of the space ; the six nerves behind the 1st perforate the pectoral muscles and become cutaneous at the side of the sternum ; the others (ten) are continued beyond the lower extremities of the intercostal spaces to be distributed in the abdominal wall.

The intercostal nerves give branches to the muscles of the same name, and about the middle of the intercostal space each gives off a large perforating branch (lateral cutaneous of man) to supply the panniculus and overlying skin.

The LUMBAR NERVES and VESSELS. The superior primary branches of these nerves (six in number) have a distribution in the loins analagous to the corresponding branches of the dorsal nerves in the back. They

supply muscular branches to the muscles over the lumbar transverse
processes, and cutaneous twigs to the skin of the loins and croup.
Branches of the lumbar arteries and veins accompany these nerves.
Each artery sends a spinal branch through the intervertebral foramen.

THE CAVITY OF THE THORAX.

Directions.—In order to expose the thoracic cavity, the chest-walls
must be in part removed ; and it is most convenient, in the first place,
to make the opening on the left side. The trunk should be allowed to
remain in the suspended position. If the diaphragm is intact, and if no
opening quite through the chest-wall has been made in the previous
dissection, then the first step should be to perforate one of the intercostal
spaces with the finger or a blunt instrument. This is to be done in
order to allow the lungs to collapse ; and a sharp instrument must not
be used, lest the surface of the lung might be injured. As soon as the
finger or instrument is withdrawn from the aperture, the air will be
heard to rush in and fill the pleural cavity, which was previously
occupied by the distended lung. This is precisely what occurs when
the chest-wall is perforated in the living animal, in which, in health, the
outer surface of the lung is closely applied to the inner surface of the
wall, following it in all its movements. The lung is kept in this distended state by the atmospheric pressure, which operates on the air
passages in the interior of the lung, but not on its exterior, where the
pressure is borne by the chest-wall ; and the lung is kept thus distended,
in opposition to a strong natural tendency to contract, which it possesses
in virtue of the large amount of elastic tissue in its structure. But
when the wall of the chest is perforated, the pressure of the atmosphere
becomes exerted on the exterior as well as the interior of the lung, and
the unopposed elasticity of the lung texture then comes into play.

By means of the saw and bone-forceps, the ribs, except the first and
a few at the end of the series, are to be removed, the upper section
being made a few inches below the head of each rib, and the lower a
little above the chondro-costal articulation.

Form and Boundaries of the Cavity (Plates 22 and 25).—If the contained organs were removed from the thorax, and a cast were taken of its
interior, it would be found to have an irregularly conical form, but the
symmetry of the cone is largely departed from. The *base* of the cone
is formed by the diaphragm, which, viewed from the thoracic side, is
markedly convex like the roof of a dome. The plane of attachment of
the diaphragm slopes downwards and forwards, so that the antero-
posterior measurement of the cavity is much less below than above.
Moreover, as the diaphragm is dome-shaped, this measurement is less
when taken from its centre than from its sides. It is in consequence of
this configuration of the diaphragm, that the liver, the stomach, and
other abdominal organs lie under cover of the ribs. The *vertex* of the

cone lies in front, and is bounded by the body of the 1st dorsal vertebra above, and at the sides by the 1st ribs, which meet below. The trachea, the œsophagus, the bloodvessels of the fore limb and head, and many important nerves are transmitted through this opening. On transverse section, the thorax is not circular, as a cone is, but gives a heart-shaped outline. It looks as if it had been squeezed laterally; and it might be described as having a *roof*, formed by the dorsal vertebræ and the ribs as far as their angles; a *floor*, much less extensive, formed by the sternum; and lateral *walls*, formed by the ribs and intercostal muscles.

Contents of the Cavity.—In point of size, the lungs are the most important organs in the thorax, the heart coming next. Besides these,

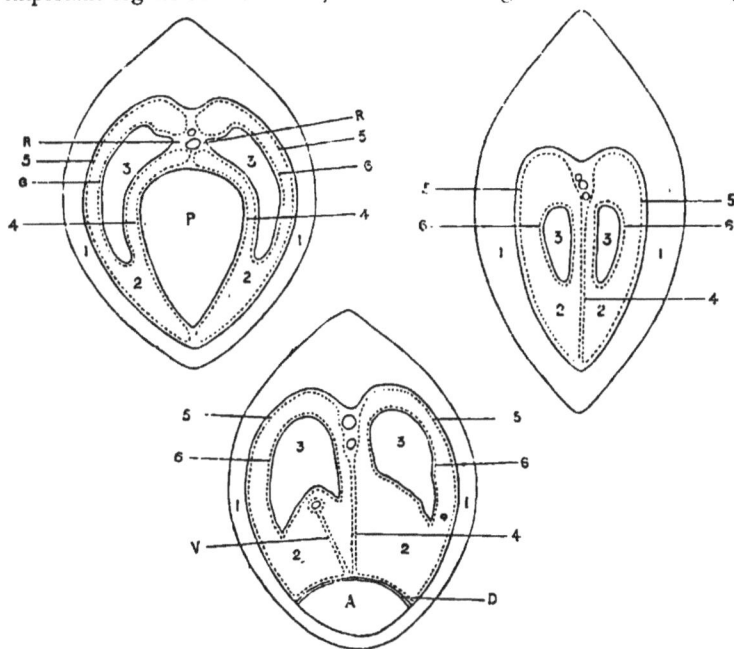

Fig. 6.

1. Chest-wall; 2. Pleural cavity or sac; 3. Lung; 4. Mediastinal pleura (parietal); 5. Costal pleura (parietal); 6. Pulmonic pleura (visceral); A. (No. 3) Abdominal cavity; D. (No. 3) Diaphragm; P. (No. 1) Pericardial sac; R. (No. 1) Root of lung; V. (No. 3) Fold of right pleural membrane enveloping posterior vena cava.

the cavity lodges the main arterial and venous trunks, the thoracic duct, the trachea, the œsophagus, and many important nerves, all of which will be examined in due course.

The PLEURÆ. Each half of the thorax possesses a serous membrane termed the pleura. Like other serous membranes, the pleura is arranged in the form of a shut sac, and consists of a visceral and a parietal portion. The *visceral* pleura is that which invests the lung, and it is therefore termed the *pulmonic* pleura. Around the root of the lung it is continuous with the parietal portion. The *parietal* pleura lines the walls of the chest on the side to which it belongs. It covers the inner surface of the ribs and intercostal muscles, forming the *costal* pleura ; it is spread over the anterior surface of the diaphragm, constituting the *diaphragmatic* pleura ; and towards the middle line of the cavity it, together with the corresponding layer of the opposite side, forms a vertical septum termed the mediastinum. This is the *mediastinal* pleura. Behind the root of the lung a double fold of pleura, termed the *ligamentum latum pulmonis*, is prolonged along the mediastinum to the diaphragm. On the right side of the chest the pleura forms a special fold that includes between its two layers the posterior vena cava and the right phrenic nerve. All these differently named divisions of the pleura are continuous the one with the other ; and they unite to form a close sac termed the pleural cavity. This disposition of the pleura will be more readily understood by reference to the accompanying diagrams (page 101), the first of which represents the arrangement of the membrane at the root of the lung, the second in front of, and the third behind, that point.

These diagrams, it is to be observed, however, are not true to nature ; for, whereas they show a distinct pleural *cavity*, in the living healthy animal that cavity has only a potential existence, the pulmonic, being everywhere in contact with the visceral, pleura. But when air is admitted to the cavity, or when inflammatory or other effusions are poured out from the surface of the membrane, the parietal and the visceral pleura become separated, and the cavity comes to have an actual existence. The free surface of the healthy pleura is exquisitely smooth, and is lubricated by a sparing amount of serous fluid, which gives it a glistening aspect. Its function is to facilitate the movements of the lung on the walls of the chest during respiration. When, in inflammation of the membrane (pleurisy), it loses its smoothness and becomes dry, these movements, which normally give rise to no sensation, are attended with the most acute pain. In structure, the pleura, like other serous membranes, comprises a single layer of endothelial cells forming the free surface of the membrane, and a sub-endothelial layer of fibrous connective-tissue supporting the bloodvessels, nerves, and lymphatics.

The MEDIASTINUM. This, as has already been said, is a septum formed towards the mesial plane of the chest by the approximation of two layers of pleura, one from each sac. At some points the right and left layers are in close contact, as, for example, in front of the heart in a lean subject ; but at other points the layers are pushed apart by organs included between them. The largest of these organs is the heart, opposite which the right and left layers of the mediastinum are distant four or five inches from one another. In the fœtus of the horse, and throughout adult life in some animals, the mediastinum is a complete imperforate septum, there being no communication between the right and left pleural sacs; but in the adult horse the mediastinum immediately behind the heart is cribriform or lace-like, and through the apertures which exist here, a pleural effusion formed on one side passes readily through to the other.

The heart, contained within its pericardial sac, is, as has already been stated, the largest organ in the mediastinum, and it is situated about the centre of that septum. For convenience of description, this division of the mediastinum and the part vertically over it may be termed the *middle mediastinum;* and the portions before and behind this, the *anterior mediastinum* and the *posterior mediastinum* respectively. Adopting this arbitrary division of the mediastinum, the organs included in it may be tabulated thus :—

In the *Anterior Mediastinum.*—The trachea ; the œsophagus ; the axillary and innominate arteries and their collateral branches; the anterior vena cava and its tributaries ; the thoracic duct ; the pneumogastric, recurrent, phrenic, and cardiac nerves ; the tracheal lymphatic glands ; and, in the fœtus and young animal, the thymus gland.

In the *Middle Mediastinum.*—The pericardium and the heart ; the common aorta and its bifurcation into anterior and posterior aortæ ; the terminations of the anterior vena cava and vena azygos ; the pulmonary vessels ; the thoracic duct ; the trachea and its bifurcation into the bronchi ; the œsophagus ; the pneumogastric, phrenic, and left recurrent nerves ; and the bronchial lymphatic glands.

In the *Posterior Mediastinum.*—The posterior aorta, the vena azygos, the thoracic duct, the œsophagus, the œsophageal continuations of the pneumogastric nerves, the left phrenic nerve, and the œsophageal lymphatic glands.

The posterior vena cava, and the right phrenic nerve in the latter part of its course are not in the mediastinum, being included in a special doubling belonging to the right pleural membrane.

THE LUNGS (PLATES 22 AND 25).

The lungs are two in number, and they occupy the greater part of the cavity of the thorax. As now seen, however, they are collapsed, and occupy but a small moiety of the cavity, a condition which makes their examination more easy. Each lung appears to lie somewhat loosely in the chest ; but if it be grasped, and an attempt be made to remove it bodily, it will be found to be attached at a point on its inner surface. This, which is termed the *root* of the lung, is the point where the bronchi and vessels enter it. Each lung presents for examination two surfaces, three borders, a base, and an apex.

The *External* (or *costal*) *surface* is much the larger of the two. It is

smooth and convex, and in health it is closely applied to the chest-wall.
The *internal* (or *mediastinal*) *surface* is moulded on the mediastinum
and the organs contained in it. Thus, it presents opposite the heart
a depression for the lodgment of that organ ; behind that point, and
near the upper limit of the surface, a longitudinal groove for the posterior
aorta ; and beneath that again a second furrow parallel to the first but
not so deep, which is the impress left by the œsophagus. This last
impression is very faint on the right lung. This surface also presents
the *root* of the lung, which is situated close behind and above the
depression for the heart ; and the *broad ligament* of the lung (or *ligamentum latum pulmonis*) already mentioned. In front of the heart, where
this surface is applied to the anterior mediastinum, it is narrow and flat.
The inner surface of the right lung presents posteriorly a small, semi-
detached lobule, not present on the left. The *base* (or *diaphragmatic
surface*) is concave and moulded on the diaphragm. This surface on the
right lung shows the base of the small, semi-detached lobule, and the
posterior vena cava disappearing into the fissure between that lobule
and the main mass of the lung. The *apex* of the lung is pointed, and
lies at the entrance to the chest. The *superior* (or *vertebral*) *border* is
long, thick, and rounded, and it is lodged in the costo-vertebral groove
at the roof of the cavity. The *inferior* (or *sternal*) *border* is short and
sharp; and opposite the heart it is widely notched, a circumstance
which allows the pericardium to be tapped at this point without danger
of wounding the lung. The notch is smaller on the right side. The
posterior (or *diaphragmatic*) *border* circumscribes the base, and the greater
part of it is included between the periphery of the diaphragm and the
chest-wall.

Directions.—The student should now attempt by the following method
to restore the lung as nearly as possible to its natural dimensions
and relations. The nozzle of a pair of bellows should be wrapped
firmly round with a strip of wet cloth until it is made of a convenient
size to fit the trachea, which is to be cut across about the middle of the
neck for its reception. The nozzle is then to be tied tightly into the
trachea with a thick piece of string carried several times round, and the
lung is to be gradually inflated while an assistant guides it into position,
and guards it from being wounded by the cut ends of the ribs. Provided
the lung has not been injured, it can by this means be restored to its
natural position, and the student should then observe the area of
pericardium which is left uncovered at the notch in the lower border.
The right side of the chest may next be opened, making the same
incisions as on the left. On raising the base of the right lung from
the diaphragm, its supernumerary lobule will be seen, and also the
posterior vena cava and right phrenic nerve invested by the special fold
of pleura. The right lung may then be inflated, and the extent of

pericardium left uncovered by lung on this side should be observed. Thereafter, the lung should be reflected towards the spine, and the pleura should be stripped off its root. The vessels, nerves, and bronchus should be isolated by teasing and scraping, rather than by cutting.

The *Root of the Lung*, it will be observed, is placed behind the upper part of the heart; and it is composed of the bronchus, bloodvessels, lymph-vessels, and nerves of the lung, with some connective-tissue. The bronchus enters each root in front and above; the pulmonary veins enter behind; and the pulmonary artery enters in front of the veins.

VESSELS. Two sets of vessels pass to and from the lung at the root, viz., the pulmonary artery and veins, and the bronchial artery and vein.

The *Pulmonary Artery* is the enormous vessel carrying impure (venous) blood from the right ventricle to be purified in the capillaries on the air-cells of the lung. It will be recognised by the thickness of its wall. The *pulmonary veins* bring the purified (arterial) blood back from the air-cells, and discharge it into the left auricle. They form at the root of the lung from two to four trunks, which are extremely short, especially on the left side. The pulmonary vessels are the *functional* vessels of the lung.

The *Bronchial Artery* is a slender vessel entering the lung on the bronchus. It carries nutritive or pure blood to the lung structure, and may therefore be termed the *nutrient* artery of the lung. The blood which it carries is led out of the lung by the *bronchial vein*, which joins the coronary sinus of the heart.

The NERVES of the lung are derived from the vagus, as will be seen at a later stage. They form a plexus at its root, and pass along the bronchi into its interior.

Directions.—Both lungs are to be left until the heart and the thoracic vessels and nerves have been examined. Proceed now to the examination of the pericardium. It is best examined from the left side, and will be sufficiently exposed by hooking the left lung towards the spine.

THE PERICARDIUM (PLATES 22 AND 26).

The pericardium is the bag that contains the heart. It occupies a position about the centre of the thorax, and between the right and left layers of the mediastinum. The sac is fibrous in structure, and is lined internally by a serous membrane. Like the organ which it encloses, the pericardium has a conical form, the point of the cone being fixed to the floor of the sternum from about the third chondro-sternal joint to within an inch of the insertion of the diaphragm across the ensiform cartilage. Above the sac is pierced by the large vessels of the heart, and there its fibrous texture blends with the outer coat of the vessels. Its outer surface is overspread by the mediastinal pleura, which can easily be stripped off.

Directions.—The pericardium should be pinched up, and slit from its apex to near the base of the heart.

The pericardium is considerably larger than the heart which it contains, but this disposition is not very evident until it is opened, when the sac can be pulled away from the heart, and a considerable cavity left between them. The inner surface of the bag and the outer surface of the heart are overspread by a serous covering—the *serous membrane of the pericardium.* The *parietal* division of this membrane is that which lines the sac; the *visceral* division covers the heart and the roots of the great vessels at its base, investing the aorta and pulmonary artery in a common tube. The *visceral* portion is also known as the *epicardium,* and around the base of the heart it is continuous with the parietal division. The free surface of this, as of other serous membranes, is exquisitely smooth, and is formed by a single layer of endothelial cells. Its object is to facilitate the movements of the heart in its sac; and for this purpose, it is kept moist by a minute quantity of serous fluid. As with the pleura, the cavity of the serous sac is only a potential one; but when inflammatory or other effusions are poured out by the membrane, it becomes an actual cavity, and the parietal and visceral layers of the membrane may be pushed widely apart. In old and emaciated subjects this is not infrequently the case, the cavity containing a considerable quantity of watery, dropsical fluid.

THE HEART (PLATES 23 AND 24).

Directions.—The pericardium may now be slit transversely, and the heart should be tilted out by introducing the hand beneath its apex. This will permit the examination of the exterior of the heart without destroying any of its connections; and afterwards, in order to observe accurately its position, it should be restored within the pericardium.

EXTERIOR OF THE HEART. The heart is a hollow muscular organ, and acts the part of a force-pump in maintaining the circulation of the blood. In its interior there are four cavities—two auricles and two ventricles, the auricle of each side being placed above the ventricle. This subdivision of the interior of the heart into cavities is indicated on its exterior by certain grooves. Thus, the *auriculo-ventricular groove* runs around the heart like a belt, and marks off the auricles from the ventricles. Although this groove is carried quite round the heart, it is not very evident in front, being concealed there by the origins of the aorta and pulmonary artery. Two other grooves, one on the right side, the other on the left, descend from the base of the heart, and become continuous a little in front of the apex. These grooves correspond to the edges of the septum which separates the cavities of the right side from those of the left. They are much more distinctly marked on the ventricular portion of the heart, where they are termed the *ventricular grooves.* The grooves of the heart lodge the coronary vessels, and a quantity of fat which is present in all but the most emaciated subjects.

In form the heart resembles a cone compressed from side to side; and its exterior may be described as presenting two surfaces, two borders, a base, and an apex. The surfaces of the heart are formed by the ventricles; and when these cavities are distended, both surfaces are convex. The *right side* of the heart is formed principally by the right ventricle, but partly also by the left. The right ventricular groove descends on it, and crosses round the anterior border a little above the apex. The *left side* belongs chiefly to the left ventricle, but partly in front to the right. It shows the left ventricular furrow crossing towards the anterior border, where it joins the furrow of the opposite side. The *anterior border* of the heart is slightly convex, and has an oblique direction downwards and backwards when the heart is in position. It belongs nearly altogether to the right ventricle, the two ventricular furrows, which denote the position of the septum between the two cavities, becoming continuous round this border a little above the apex. The *posterior border* is thicker and less flaccid than the anterior. It is nearly straight, and is disposed almost vertically when the heart is in position. It belongs entirely to the left ventricle. The *base* of the heart lies above, and is formed by the auricles. At this point the large vessels pass to and from the heart, and form the principal means of its suspension. The left auricle forms the left posterior part of the base, and consists of a sinus venosus into which the pulmonary veins open, and an ear-shaped appendix—the auricula, the latter being most posterior. The right auricle forms the right anterior part of the base, and also consists of a sinus venosus and an auricula, the latter lying in front. The large systemic veins discharge themselves into the sinus venosus of the right auricle. The *apex* of the heart is blunt and firm, and belongs to the left ventricle.

Position and relations.—In order to study these, the heart should now be restored to its natural position within the pericardium.

The position of the heart may be expressed with regard to the skeleton as follows :—It lies beneath the bodies of the dorsal vertebræ from the 4th to the 10th inclusive ; it responds to the four ribs behind the 2nd ; and it is placed above the sternum from about the 3rd chondro-sternal joint to within an inch of its posterior extremity.

In a medium-sized animal the most anterior part of the heart (right auricula) is distant about four or five inches from the entrance to the chest ; the posterior border at its upper part is separated by about the same interval from the tendinous centre of the diaphragm, but at its lower part it is only about an inch in front of the insertion of the rim of the diaphragm across the ensiform cartilage ; and during great distension of the abdominal viscera, the diaphragm may be driven forwards so as to entirely obliterate the interval between it and the heart, a condition which interferes not only with respiration but also with the movements of the heart.

The base of the heart has its mid point a little to the right of the mesial plane of the body, and is distant about six inches from the spine, to which it is suspended by the great systemic vessels.

The apex of the heart lies over the posterior extremity of the sternum, and slightly to the left of the mesial plane.

THE NERVES AND VESSELS OF THE LEFT SIDE OF THE THORAX (PLATE 22).

Position.—It will be most convenient to lower the trunk from its suspended position, and lay the thorax flat on a table, with the left side upwards.

Directions.—Sever the insertion of the scalenus into the 1st rib, and then remove that bone by sawing through its lower extremity and disarticulating its costo-vertebral joints. In these operations take care not to cut the vessels or nerves to the inner side of the rib. In order to follow many of the nerves and vessels of the thorax, but little dissection is necessary, as they show distinctly through the transparent pleura which covers them. The phrenic, pneumogastric, and cardiac nerves should be found in the anterior mediastinum, and traced backwards; the sympathetic chain will be seen at the roof of the cavity, extending under the costo-vertebral articulations.

The Left PHRENIC or DIAPHRAGMATIC NERVE is formed at the root of the neck by the union of three branches (Plate 3), the smallest of which is not constantly present. The inconstant branch is from the inferior primary branch of the 5th cervical nerve, the others are furnished by the corresponding branches of the 6th and 7th cervical nerves. The nerve, as thus formed, enters the chest between the first pair of ribs, passing between the axillary artery and the origin of its inferior cervical branch. Continuing backwards between the layers of the mediastinum, it crosses the common trunk of the dorsal, superior cervical, and vertebral veins, and the pericardium at the level of the common aorta. Behind the heart it passes under the root of the lung, through the posterior mediastinum, and is distributed to the left half of the diaphragm (muscular rim and pillar), of which it is the motor nerve.

The Left PNEUMOGASTRIC, VAGUS, or 10TH CRANIAL NERVE. At the entrance to the chest this nerve lies on the trachea, at the upper edge of the cephalic trunk, and a little below the sympathetic. It crosses in beneath the arch of the left axillary, in company with a cardiac nerve. It is continued backwards across the angle of separation between the anterior and posterior aortæ; and crossing the root of the latter vessel, it reaches the root of the lung, where it divides. The upper division is continued backwards to fuse above the œsophagus with the corresponding branch from the right vagus, this fusion taking place about the middle of the posterior mediastinum. The lower division unites in the same way with a branch from the nerve of the opposite side, the fusion taking place on the left bronchus. The resulting nerves are termed the *superior* and *inferior œsophageal nerves*, and they are continued backwards, the one above, and the other below, the gullet, giving branches to it and accompanying it through the foramen sinistrum of the diaphragm.*

In this part of its course the vagus detaches the following branches :—

1. A *Branch of Communication* with the middle cervical ganglion of the sympathetic (or with the inferior ganglion when the middle is not developed). It is given off within the 1st rib.

2. The left *Inferior (recurrent) Laryngeal Nerve.*—This is detached at the root of the posterior aorta ; and turning round behind the vessel at that point, it gains its inner side, to be included between the artery and the left bronchus, where it receives twigs from the cardiac nerves. It then passes forwards along the lower face of the trachea, in company with a cardiac nerve ; and issuing from the chest, it is continued up the neck to the larynx. As it is included between the aorta and left bronchus, it is related to the bronchial lymphatic glands. Within the thorax the nerve gives branches that pass upwards and forwards to the trachea and œsophagus. The left recurrent is the nerve implicated in "roaring."

3. *Pulmonary Branches.*—These form at the root of the lung a plexus from which filaments are continued into the lung along the ramifications of the air tube.

DORSAL ROOTS of the BRACHIAL PLEXUS. These are two branches of the 1st and 2nd dorsal nerves respectively. They will be found at the upper part of the 1st and 2nd intercostal spaces. After giving branches to the inferior cervical ganglion, they turn round the inner surface of the 1st rib, close to its upper extremity.

The SYMPATHETIC NERVE. The cervical cord of the sympathetic, which in the neck is fused with the vagus, separates from it at the entrance to the chest, and terminates in a stellate greyish ganglion—the middle cervical ganglion.

The *Middle Cervical Ganglion.*—This will be found within the 1st rib, or in front of it, at the line of contact of the trachea and œsophagus. A thick connecting branch continues it up to another enlargement—the inferior cervical ganglion. The middle cervical ganglion has a branch of communication with the vagus, and gives off two or three cardiac nerves.

---

* The superior œsophageal nerve is generally, if not always, larger than the inferior ; and in most cases I have found that the upper nerve is formed in greater proportion by the left vagus than by the right, while the lower is formed about equally from each. Chauveau, on the other hand, describes and figures the upper nerve as being formed mainly by the right vagus, and the lower by the left vagus.

The *Inferior Cervical Ganglion* is placed a little above the preceding, to which it is connected by a short thick nerve. It rests on the longus colli, between the vertebral and superior cervical arteries. It is joined by the vertebral nerve, and by short branches from the inferior primary divisions of the 8th cervical and first two dorsal nerves. By its posterior extremity it is continued into the dorsal cord of the sympathetic. It gives off a cardiac filament.

Cardiac Nerves of the left side.—These nerves, like many others, have a variable disposition, but the following is what I have found to be the most common arrangement.

The middle cervical ganglion detaches two cardiac nerves : (1) One of these (which may be double at its origin), the smaller of the two, is distributed to the great arteries in the anterior mediastinum. (2) The other immediately divides into two branches—a lower and an upper. (*a*) The lower branch, joined by a filament from the vagus, passes beneath the arch of the left axillary in company with the vagus, and reaching the angle of bifurcation of the common aorta, it divides, one branch continuing backwards on the posterior aorta to dip down between the right and left divisions of the pulmonary artery and gain the left auricle, while the other descends along the common aorta, and uniting at the origin of the right coronary artery with a cardiac branch of the right side, is distributed to the roots of the great arteries and to the ventricles, the largest branches following the right coronary artery. (*b*) The upper division of the second nerve, passing to the inner side of the left axillary, and along the lower face of the trachea, unites with a right cardiac filament, crosses to the right of the common aorta, and is reflected round that trunk to gain the left coronary artery, its divisions following the main branches of that vessel.

The inferior cervical ganglion detaches a slender cardiac nerve which, after throwing off some twigs to the arteries in the anterior mediastinum, passes downwards and backwards to the left auricle.

[" The cardiac nerves of the horse (left side) ordinarily have the following disposition : There are found four nerves, two of which, very slender, proceed from the middle cervical ganglion and lose themselves on the vessels arising from the convexity of the brachial trunk. The two others are the one superficial, the other deep. The superficial nerve, the more voluminous, commences by a filament which springs from the middle cervical ganglion, passes backward and downwards, contracts beneath the brachial trunk an anastomosis *en arcade* with a branch detached from the inferior cervical ganglion, and then places itself alongside of the following. The deep nerve is formed at first by three elements : (1) of medullary fibres furnished by the spinal pairs ; (2) of a ramuscule furnished by the cervical cord of the sympathetic ; (3) of a slender filament which proceeds from the left pneumogastric at the entrance to the chest. It places itself in the direction of the heart, adheres to the superficial nerve, is inflected on the concavity of the brachial artery, margins this vessel to the left and insinuates itself between the aorta and the pulmonary artery. At this point these nerves are distributed to the heart and to the great vessels, a branch passes under the right auricle and plunges into the cardiac muscle; a second spreads itself over the origin of the pulmonary artery and over the right ventricle; two other branches, grayish, plexiform, anastomose more or less between the aorta and the pulmonary artery, unite under the aortic root with a nerve which comes from the right side, then descend in the vertical furrow of the heart, and are expended in the left ventricle; finally, some other ramuscules, parallel to the pneumo-gastric, are expended on the pulmonary artery and on the aorta."—*Chauveau*. "Traité d'anatomie comparée des animaux domestiques."]

The left *Dorsal Cord of the Sympathetic.*—This will be seen through the transparent pleura, extending beneath the costo-vertebral articulations. The first portion of the cord is concealed at the outer edge of the longus colli muscle, where it joins the inferior cervical ganglion. It crosses the intercostal vessels superficially; and in company with it, from the 6th intercostal space backwards, is the great splanchnic nerve. Poster-iorly it passes between the psoas parvus and the left crus of the diaphragm, and is continued as the lumbar cord. The cord is studded with ganglia of a flattened form and greyish colour, there being a gang-lion for each intercostal space. Each ganglion is placed at the posterior part of the space to which it belongs, and partly on the posterior rib. It is connected by an *afferent* filament to the intercostal nerve of the same space, and from it proceed other branches, which are sometimes named *efferent*. The efferent branches from the first five or six ganglia are very small, and pass to the adjacent arteries, ligaments, or vertebræ. The efferent branches from the succeeding ganglia unite to form the splanchnic nerves.

The *Great Splanchnic Nerve* lies to the inner side of the gangliated cord, as far as the 15th intercostal space. There it crosses to the outer side, and is continued backwards to enter the abdomen by passing between the psoas parvus and the rim of the diaphragm. In the abdomen it joins the semilunar ganglion. The first efferent filament contributing to the formation of the nerve comes usually from the 6th ganglion, and the last from the 16th. The intermediate ganglia con-tribute irregularly, some sending no branch, in which case the next ganglion contributing sends a branch of more than the usual size.

The *Small Splanchnic Nerve* is either the efferent filament from the 17th ganglion, or it is formed by the union of that and the filament from the 16th. It passes directly to the solar, the renal, or the supra-renal plexus.

The PULMONARY ARTERY. This is a short vessel of enormous calibre. It springs from the conus arteriosus of the right ventricle; and passing in front of the common aorta, it gains its left side, crosses the root of the posterior aorta, and divides behind it into a right and a left branch, one for each lung. Each of these enters the root of the lung and divides. As the trunk of the pulmonary artery rests on the root of the posterior aorta, it is connected to it, in the adult, by a fibrous cord which is the remains of the *ductus arteriosus*—a vessel which in the

fœtus brings the two arteries into communication. The pulmonary
artery conveys venous blood to the lungs to be purified.

The COMMON AORTA. This is the primary trunk of the *systemic*
arteries. It is of great calibre, but not more than three inches in length.
It springs from the left ventricle, and divides into two unequal vessels—
the anterior and the posterior aorta. Where the vessel springs from
the ventricle, it shows, when injected, three bulgings, each corresponding
to a *sinus of Valsalva*. From two of these sinuses spring the right and
left coronary arteries of the heart. These, which are the first collateral
branches of the arterial tree, are described with the heart.

The POSTERIOR AORTA is by far the longer of the two terminal branches
of the common trunk, and it has also the greater calibre. It passes
backwards and upwards, describing a curve—the *arch* of the aorta, and
reaches the spine at the 10th dorsal vertebra. From that point it is
continued backwards along the vertebral bodies, being at first a little to
the left of the middle line; but it gradually inclines to the right, until,
at the 14th dorsal vertebra, it lies almost entirely to the right of the
median plane of the body. It passes into the abdominal cavity through
the hiatus aorticus—an opening between the pillars of the diaphragm.
The arch of the vessel is crossed to the right by the œsophagus, and by
the termination of the trachea. The remaining portion of the artery
is related on its right to the thoracic duct and vena azygos, the duct
being usually between the vein and artery, but sometimes to the left of
the latter. The thoracic branches of the posterior aorta are :—

1. The *Broncho-œsophageal Artery.*—This vessel will be more conveniently dissected with the right side of the chest. It is described at
page 118.

2. *Intercostal Arteries.*—The last thirteen of these generally have this
origin. They spring from the upper aspect of the artery, and pass over the
vertebral bodies, crossing beneath the dorsal cord of the sympathetic to
gain the upper end of an intercostal space. Here each gives off a large
dorso-spinal branch, and places itself at the posterior border of a rib,
along which it descends. The latter part of the intercostals and their
dorso-spinal branches have already been followed in the dissection of the
chest-wall and back.

The ANTERIOR AORTA. This vessel, after a course of not more than
three inches, divides into two vessels of unequal size. The left and smaller
of the two is the left axillary artery; the other is the arteria innominata. The direction of the anterior aorta is oblique upwards and forwards, and it is in great part included within the pericardial sac. It has
no collateral branches of a size meriting description. Of its terminal
branches only the left axillary will be followed now The left axillary
is the vessel for the supply of the neck, the fore limb, and the subjacent
part of the chest-wall on the left side; while the arteries innominata,

besides supplying the corresponding parts on the opposite side, carries blood for the head.

The LEFT AXILLARY ARTERY is smaller than the other division of the anterior aorta, and placed at a higher level. It passes forwards in the anterior mediastinum, describing a curve which has its convexity directed upwards and forwards. It leaves the chest by passing to the inner side of the 1st rib, and turns round the anterior border of the bone, where it leaves a smooth impression below the lowest fibres of the scalenus. From this point it is directed downwards and backwards across the inner aspect of the shoulder, beyond which it is continued as the brachial or humeral artery. The vessel has thus a part within the thorax, and another in the axilla; but only the former presents itself now. In the human subject the artery passes beneath the clavicle and is termed the *Subclavian*. The arch formed by the thoracic part of the left axillary rests to its right on the trachea, and touches at its highest point the œsophagus. Beneath the arch the vagus and phrenic nerves, and one of the left cardiac nerves pass backwards. The common trunk of the left dorsal, superior cervical, and vertebral veins crosses it on the left, in passing down to the anterior vena cava. The collateral branches of the artery arising within the chest are four in number, three of them, viz., the dorsal, superior cervical, and vertebral arteries, arise from the summit of the arch ; the other, the internal thoracic or mammary artery, takes origin from its lower aspect, at the hinder edge of the 1st rib.

The DORSAL ARTERY passes upwards and forwards across the œsophagus and longus colli muscle, and disappears at the upper end of the 2nd intercostal space. At the outer edge of the longus colli it gives off the *subcostal* artery, a vessel which furnishes the 2nd, 3rd, and 4th intercostal arteries. The 2nd intercostal artery may arise directly from the trunk of the dorsal. External to the chest the dorsal artery is distributed to the parts beneath the scapula, and to the upper part of the neck.

The SUPERIOR CERVICAL ARTERY arises a little in advance of the preceding. It crosses the œsophagus and longus colli, and perforates the upper end of the 1st intercostal space. It supplies the 1st intercostal artery ; and, external to the chest, it is distributed in the neck.

The VERTEBRAL ARTERY has its origin a little in front of the preceding. Passing obliquely upwards and forwards, it crosses the inner side of the 1st rib near its upper extremity, and enters the root of the neck to pass in succession through the series of vertebral foramina.

The INTERNAL THORACIC ARTERY arises at the inner side or hinder edge of the 1st rib. It descends along the inner face of the rib to the floor of the chest, where it will be followed at a later stage.

The SUPERIOR CERVICAL and EXTERNAL THORACIC ARTERIES are given off from the axillary at the anterior edge of the 1st rib, and their roots may be seen now, but they are distributed to parts without the thorax.

I

The ANTERIOR VENA CAVA. This large vessel will be seen below the large arteries in the anterior mediastinum. It is formed at the entrance to the chest by the union of the axillary and jugular veins of both sides, and it terminates in the right auricle. It is better seen on the right side of the thorax, and will be more fully described in that connection. It receives the following branches on this side :—

1. The *Internal Thoracic*, which accompanies, and exactly corresponds to, the homonymous artery.

2. A large venous trunk formed by the union of the *vertebral, superior cervical*, and *dorsal veins.* It crosses to the left of the axillary artery to reach the vena cava. Very exceptionally, as in Plate 23, the vertebral vein may join the cava independently. The vertebral and superior cervical veins exactly correspond to the arteries of the same names, but the *subcostal* root of the dorsal vein is of greater extent than the corresponding artery, for it drains the intercostal spaces from the 3rd to the 11th or 12th. The left dorsal vein is also called the *small vena azygos*. These veins are superficially placed to the corresponding arteries as they lie on the œsophagus and longus colli.

Intercostal Veins.—The last five or six of the left side join the *great vena azygos*, a vessel of the right side of the chest.

The THORACIC DUCT. This is the largest lymphatic vessel in the body, and has a calibre about twice that of a goose quill. It will be most readily found in the angle of separation of the anterior and posterior aortæ, resting on the trachea, at the lower edge of the œsophagus. It will be recognised as a very thin-walled vessel, empty or with a small amount of coloured contents, so that it might be mistaken for a vein. There is not, however, any vein of so large a size in this situation. Open it, and pass a blunt probe along it towards the entrance of the chest. Most commonly the duct has the following course :—Entering the chest by the hiatus aorticus, to the right of the posterior aorta, it extends forwards along the spine, having the aorta on its left and the great vena azygos on its right. It descends from the spine on the right side of the aortic arch, crosses the before-mentioned angle, where it rests on the trachea. It then passes to the right side of the left axillary artery, and dips down between that vessel and the arteria innominata. It terminates at the anterior edge of the 1st rib, in the angle of junction of the left jugular with its fellow or with the left axillary vein, that is, at the beginning of the anterior vena cava. At its termination it is slightly dilated, and furnished with a valve. The duct may be found to the left of the posterior aorta, or it may be double as far as the heart, there being a branch on each side of the aorta. The thoracic duct discharges into the venous system the lymph collected throughout the whole animal except the right fore limb, and the right side of the head, neck, chest-wall, and diaphragm.

It is by this channel, also, that the chyle absorbed from the intestine enters the red-blood vessels. The before-mentioned exceptional areas are drained by the right lymphatic duct, a short vessel to be sought afterwards on the right side.

The TRACHEA. The thoracic portion of the windpipe is situated in the middle plane of the cavity. Entering between the first pair of ribs, it passes backwards through the anterior mediastinum ; and over the base of the heart it bifurcates to form the right and left bronchi. The angle of bifurcation is under the 6th dorsal vertebra. It is related above to the œsophagus and right longus colli ; and below to the cephalic trunk, arteria innominata, anterior vena cava, and right auricle. On its left side are the arch of the axillary, the thoracic duct, and the arch of the posterior aorta. On the right side it is crossed near its termination by the great vena azygos, as will be seen at a later stage. It is also related to the vagus, recurrent, sympathetic, and cardiac nerves ; and to the prepectoral, tracheal, and bronchial lymphatic glands.

The ŒSOPHAGUS. At the entrance to the chest the gullet lies above the trachea, and a little to its left side ; but as it passes backwards beneath the longus colli muscle, it mounts on to the middle of the upper face of the trachea, and passes directly over its bifurcation, having the arch of the aorta to its left. Beyond that it enters the posterior mediastinum, between whose layers it passes, a few inches below the spine, to perforate the diaphragm by the foramen sinistrum. At its entrance into the chest, and for some distance beyond that point, the muscular wall of the tube is red, but behind the heart it is pale.

The *Structure* of the œsophagus is described at page 150,·that of the trachea at page 149.

LYMPHATIC GLANDS of the thorax. The following groups of glands may be seen at this stage :—

1. *Œsophageal* glands of small size, along the œsophagus, between the layers of the posterior mediastinum.

2. *Bronchial* glands, situated at the bifurcation of the trachea, and extending along the bronchi. The lymphatic vessels of the lung join these.

3. *Tracheal* or *Cardiac* glands, a double chain of glands on the lower face of the trachea, in the anterior mediastinum, and placed on the course of the lymphatic vessels from the heart.

4. A series of small glands beneath the pleura, at the upper extremities of the intercostal spaces.

5. *Prepectoral* glands.—These belong to the neck rather than to the thorax, but some of them may have been left by the dissector of the former region. They are situated at the entrance to the chest, beneath the great vessels.

The THYMUS GLAND. In the fœtus this is a considerable organ, composed of lymphoid tissue, and included between the layers of the anterior mediastinum. It steadily atrophies after birth, and in the adult only the shrivelled remains of it will be found.

THE NERVES AND VESSELS OF THE RIGHT SIDE OF THE THORAX
(PLATES 25 AND 26).

Directions.—Reverse the position of the thorax, turning the right side upwards, and proceed as already directed for the display of the structures on the left side (page 108).

The RIGHT PHRENIC NERVE. This nerve enters the chest by passing beneath the right axillary artery, being included between that vessel and the anterior vena cava. In the anterior mediastinum it lies on the side of the anterior vena cava. It crosses the pericardium as on the left side, and behind the heart it passes across or below the posterior vena cava to reach the diaphragm, where it terminates. Behind the heart the nerve and the vena cava are included between the layers of a special fold of pleura which passes upwards from the diaphragm and floor of the chest to envelop them.

The RIGHT VAGUS. Separating from the cervical cord of the sympathetic, the right vagus enters the chest by passing under the arch of the right axillary in company with a cardiac nerve, having the anterior vena cava below. It is then directed obliquely backwards and upwards across the trachea; and crossing to the inner side of the great vena azygos, it divides at the line of contact of the gullet and windpipe. Each branch unites, as already described, with the corresponding branch of the left vagus, thus forming the superior and inferior œsophageal nerves. The thoracic branches of the right vagus are :—

1. *Branches of Communication* with the middle and inferior cervical ganglia of the sympathetic.

2. The right *Inferior (recurrent) Laryngeal.*—This nerve differs from the left in its relations and point of origin. It is given off from the vagus at the origin of the dorso-cervical artery. Turning round behind the root of this trunk, between it and the trachea, it passes forwards on the lower face of the windpipe, above the cephalic artery, and internal to the middle cervical ganglion of the sympathetic. Reaching the root of the neck, it crosses between the carotid artery and the trachea, and is continued up the neck below the artery. In the larynx it is distributed in the same manner as the left. In the chest it communicates with the cardiac nerves and with the middle cervical ganglion of the sympathetic, and emits tracheal and œsophageal filaments as on the left side. The right recurrent nerve, it will be observed, is considerably shorter than the left, having its origin at the posterior edge of the 1st rib, while the left has its origin at the base of the heart. Moreover, the

right is reflected round a comparatively small artery, while the left is reflected round the great aorta. The right recurrent nerve is not implicated in "roaring."

3. *Cardiac Branches*, variable in number, pass downwards and backwards to reach the lower face of the trachea, whence, after anastomosing intricately with the sympathetic cardiac nerves, they pass on to the right auricle.

4. *Pulmonary Branches* as on the left side.

DORSAL ROOTS of the BRACHIAL PLEXUS. These do not differ from those of the left side (page 109).

The SYMPATHETIC NERVE.

The *Middle Cervical Ganglion.*—This resembles that of the left side. It is placed on the trachea, internal to the insertion of the scalenus muscle into the 1st rib. It receives the cervical cord of the sympathetic in front, and behind it is prolonged by a short cord connecting it to the inferior cervical ganglion. It communicates with the vagus and recurrent nerves, and gives off the cardiac nerve accompanying the vagus beneath the axillary artery.

The *Inferior Cervical Ganglion* is situated on the longus colli, at the upper edge of the trachea, and between the vertebral and superior cervical arteries. It receives the vertebral nerve and branches from the inferior primary divisions of the 8th cervical and first two dorsal nerves, and is continued into the dorsal cord of the sympathetic as on the left side. It emits a cardiac nerve.

Cardiac Nerves of the right side.—(1) The middle cervical ganglion detaches a considerable cardiac nerve which accompanies the right vagus in passing back beneath the arch of the axillary artery. Reaching the lower face of the trachea, it unites with one of the cardiac nerves of the left side, and is reflected behind the common aorta to be distributed to the left side of the heart, as already described. This nerve emits a branch to unite with another cardiac nerve of the left side—that which follows the right coronary artery. (2) The inferior cervical ganglion gives origin to a cardiac nerve, smaller than the preceding, which it joins after giving fibres to the right vagus and recurrent. (3) The cardiac branches of the right vagus have already been seen.

["On the right side we reckon two principal cardiac nerves and four secondary filaments. The first cardiac nerve is a long branch which takes origin at the level of the middle cervical ganglion. It is formed by fibres from the sympathetic and by a fasciculus furnished by the right pneumogastric, at the entrance to the chest; it receives probably also some medullary fibres through the intermediation of a branch of communication thrown between the middle ganglion and the inferior ganglion. This nerve is reinforced by two filaments which proceed from the inferior cervical ganglion, and sometimes from the second middle ganglion, of which one, the posterior, is reinforced in the same way by a left sympathetic filament which gains its destination in passing alongside of the recurrent nerve. When it is fully constituted the first right nerve creeps over the base of the heart, turns round the root of the aorta, and mixes its terminal filaments with those of the left cardiac nerves. The second right cardiac nerve is formed by the union of three branches which take origin in succession from the corresponding pneumogastric, behind the dorsal artery, along the right side of the trachea. This nerve is in communication with the sympathetic of the dorsal region by three branches which approach the last beneath the first, fourth, and sixth ribs.

When the second right nerve arrives above the termination of the anterior vena cava, it divides into two branches : the one is thrown into the roof of the auricles ; the other, reinforced by a filament coming from the pneumogastric, is expended, by numerous filaments, on the surface of the left ventricle ; some reaching as far as the right ventricle.

The four secondary filaments are arranged like the steps of a ladder on the portion of the pneumogastric comprised between the entrance of the chest and the division of the bronchi. These filaments are expended in the great vessels and in the walls of the heart." —*Chauveau*. "Traité d'anatomie comparée des animaux domestiques."]

The *Right Dorsal Cord of the Sympathetic* does not differ materially from the left.

The POSTERIOR AORTA is here seen in a large part of its course, but it has already been fully described in connection with the left side. It detaches to this side thirteen intercostals, exactly similar to those of the left.

The BRONCHO-ŒSOPHAGEAL ARTERY. This artery arises from the convexity of the aortic arch, a little anterior to the bifurcation of the trachea. Generally, as in Plate 24, it arises not independently but as a division of a short vessel which is at the same time the common trunk for the 1st and 2nd pairs of aortic intercostals. It is a small vessel, not larger than an intercostal. It is reflected downwards and backwards on the right side of the aorta, and divides into the *bronchial* trunk and the *œsophageal* artery. The *œsophageal* artery, which is the smaller of the two, is continued backwards above the gullet, through the posterior mediastinum, extending sometimes to near the foramen sinistrum, and anastomosing with the pleuro-œsophageal branch of the gastric artery. Sometimes there is an analogous vessel in the mediastinum below the œsophagus (inferior œsophageal), but when present this is a very slender artery. The *bronchial* trunk dips down between the aorta and the gullet, and bifurcates to form the right and left bronchial arteries, each of which enters the root of the lung on the bronchus. It is the nutrient vessel to the lung. The above-mentioned inferior œsophageal may be a branch of one of the bronchial arteries.

The ARTERIA INNOMINATA is the right division of the anterior aorta, the left axillary artery being the other division. In calibre it is greater than the left axillary, and it is placed on a lower level. It is related to the trachea above ; and to the anterior vena cava below and to the left. After a course of about two inches, it divides to form the cephalic trunk and the right axillary artery, and immediately in front of its point of division it detaches the dorso-cervical artery.

The DORSO-CERVICAL ARTERY. This is a short trunk which passes upwards on the trachea, and divides to form the dorsal and superior cervical arteries, which have precisely the same course and distribution as those of the left side. They have also the same connections, save that they do not touch the œsophagus.

The CEPHALIC ARTERY. This vessel, which has a length of about two or three inches, passes directly forwards at the lower face of the trachea, and bifurcates at the entrance to the chest, forming the common carotid arteries (right and left).

The RIGHT AXILLARY ARTERY. This vessel in its intrathoracic course forms a continuous curve, or arch, with the arteria innominata; this arch being, however, on a lower level, and less abrupt, than that of the left axillary. It gives off here the vertebral and internal thoracic arteries.

The VERTEBRAL ARTERY and the INTERNAL THORACIC ARTERY do not differ from the homonymous vessels of the left side.

The ANTERIOR VENA CAVA. This large vessel, already referred to, is best seen from the right side of the chest. It is formed at the entrance to the chest by the union of the jugular and axillary veins of both sides, and its initial portion is fixed by fibrous processes to the inner surfaces of the first pair of ribs. It passes backwards through the anterior mediastinum, being there related to the great arteries, beneath and to the right of which it is placed.* It enters the roof of the right auricle. Besides the vessels already seen entering it on the left side (page 14), it receives—

1. The *Internal Thoracic Vein.*

2. The *Vertebral Vein.*

3. A trunk formed by the union of the *dorsal* and *superior cervical veins.* (These veins may enter independently.)

4. The *Great Vena Azygos* (sometimes). This large vein begins behind the hiatus aorticus, where it receives the first pair of lumbar veins. Passing through the hiatus, it extends along the dorsal portion of the spine to the right of the posterior aorta, the thoracic duct being usually between the two vessels. At the 6th or 7th dorsal vertebra it leaves the spine and curves downwards to the right of the aortic arch, the œsophagus, and the trachea; terminating either in the anterior vena cava, or in the auricle immediately behind the opening of that vein. Besides the first pair of *lumbar* veins, it receives the last thirteen *intercostal veins* of the right side and the last five or six of the left.

The POSTERIOR VENA CAVA. This great vein enters the thorax through the foramen dextrum of the diaphragm. It passes directly forwards to terminate in the right auricle, being included between the main mass of the right lung and its internal lobule, and placed at the upper edge of a double serous fold belonging to the right pleural membrane. The right phrenic nerve is in company with it.

The RIGHT LYMPHATIC DUCT. This is a short lymphatic vessel (not

---

* The vein is in the natural position in Plate 25. In Plates 24 and 26 it is represented as smaller than natural, and slightly lowered in position, in order to expose the arteries.

more than two inches) which empties itself into the initial part of the
anterior vena cava, at the angle of junction of the jugular and axillary

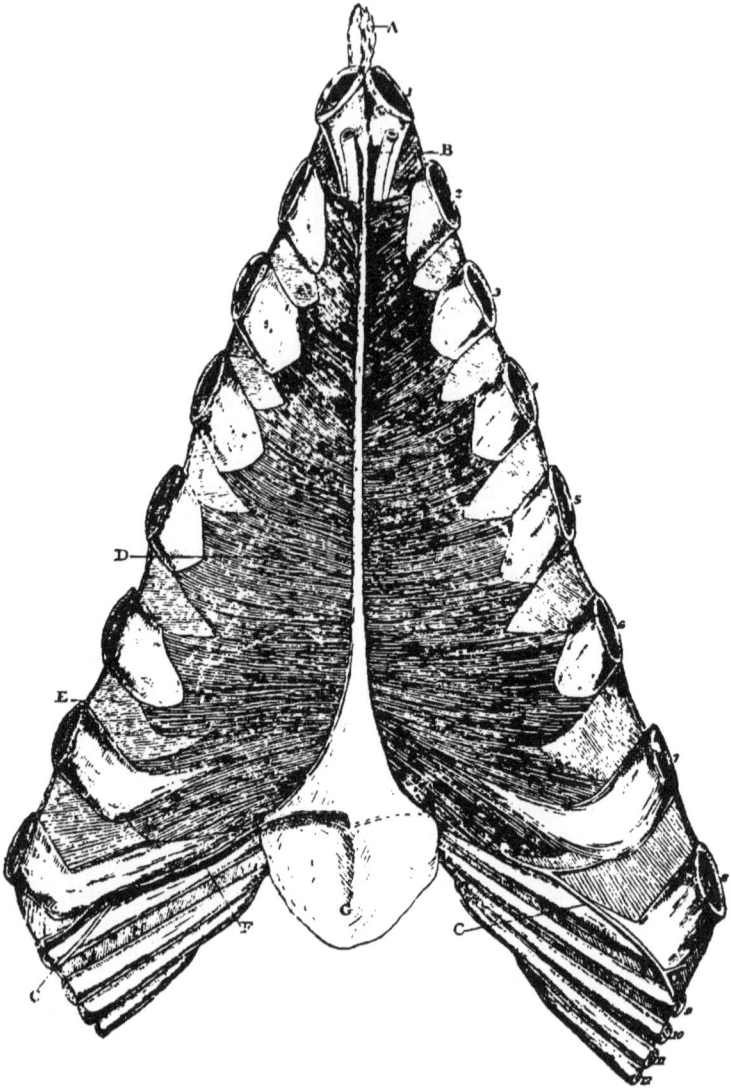

FIG. 7.

FLOOR OF THE THORAX.

Nos. 1 to 8 indicate the corresponding ribs. Nos. 9 to 12, the cartilages of the corresponding ribs.
A. Cariniform cartilage; B. Internal thoracic artery; C. Asternal artery; D. Triangularis sterni;
E. An internal intercostal muscle; F. Rim of diaphragm; G. Ensiform cartilage.

veins of the right side. Its opening is provided with a valve. It discharges into the venous system the lymph collected in the right anterior half of the animal, viz., the right fore limb, and the right side of the head, neck, chest-wall, and diaphragm.

Directions.—Cut out the trachea and lungs with the heart and great vessels. Sever the lungs from the heart by cutting the great vessels at the root, and set both lungs and heart aside in carbolic or other preservative solution to serve in the examination of the structure of these organs. Or, since they are likely to be much decomposed, it will be better to discard them if fresh organs can be obtained. In the meantime separate the sternum and costal cartilages as shown in Fig. 7, and dissect the triangularis sterni muscle and the internal thoracic vessels. Portions of the longus colli and psoas muscles which are attached to the lower face of the dorsal vertebræ should be noticed. The longus colli is described at page 156, and the psoas muscles at page 325.

The TRIANGULARIS STERNI (Fig. 7). This muscle *arises* from the lateral margin of the thoracic surface of the sternum, beginning at a point opposite the 2nd costal cartilage, and extending backwards to the ensiform cartilage, from the edge of which the last few fibres arise. It is *inserted* into the costal cartilages from the 2nd to the 8th inclusive, and into an aponeurosis on the internal intercostal muscles. Its outer edge is strongly serrated. It covers the internal thoracic vessels, and is lined by pleura on its upper face.

Action.—It pulls the cartilages to which it is attached inwards and backwards, and thus assists in expiration.

The INTERNAL THORACIC (MAMMARY) ARTERY (Fig. 7). This vessel, detached from the axillary artery at the 1st rib, descends on the inner face of that bone, and disappears beneath the triangularis sterni muscle. When the muscle is removed, the artery is seen to pass backwards at the edge of the sternum, crossing the chondro-sternal joints. Over or about the 8th of these joints it divides into the asternal and anterior abdominal arteries. The *asternal* branch emerges from under cover of the triangularis sterni, and runs up the cartilage of the 9th rib, on the thoracic side of the origin of the diaphragm. About the upper end of the cartilage it passes through the edge of the diaphragm to its abdominal side. The *anterior abdominal artery* dips down between the cartilage of the 9th rib and the edge of the ensiform cartilage, and enters the abdominal wall. The collateral branches of the internal thoracic are :—

1. Branches to the mediastinum and pericardium.

2. Pectoral branches, perforating the intercostal space and anastomosing with the external thoracic artery.

3. Intercostal branches, which ascend to anastomose with the inter-

costal arteries. This series is continued by branches of the asternal artery.

The INTERNAL THORACIC VEIN runs in company with the artery, and internal to it. Beneath the triangularis sterni it is placed between the artery and the fibrous cord that traverse the edge of the sternum.

SUPRASTERNAL LYMPHATIC GLANDS. These include (1) a group of glands on the thoracic side of the insertion of the diaphragm across the ensiform cartilage, and (2) some small scattered glands along the course of the internal thoracic vessels.

EXAMINATION OF THE LUNG.

Physical Characters.—The exterior of the lung is exquisitely smooth in virtue of its pleural covering. Through this thin, transparent covering, the surface is seen, especially when the lung is distended, to be divided by intersecting lines into small areas, each of which corresponds to a lobule of the lung. The lines are formed by the interlobular connective-tissue.

The colour of the lung varies with the age of the animal. In the young subject it is pale pink, but in old animals it is of a grayish or slaty hue. In the fœtus it is a bright pink.

The lung is spongy to the touch, and its cut surface has the same appearance. It is also markedly elastic, this quality being best illustrated by the rapidity with which the inflated lung collapses when the distending force is removed. It crepitates on pressure with the fingers, and it floats on water. The fœtal lung is non-crepitant, and sinks in water.

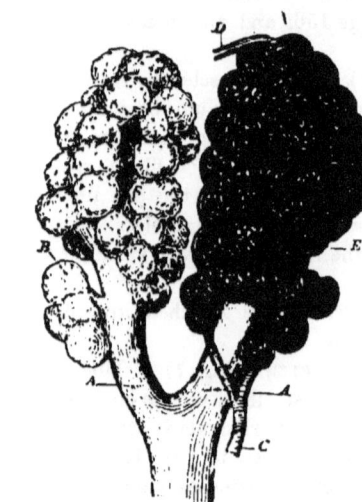

FIG. 8.

TERMINATION OF THE AIR PASSAGES IN THE LUNG (modified from *Turner*).

A, A. Terminal bronchioles; B. An infundibulum, showing the air-cells on its surface; C. Pulmonary artery; D. Pulmonary vein; E. Pulmonary capillaries.

Structure.—When the bronchus enters the lung, it divides again and again until there results a remarkable tree-like arrangement of bronchial tubes. Of this tree, the bronchus entering the root of the lung forms the main stem; and as the division is traced onwards, the bronchial tubes, representing the branches, become smaller and smaller, until there is reached a tube of comparatively small calibre which belongs exclusively to one lobule, and is therefore termed a *lobular* or *terminal* bronchus. The left bronchus has a length of three or four inches before dividing, but the right immediately gives off from its outer side a considerable branch (Plate 26). Within each lobule the terminal bronchus ramifies, forming smaller tubes or *bronchioles*, the last and smallest of which lead into recesses or dilatations. Each such dilatation is termed an *alveolar passage*, and it is bounded by delicate sacculated walls, each sacculation being an *infundibulum*. The infundibula are themselves sacculated, the minute recesses of their walls being termed *air-cells*. The air-cell is thus the ultimate part of the air passages within the lung, and a group of air-cells forms an infundibulum. The wall of an air-cell consists of a delicate membrane supporting the capillary plexus of the pulmonary vessels, and lined towards the air passage by a single layer of squamous cells. The bronchial tubes comprise in their walls: (1) an outer fibro-elastic coat sustaining segmented rings of cartilage; (2) within the preceding, a complete coat of non-striped muscular fibres circularly arranged; (3) an inner fibro-elastic coat; (4) a mucous membrane with a ciliated epithelium on its free surface. Numerous mucous glands lie in the outer fibrous coat, and discharge their secretion into the bronchus. The bronchi in their ramifications are accompanied by divisions of the pulmonary artery and veins, these two sets of vessels being connected by the capillary plexus on the air-cells. Along the bronchi run also the much smaller branches of the bronchial vessels, as well as nerves and lymphatics.

Connective-tissue forms a framework for the lung. It surrounds and connects the bronchi and vessels as they run together in the lung substance; it connects and isolates the adjacent lobules; and beneath the pleura it forms a fibrous capsule for the lung. Lymphatic vessels are abundantly distributed in it, and form three principal sets, viz., sub-pleural, perivascular (around the pulmonary vessels), and peribronchial.

DISSECTION OF THE HEART.

The VESSELS of the HEART.

The *Coronary Arteries* (Plates 23 and 24) carry arterial blood to nourish the heart-wall. They are two in number, distinguished as right and left. Each arises from the common aorta, and has its mouth in one of the sinuses of Valsalva.

The *Right Coronary Artery* passes forwards to the right of the pulmonary artery at its root; and encircling the right auricular appendix, it places itself in the auriculo-ventricular furrow, in which it passes to the right side of the heart. On reaching the origin of the right ventricular furrow, it divides, one branch descending in that furrow, while the other continues the course of the main trunk in the auriculo-ventricular

groove. The terminal twigs of the vertical branch enter the heart
a little above the apex; the horizontal branch reaches as far as the
posterior border of the heart.

The *Left Coronary Artery* passes outwards and to the left, between
the pulmonary artery and the left auricular appendix. Reaching the
auriculo-ventricular furrow at this point, it divides into a vertical and a
horizontal branch. The former descends in the left ventricular furrow;
the latter turns backwards along the auriculo-ventricular furrow. The
terminal portion of the vertical branch turns round the anterior border
of the heart, and ends in twigs that enter the ventricular wall on the
right of its apex; the horizontal branch terminates in the same way
near the posterior border of the heart.

The corresponding branches of the right and left arteries, thus, approach
each other at their terminations, but they do not anastomose; nor is
there any anastomosis between the arteries through their collateral
branches. Still more, there is no anastomosis between the adjacent
collateral branches of the same coronary artery.*

The *Coronary Veins.*—These arise from the capillaries of the coronary
arteries. The principal vessel of the right side ascends in the right
ventricular groove, and at the auriculo-ventricular furrow it joins the
main vein of the left side. The latter ascends at first in the left ventricular furrow, at the top of which it enters the auriculo-ventricular
groove. Along this it is reflected backwards; and turning round the
posterior border of the heart, it joins the right vein. The dilated vessel
resulting from this union is termed the *coronary venous sinus,* and it
opens into the right auricle, beneath the mouth of the posterior vena
cava.

For the most part, the veins arising in the wall of the right auricle
do not join the large coronary veins, but open into the cavity independently, by minute mouths—the *foramina Thebesii.*

The NERVES of the HEART have already been described (pages 110 and
117). In the heart of a lean subject long filaments are visible without
dissection, descending beneath the serous covering.

The INTERIOR of the HEART.—The cavities of the heart should be
studied in the order in which the blood passes through them, and
therefore the right auricle falls to be examined first. The terminations of the anterior and posterior cavæ and vena azygos (provided
that has an independent opening) should be identified; and then an
incision should be made along the wall of the sinus venosus, from
the opening of the anterior to that of the posterior cava. Another
incision should be carried from this one to the point of the auricula;

* Percivall, Leyh, Chauveau, and all the other authors that I am acquainted with state that the
coronary arteries anastomose with one another. That they do not, I have repeatedly proved by
injecting one of them, by which method none of the injection, however fine, can be driven into the
other artery.

and the clots of blood having been cleared out, the cavity will be ready for examination.

The CAVITY of the RIGHT AURICLE. The interior of this and the other cavities of the heart is smooth and glistening in virtue of an endothelial membrane termed the *endocardium*, which is here continuous with the endothelial lining of the great veins. It will be observed that the muscular wall of the auricle is thrown into parallel ridges, which from their resemblance to the teeth of a comb are termed *musculi pectinati*. The venous orifices by which the blood is poured into the cavity are all found in the sinus venosus, and are as follows:—1. The anterior vena cava empties itself into the anterior part of the roof of the sinus. 2. The posterior vena cava discharges itself at the lower and back part of the outer wall of the sinus. 3. The coronary venous sinus conveys the blood from the wall of the heart itself, and its mouth will be found under that of the posterior vena cava. 4. The vena azygos sometimes has an independent opening into the auricle, and it then discharges itself by the roof of the sinus, behind the mouth of the anterior cava. At other times it opens into the anterior cava. 5. The *venæ cordis*

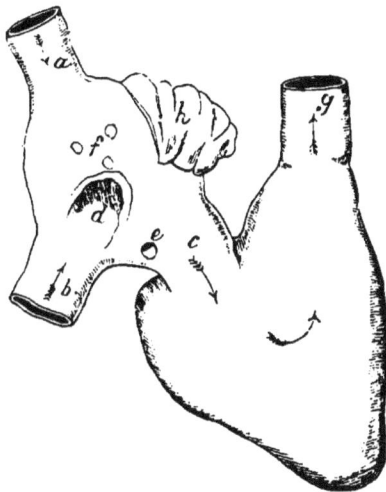

FIG. 9.

DIAGRAM OF THE TWO CAVITIES OF THE RIGHT SIDE OF THE HEART (*Ellis*).

a. Anterior cava ; *b.* Posterior cava ; *c.* Right auriculo-ventricular opening ; *d.* Fossa ovalis ; *e.* Opening of the coronary sinus ; *f.* Foramina Thebesii, the openings of veins ; *g.* Aperture of the pulmonary artery ; *h.* Auricular appendix.

minimæ are small veins of the wall of the right auricle, which, instead of discharging themselves by the coronary sinus, open directly on the wall by minute mouths named the *foramina Thebesii*. Of all these orifices, that of the coronary sinus is the only one provided with a valve. It is a thin fold of the lining membrane, termed the *valve of Thebesius*.

The inner wall of the sinus venosus is formed by the auricular septum, which is the partition between the two auricles. On this the following objects are to be noticed:—1. Between the orifices of the anterior and posterior cavæ is a muscular prominence—the *tubercle of*

Lower.   2. Above and in front of the opening of the posterior cava is a depression of the septum that looks like another venous orifice. This is the *fossa ovalis*, and it marks the former position of the *foramen ovale*—an aperture which, in the fœtus, established a communication between the right and left auricles. The raised border which surrounds the fossa is termed the *annulus ovalis.* In the fœtus of many animals, but not of the horse, a valve, termed the *Eustachian* valve, directs the blood from the posterior cava through the foramen ovale. After birth the foramen ovale in nearly every case becomes completely closed, but sometimes an oblique slit remains, which, however, does not necessarily permit any blood to pass through the septum.

The blood which passes through the right auricle is venous in character. It has been circulating among the tissues, and it is poured into the cavity at the venous orifices already enumerated. When the auricle contracts, the blood is passed into the ventricle of the same side, by a large aperture of communication between the two cavities—the *right auriculo-ventricular opening.*

Directions.—The fore and middle fingers of the left hand should be introduced through the auriculo-ventricular opening, so as to grasp, between the fingers and thumb, the wall of the right ventricle close to the angle of junction between the right ventricular and the auriculo-ventricular groove.   The scalpel should then with the right hand be passed through the wall of the ventricle at that point, and carried downwards in front of the right ventricular furrow; and following that furrow round the anterior border of the heart, the incision should be continued up in front of the left ventricular furrow, as far as the root of the pulmonary artery. This will enable nearly the entire wall of the right ventricle to be raised as a triangular flap, and will give a good view of the cavity when looked into from below.

The CAVITY of the RIGHT VENTRICLE. This cavity is widest above, and tapers to its lowest point; and its shape is such that its transverse section gives a crescentic outline, the wall of the ventricle being concave towards the cavity, while the septum is convex in the same direction. The inner surface of its wall is rendered irregular by muscular bands and prominences—the *columnæ carneæ,* of which there are several varieties: 1. Some of them have the form of bars or ridges sculptured on the wall of the heart, to which they give a sponge-like appearance. 2. Others, the *trabeculæ carneæ,* are veritable bands or strings between which and the wall of the ventricle the handle of a scalpel may be passed. Of this variety two or three very tendinous strings, sometimes more or less reticulate, stretch between the wall and the middle of the septum; and since they are believed to prevent over-distension of the ventricle, they have been named *moderator bands.* Other strings occur in the angle of junction of wall and septum, and still others stretch between different parts of the wall. 3. A third variety are blunt, nipple-shaped prominences called *musculi papillares*, of which there are commonly three in this cavity, one being placed on the wall and two on the septum. Radiating from each of these is a set of fibrous strings— the *chordæ tendineæ*, which are attached by their other ends to the segments of the valve guarding the auriculo-ventricular opening. The *right auriculo-ventricular opening* is situated at the base of the cavity, and is a very large orifice. It is provided with a valve composed of three main cusps, or segments, and hence named *tricuspid*. Each of these cusps is triangular in shape, being fixed by its base to the wall of the heart, and having its edges free and directed towards those of the adjacent cusps. When the blood stream is rushing through the opening, the segments of the valve hang down into the ventricle, and have one surface directed towards the blood stream, and the other to the wall of the ventricle. The first of these surfaces is smooth ; the other is rough, and to it and the apex and edges of the cusp, the chordæ tendineæ are attached. When, during contraction of the ventricle, the blood tends to regurgitate through the opening, the cusps are floated upwards, and, meeting each other, close the orifice. To the efficiency of this action, the chordæ tendineæ passing from the musculi papillares are essential ; for, being attached to the edges and lower surfaces of the cusps, they prevent the latter from being carried right up into the auricle. The three principal cusps generally alternate around the opening with three of much smaller size. There are three musculi papillares, each with its set of chordæ tendineæ, and three large cusps ; but it will be observed that one set of the chordæ tendineæ does not pass entirely to one cusp, but divides itself between two adjacent segments.

Directions.—A better view of the tricuspid valve may now be obtained by cutting through the auriculo-ventricular ring near the point where the first incision was begun in opening the ventricle, selecting the interval between two cusps. When some of the chordæ tendineæ have been cut, this will enable the wall of the ventricle to be thrown outwards.

When the ventricle contracts, the blood, prevented by the tricuspid valve from passing back into the auricle, is forced upwards into the left-anterior portion of the ventricle, and leaves the cavity by the pulmonary artery. This portion of the cavity, which leads up to the artery, is termed the *conus arteriosus*. The *orifice* of the *pulmonary artery* is surrounded by a valve composed of three crescentic segments, and hence termed the *semilunar* valve. The convex border of each segment is fixed to the wall of the artery where it springs from the ventricle. The concave border is free, and shows at its mid point a minute, fibro-cartilaginous thickening—the *nodulus* or *corpus Arantii*. On each side of the *corpus* a small cresentic portion near the free edge of the segment, and distinguished from the rest by its thinness, is termed the *lunula*. One surface of the valve is convex, and, during contraction of the ventricle, it is directed to the blood-stream; the other is concave, and directed to the wall of the artery, which, opposite each segment, forms a pouch—the *sinus of Valsalva*. When the ventricle has ceased to contract, the elastic recoil of the artery forces the blood against the concave side of the segments, and carries them inwards till they meet and completely close the opening. The blood is thus propelled along the pulmonary arteries to the lung, where, in the capillary plexus on the walls of the air-cells, it is purified. The purified fluid is then carried from the lungs by the pulmonary veins, which pour it into the left auricle of the heart.

Directions.—The cavity of the left auricle is to be exposed by an incision from the right to the left pulmonary veins, and by another from the first to the point of the appendix.

The CAVITY of the LEFT AURICLE is smaller than the right, but, like it, consists of a *sinus venosus* and an ear-shaped appendage—the *auricula*. The pulmonary veins open on the roof of the sinus venosus; and most commonly they have four openings—two from each lung, but they may have as many as eight. They are not provided with valves. The auricula and adjacent part of the sinus venosus show *musculi pectinati* similar to these of the right auricle. In the floor of the cavity is the *left auriculo-ventricular opening*, by which, on contraction of the auricle, the blood is passed into the left ventricle.

Directions.—The left ventricle should be opened by an incision similar to that used on the right side. The point of the scalpel should be passed through the wall of the ventricle near the upper end of the left ventricular furrow, and the incision should

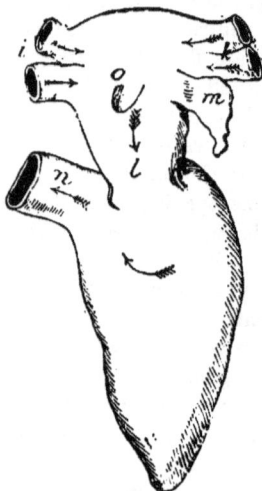

FIG. 10

DIAGRAM OF THE TWO CAVITIES OF THE LEFT SIDE OF THE HEART (*Ellis*).

k. Left pulmonary veins; *i.* Right pulmonary veins; *o.* Remains of foramen ovale; *l.* Left auriculo-ventricular opening; *m.* Auricular appendix; *n.* Aperture of the aorta.

be carried down the left side of the ventricle, round the apex, and up the right side to within a short distance of the auriculo-ventricular groove, the cut being made near the septum, to which the ventricular furrows will serve as a guide.

The CAVITY of the LEFT VENTRICLE is longer than the right, and is almost conical in shape, the base being at the auriculo-ventricular opening. On transverse section, it gives an oval or nearly circular outline, the septum, as well as the wall of the ventricle, being concave towards the cavity. It will be observed that its wall is about thrice the thickness of that of the right cavity, a circumstance which makes it easy to distinguish the right and left ventricles in the undissected heart; for, whereas the former appears flabby, the latter is firm and solid-looking. The left ventricle possesses *columnæ carneæ* like those on the right side. The *musculi papillares* are two in number, and are of very large size. They are placed on the wall, and are provided with *chordæ tendineæ* stronger than those of the right cavity. The base of the cavity shows the *left auriculo-ventricular opening*, which is somewhat smaller than the right. It is guarded by a valve with two large cusps, and hence called the *bicuspid* valve. It is also very commonly designated the *mitral* valve, from a fancied resemblance to a bishop's mitre. The cusps are stronger than those of the tricuspid valve, with which they agree in shape and disposition. Two smaller segments

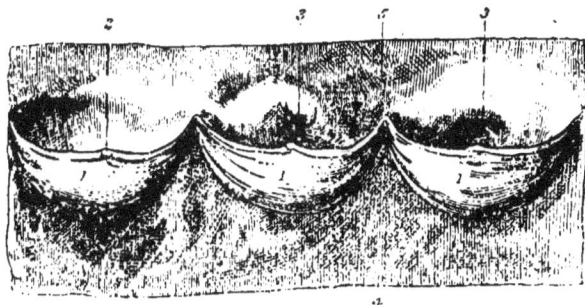

FIG. 11.

ROOT OF THE COMMON AORTA LAID OPEN.

1, 1, 1. Semilunar segments of the aortic valve; 2. Corpus Arantii; 3, 3. Orifices of right and left coronary arteries from two of the sinuses of Valsalva; 4. Ventricular wall; 5. Arterial wall.

alternate with the main ones. In mode of action the mitral exactly resembles the tricuspid valve. When the ventricle contracts, the blood, prevented from regurgitating into the auricle, is forced out of the cavity along the great systemic artery—the common aorta, which springs from the right-anterior part of the base of the ventricle. The *aortic orifice* is guarded by a three-segmented *semilunar* valve. These segments are stronger than those at the mouth of the pulmonary artery, which they otherwise exactly resemble. Opposite to each a large *sinus of Valsalva* is developed on the wall of the artery, and from two of these spring the right and left coronary arteries of the heart.

STRUCTURE OF THE HEART.

In structure the heart consists of a muscular wall, an external serous investment—the *epicardium*, and an internal serous lining—the *endocardium*. The valves are folds of the endocardium, strengthened with fibrous

K

connective-tissue, to which are added some elastic fibres. The muscular tissue is of the striped variety (although not under the control of the will), and its fibres are grouped in bundles separated by fibrous connective-tissue. Connective-tissue occurs also in large amount in the neighbourhood of the auriculo-ventricular and arterial openings, where it is aggregated in the form of rings, or zones. These rings give to the orifices that firmness which is necessary for the efficient working of the valves, and at the same time give origin to some of the muscular fibres. The tissue of which they are composed is mainly fibro-cartilaginous. In the heart of the ox, and rarely also in the horse, a bone—the *os cordis*—is developed in the angle between the aortic ring and the two auriculo-ventricular rings.

Directions.—The arrangement of the muscular tissue in the wall of the heart is exceedingly complex, and cannot be studied except in a heart specially prepared. A heart from any of the domestic animals, but preferably from the horse, should be procured, and boiled for about an hour. This will favour the dissection of the fibres, by making them firm and softening the connective-tissue between them. The epicardium, fat, and vessels having been cleaned off the surface of the heart, the auricles should be first examined.

The auricles have the muscular fibres of their walls distinct from those of the ventricles. Moreover, the fibres are arranged in two layers— a deep set proper to each auricle, and a superficial set common to both, some of the fibres of the latter stratum being carried into the auricular septum. In the deep stratum some of the fibres run obliquely in the wall, while others are arranged as rings around the auricula and the different venous orifices, the latter playing an important part in preventing regurgitation into the veins when the auricle contracts.

Directions.—Separation of the auricles from the ventricles should next be effected by cutting the auriculo-ventricular fibrous rings, which form the bond of connection between the auricular and ventricular fibres. By combined cutting and teasing the following facts may be observed.

Over the whole exterior of the ventricles the fibres have an oblique direction. Thus, on the left side the fibres pass obliquely downwards and backwards, and on the right side downwards and forwards. At the left ventricular furrow many of the fibres dip into the septum; but on the right side the fibres of the left ventricle pass across the furrow, and are directly continued on the right ventricle. At the apex of the heart the fibres turn inwards in a whorl-like manner and disappear from view. If a thin stratum of these superficial fibres be now removed, they will be found to cover others having a less oblique course ; and further dissection will show that the fibres become less and less oblique until the centre of the wall is reached, where the fibres are approximately horizontal. On peeling off these horizontal fibres, a deeper set will be exposed which are slightly inclined, but in a direction opposite to the superficial fibres; and the obliquity of these increases until the inner surface of the wall is reached, where the fibres have a degree of obliquity equal to the most external fibres. So much the student will probably be able to make out without much difficulty, but according to Pettigrew

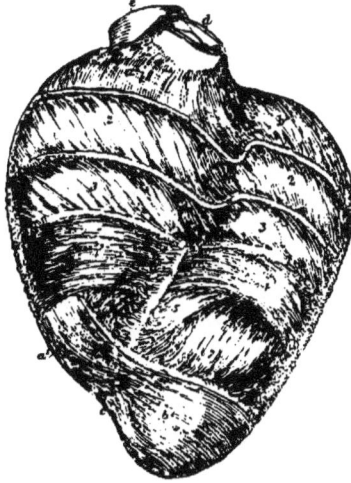

FIG. 12.

VIEW OF A PARTIAL DISSECTION OF THE FIBRES OF THE LEFT WALL OF THE VENTRICLES IN A SHEEP'S HEART, DESIGNED TO SHOW THE DIFFERENT DEGREES OF OBLIQUITY OF THE FIBRES (*Allen Thomson*).

At the base and apex the superficial layer of fibres is displayed : in the intervening space, more and more of the fibres have been removed from above downwards, reaching to a greater depth on the left than on the right side. a^1. a^1. The superficial layer of the right ventricle ; b^1. b^1. The same of the left ventricle ; at 2 this superficial layer has been removed so as to expose the fibres underneath, which are seen to have the same direction as the superficial ones over the left ventricle, but different over the right ; at 3 some of these have been removed, but the direction is only slightly different ; 4. Transverse or annular fibres occupying the middle of the thickness of the ventricular walls ; 5, 6, 7. Internal fibres passing downwards towards the apex to emerge at the whorl ; between *c. c.* the left ventricular groove, over which the fibres of the superficial layer are seen crossing ; in the remaining part of the groove, some of the deeper fibres turn backwards towards the septum ; *d.* The pulmonary artery ; *e.* The aorta.

the fibres are arranged in seven determinate layers—three external which are oblique, three internal, also oblique but in the opposite direction, and a central which is horizontal. Further, he describes the fibres of the most external layer as turning in at the auriculo-ventricular orifices and at the apex of the heart to become continuous with the layer beneath the endocardium. In like manner, he supposes that the second layer is continuous with the sixth, and the third with the fifth, while the fourth or central layer has a zone-like arrangement. In truth, however, the fibres of the same stratum anastomose not only with one another, but also with the fibres of adjacent strata, as is shown by the rough surface which is left when one set of fibres is removed from the underlying set.

Directions.—The joints and ligaments of the dorso-lumbar part of the spinal column and of the ribs must now be dissected. The ligaments of the lumbar region will be exposed by carefully removing from the surface of the bones the remains of muscles and other textures already examined. The whole of the dorsal region need not be dissected in order to expose the ligaments, but it will suffice to take a segment containing four or five vertebræ with their costal articulations intact. The articulations of the costal cartilages to the sternum are to be examined on the part of the thorax removed for the display of the triangularis sterni muscle.

ARTICULATIONS OF THE RIBS.

Each rib is articulated to the spinal column at two points, viz., by its head, and by its tubercle. The head is received into a cup-like cavity formed by two adjacent vertebral bodies and the disc that unites them. This is the *costo-central* joint. The tubercle articulates with the flat facet on the transverse process belonging to the posterior of the vertebræ to which the head is articulated. This is the *costo-transverse* joint.

COSTO-CENTRAL JOINT. This possesses two ligaments—the costo-vertebral and interarticular, and two synovial sacs.

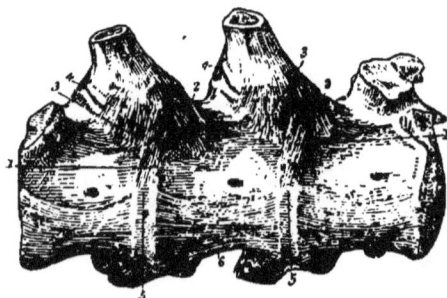

FIG. 13.

TWO COSTO-VERTEBRAL, AND TWO INTERVERTEBRAL JOINTS, VIEWED FROM BELOW.

1. Attachment of costo-vertebral (stellate) ligament to intervertebral disc; 2. and 3. Attachments of the same ligament to the anterior and posterior vertebral bodies; 4. Posterior costo-transverse ligament; 5. Intervertebral disc, covered by 6. the inferior common ligament.

The *Costo-vertebral* or *Stellate Ligament* is placed beneath the joint. Its fibres radiate from the rib just below its articular head, and become attached to the body of the vertebra in front, to the body of the vertebra behind, and to the intermediate disc.

The *Interarticular Ligament* is fixed to the groove dividing the articular head of the rib into two facets. It passes inwards across the floor of the spinal canal, being united to the upper edge of the intervertebral disc; and on the middle line it becomes continuous with the corresponding ligament of the opposite rib. It is not present in the 1st rib. It should be displayed by disarticulating the costo-transverse joints of the first vertebra in the segment, and then removing the arch. On removing the superior common ligament, it will be found in the interval between the two vertebral bodies.

Synovial Sacs.—There is one sac on each side of the interarticular ligament. There is only one sac for the first costo-central joint.

COSTO-TRANSVERSE JOINT. This is maintained by two ligaments—an anterior and a posterior costo-transverse ligament, and it possesses a synovial sac.

FIG. 14.

TWO COSTO-VERTEBRAL, AND TWO INTERVERTEBRAL JOINTS, VIEWED FROM ABOVE. THE LAMINÆ OF THE VERTEBRÆ HAVE BEEN REMOVED.

1. Posterior costo-transverse ligament; 2. Anterior (interosseous) costo-transverse ligament 3, 3. Superior common ligament.

The *Posterior Costo-transverse Ligament* is composed of a band of fibres stretching across the joint behind, and fixed by its extremities to the rib and transverse process.

The *Anterior (Interosseous) Costo-transverse Ligament* stretches between the antero-inferior aspect of the transverse process and the neck of the rib. It is partly concealed by the transverse process, and is best seen when viewed from above and in front.

Synovial Sac.—This will be exposed by removing the posterior costo-transverse ligament. In the last two or three ribs there is no separate synovial sac for the costo-transverse joint; but the posterior costo-central sac is extended over it, the two articular surfaces being in these ribs continuous.

CHONDRO-COSTAL JOINT. The inferior extremity of the rib is slightly excavated, and receives the extremity of the costal cartilage. The periosteum passes from the rib to the cartilage, and serves to consolidate the union.

CHONDRO-STERNAL JOINT. This is the joint by which the costal cartilage of each of the first eight ribs is articulated to the sternum. Peripheral fibres envelop the joint and form a *capsular ligament*. Above and below the joint the capsule is somewhat thickened, forming the *superior* and *inferior costo-sternal ligaments*.

Synovial Sac.—The joint possesses a synovial membrane. The cartilages of the first pair of ribs meet in a common joint on the middle line, and there is a single synovial sac common to these chondro-sternal articulations, and to the facet between the two cartilages.

UNION OF THE COSTAL CARTILAGES TO ONE ANOTHER. The cartilage of the first asternal rib (9th) is firmly bound to the preceding carti- lage by short fibrous bands. It is further bound to the lower face of the xiphoid cartilage by a small band—the *chondro-xiphoid ligament*. From the tip of each succeeding cartilage, a yellow elastic band is carried to the posterior edge of the cartilage in front of it.

MOVEMENTS OF THE RIBS. Each rib with its cartilage moves around an imaginary axis joining the head of the rib and the sternal end of the cartilage. In inspiration the rib moves forwards and outwards round this axis, so as to bring the middle portion of the rib towards the position occupied by the preceding rib at the end of expiration. This movement lengthens the line joining the mid point of each rib to the corresponding point on the opposite rib, and thus increases the capacity of the chest by increasing its transverse diameter. During expiration the rib falls into its original position by moving in the opposite direction. In these movements the head of the rib and the extremity of the costal cartilage rotate slightly in their cavities, but without change of place. The tubercle of the rib glides on the facet of the transverse process, moving in a circle whose centre is the costo-central joint.

The STERNUM. There are no joints in the sternum of the horse, in which the osseous segments are simply united by persisting portions of the original cartilaginous mass. In this connection, however, there may be noticed the two suprasternal fibrous cords which pass, one at each side of the thoracic surface of the sternum, internal to the mammary vessels.

INTER-VERTEBRAL JOINTS AND LIGAMENTS.

In the dorsal and lumbar regions adjacent vertebræ are connected (1) by an amphiarthrodial joint between their bodies, and (2) by synovial joints between their articular processes. These same joints are formed between the last lumbar vertebra and the sacrum ; and, in addition, there is a synovial joint between the last lumbar transverse process and the base of the sacrum. Inter-transverse joints are also developed between the 4th and 5th, and 5th and 6th lumbar transverse processes. The ligaments may be classified into :—1. Those connecting the processes and neural arches of adjacent bones. 2. Those connecting adjacent vertebral bodies.

LIGAMENTS of the PROCESSES and NEURAL ARCHES :—

The *Supraspinous Ligament* is a strong longitudinal band, or cord, extending along the tips of the spinous processes. It is continued backwards on the sacral spines ; and in the anterior part of the dorsal region its texture changes from white fibrous to yellow elastic tissue, and is continued forwards as the funicular portion of the ligamentum nuchæ. It not only tends to maintain the union of the vertebræ, but also affords a point of origin to muscles of the back and loins.

The *Interspinous Ligaments* occupy the interspaces of the spinous processes. In each space the ligament consists of a right and a left layer whose fibres have a downward and backward direction. This oblique direction of the fibres favours the separation of the spines during flexion.

Capsular Ligaments of the Articular Processes.—These complete the diarthrodial joint formed between the articular processes of adjacent vertebræ, and support the synovial sac of the joint. One of the ligaments should be slit open to display the synovial membrane lining its inner surface.

The *Ligamenta subflava*, or *Ligaments of the Arches.*—These pass between the edges of adjacent neural arches. They are best seen by sawing horizontally through the pedicles of two vertebræ, close to the body, and then pulling the arches apart while they are viewed from below.

Capsular Ligaments of the Transverse Processes.—These surround the joints developed between the transverse processes of the 4th and 5th, and 5th and 6th lumbar transverse processes, and between the last of these processes and the base of the sacrum. On removing them, the joints will be found to possess a synovial membrane. In old subjects, however, these joints are generally obliterated by anchylosis.

LIGAMENTS of the BODIES :—

The *Inferior Common Ligament* is a thin stratum of fibres covering the lower face of the vertebral bodies and the intervertebral discs. It is continued backwards beneath the sacrum, but it is not traceable as a distinct ligament farther forwards than the 6th dorsal vertebra.

The *Superior Common Ligament* lies on the floor of the spinal canal, and must be exposed by the removal of the neural arches. It is a riband-like structure adherent to the vertebral bodies, and to the inter-articular ligament of the ribs or to the upper edge of the intervertebral disc. The edges of the ligament are scalloped, the ligament being widest where it passes over the intervertebral discs, and narrowest at the middle of the vertebral bodies.   It is continued into the sacral and coccygeal regions.

The *Intervertebral Substance.*—Between every two adjacent vertebral bodies there is interposed a disc of fibro-cartilage.   This will be best seen on making a vertical mesial section of two centra.   The disc is thinner in the back than in the loins or neck.   The disc between the last lumbar body and the sacrum is especially thick.   In the dorsal region they concur in forming the cavity for the head of a rib.   The disc is not of uniform texture throughout.   The peripheral part of each is composed of alternating layers of fibrous tissue and fibro-cartilage. In each layer the fibres pass in an oblique direction between the two bones, and in successive layers the fibres are alternately oblique in opposite directions.   The central portion of the disc is pulpy, soft, and elastic ; and is interesting as being a persistent portion of the foetal *chorda dorsalis.*   The peripheral part of the disc constitutes an extremely resistant bond of union between the two vertebræ, while the central pulpy portion permits rotation of the one bone on the other.

MOVEMENTS of the dorso-lumbar part of the spinal column.   These are *flexion, extension, lateral inclination,* and *rotation.   Flexion* and *extension* are opposite movements taking place in a vertical plane.   In *flexion* the downward concavity of the column is increased, in *extension* it is diminished.   These movements are much more restricted here than in the neck, owing to the thinness of the intervertebral discs and the large size of the spinous processes.   They have a greater range in the loins than in the back.   *Lateral bending* is also much less free than in the neck, being impeded by the thinness of the intervertebral substance, and by the ribs and lumbar transverse processes.   Its greatest range of movement is in the anterior part of the lumbar region.   *Rotation* is the twisting or turning of a vertebra round a longitudinal axis passing through its body.   It is not permitted in the lumbar region, owing to the form of the articular processes ; and even in the back, it is scarcely appreciable.

THE SPINAL CORD.

Directions.—To expose the spinal cord of the horse in the whole of its extent is a tedious and difficult operation.   Moreover, where the dissection of the parts surrounding the vertebral column is apportioned between the dissectors of the neck, thorax, abdomen, and pelvis, it is quite impossible, without unduly interfering with what is otherwise the most convenient course of dissection, to expose at once the entire cord. This, however, is not a matter of much importance, since, in all the main features of its structure, the spinal cord of the horse is identical with that of any other mammal.   The student is therefore advised to study the cord of a dog or a cat, which may be exposed without difficulty. One of these animals having been secured (and preferably a dog), it should be fastened to a table in the prone position, and a mesial incision through the skin and muscles, down to the vertebræ, should be made from the occiput to the root of the tail. With the knife the muscles are to be reflected so as to expose the vertebræ as far as the junction of the arch with the body. The spinal canal is then to be opened by removing the arches with a chisel and mallet, or with bone-forceps. In the dorsal region each arch must be disarticulated from its connection with the ribs. The spinal cord enclosed within its membranes will now be exposed, and between the outer membrane and the bones are some veins and a quantity of fat.

Membranes, or Meninges, of the Spinal Cord.

The DURA MATER. This is the most external of the membranes. It is the protective envelope of the cord, and has the form of tubular membrane of fibrous connective-tissue, extending from the foramen magnum, where it is continuous with the corresponding envelope of the brain, to the posterior end of the spinal canal in the coccygeal region. It does not form a tight-fitting covering to the cord, but invests it somewhat loosely. Its outer surface, it is to be observed, is smooth, and does not line the vertebræ, which have the ordinary periosteal covering. It is connected by some slender fibrous processes to the superior common ligament. The capacity of the tube varies with the thickness of the cord, being greater at the atlas, lower part of the neck, and lumbar region than at the intermediate points. The spinal cord does not extend beyond the middle of the sacrum, but the dura mater is prolonged a few inches beyond that as an impervious, tapering process. On each side the dura mater is perforated by the roots of the spinal nerves, and along these it sends offsets as far as the intervertebral foramina.

Directions.—A small piece of the dura mater should be pinched up with forceps and snipped through. Beginning at the slit thus formed, it should be laid open backwards and forwards along the middle line. As this is being done, the membrane should be pinched up, so as to prevent injury to the cord.

The ARACHNOID is the second of the membranes of the cord. It is much more delicate than the dura mater, and in disposition and structure it is comparable to a serous membrane. Like such membranes, it encloses a cavity, or sac, and consists of a *parietal* and a *visceral* portion. The sac is known as the *arachnoid cavity*, or *sub-dural space*, receiving the latter designation from its relation to the dura mater. The *parietal* division of the membrane is represented by a layer of endothelium lining the inner surface of the dura mater, to which it gives a smooth and glistening aspect, but from which it is not separable by dissection. The *visceral* division invests the cord and pia mater as a thin transparent membrane, but it does so loosely, leaving a space between it and the outer surface of the pia mater.   This, which is the *sub-arachnoid space*, contains a variable amount of an alkaline fluid—the sub-arachnoid fluid, which acts as a water-bed to the cord.   As the roots of the spinal nerves

extend outwards, they take with them a covering from the visceral arachnoid ; and where they pierce the dura mater, this covering becomes continuous with the parietal layer.

The PIA MATER is the vascular membrane of the cord.   It consists of areolar connective-tissue in which the vessels subdivide before entering the cord.   It invests the cord closely, and is intimately connected to it; sending a considerable process into the inferior median fissure, and numerous other slender filaments which penetrate the substance of the cord.   On each side it is connected to the inner surface of the dura mater by a series of pointed processes constituting the *ligamentum denticulatum.*   Each of these processes of pia mater passes outwards from the side of the cord, and,

FIG. 15.

VIEW OF THE MEMBRANES OF THE SPINAL CORD (*Ellis*).

a. Dura mater cut open and reflected ; *b.* Small part of the translucent arachnoid, left ; *h.* Pia mater closely investing the spinal cord ; *c.* Ligamentum denticulatum on the side of the cord, shown by cutting through the inferior roots of the nerves ; *d.* One of the processes joining it to the dura mater ; *g.* Middle spinal artery ; *e.* Inferior roots of the nerves, cut ; *f.* Superior roots.

carrying the arachnoid with it, becomes attached to the dura mater, midway between the points of perforation of the superior and the inferior nerve-roots. Behind the point in the sacral region at which the spinal cord stops, the pia mater is prolonged as an attenuated thread—the *filum terminale*—which is enclosed by, and blends with, the tapering end of the dura mater.

Directions.—Before the removal of the cord the student should observe its varying thickness at different points, and the disposition of the spinal nerves within the spinal canal.

The spinal cord begins at the foramen magnum by continuity with the medulla oblongata, and it is here of considerable thickness.   Tracing it backwards, it is seen to become thicker behind the middle of the cervical region, forming the *cervical enlargement*, which extends as far as the 2nd dorsal vertebra.   It is from this enlargement that the nerves which supply the fore limb are given off.   Beyond the 2nd dorsal vertebra the cord contracts slightly, so as to become about the middle of the back smaller even than in its initial portion.   Preserving this diminished thickness throughout the dorsal region, it again expands in the lumbar region, forming a second swelling—the *lumbar enlargement*, from which the nerves for the supply of the hind limb are detached. Beyond the lumbar enlargement the cord rapidly becomes reduced in volume, and tapers to a point about the 2nd sacral segment. This tapering extremity of the cord—the *conus medullaris*—is prolonged backwards by the filum terminale, into which its nervous structure is continued for a little distance.

The SPINAL NERVES of the horse number forty-two or forty-three pairs, and their number in the different regions of the vertebral column is expressed in the following formula :—

$$C_8 D_{18} L_6 S_5 C_{5 \text{ or } 6}.$$

The 1st cervical nerve leaves the canal by the antero-internal foramen of the atlas, the 2nd by the foramen in the front of the arch of the axis, and the others in succession pass out by the intervertebral foramina.

In the other regions the nerves are numbered according to the vertebræ behind which they emerge; thus, the 1st dorsal nerve emerges by the intervertebral foramen behind the 1st dorsal vertebra, and so on with the others.

In the cervical region the nerves pass nearly directly outwards from the cord to their points of exit from the canal. In the dorsal region, however, it will be observed that each nerve is slightly inclined backwards from the side of the cord to the foramen by which it emerges. In the lumbar region this backward inclination of the nerves becomes

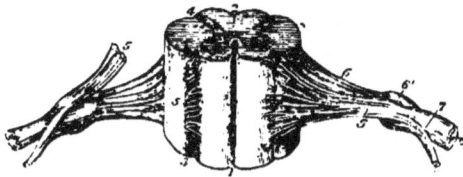

FIG. 16.

PORTION OF SPINAL CORD WITH THE ROOTS OF THE NERVES (*Quain*).

1. Inferior median fissure ; 2. Superior median fissure ; 3. Infero-lateral fissure (exaggerated) ;
4. Supero-lateral fissure ; 5. Inferior roots, passing under the ganglion (on the left side these are cut);
6. Superior roots, the fibres of which pass into the ganglion—6' ; 7. The united or compound nerve ;
8. Superior primary branch ; 9. Inferior primary branch.

augmented, and it continues to increase in the same way in each nerve of the sacral and coccygeal regions. The sacral nerves thus have their roots detached from the lumbar part of the cord, while the coccygeal nerves are given off by the terminal part of the cord, which, as already stated, does not extend beyond the middle of the sacrum. These last nerves have therefore a length of several inches within the spinal canal; and as they pass back together, each to reach its aperture of exit, they have an arrangement which resembles the hairs of a horse's tail, and is therefore termed the *cauda equina*.

ROOTS of the NERVES. Each spinal nerve has two roots connecting it with the spinal cord—a superior and an inferior. The *superior, sensory*, or *ganglionic* root consists of filaments which arise from along the supero-lateral fissure of the cord. These filaments perforate the dura mater, and converge towards the intervertebral foramen, where they form a cord on which there is superposed a reddish oval ganglion. Immediately beyond the ganglion the cord mixes its fibres with the inferior root.

The *inferior, motor*, or *aganglionic* root consists of fibres detached from the cord along its infero-lateral fissure. These, which are fewer and smaller than those of the superior root, perforate the dura mater by openings distinct from those for the superior root; and, converging towards the intervertebral foramen, they join the superior root immediately external to the point at which the ganglion is placed on it. The fibres of the inferior root, thus, have no connection with the ganglion. Where the superior and inferior roots meet in the intervertebral foramen, they mix their fibres and form a short common cord, which almost immediately divides into two—the *superior* and the *inferior primary branch;* and each of these contains fibres from both roots. Both branches emerge by an intervertebral foramen, and, roughly speaking, the series of superior primary branches supply the skin and muscles above their points of emergence, while the inferior primary branches are distributed to the skin and muscles below their points of emergence, including the limbs. From the common trunk formed by the union of the two roots, a filament re-enters the spinal canal to be distributed to the bones and vessels.

In the region of the neck the *spinal accessory nerve* (page 255) passes along each side of the cord, between the superior and inferior roots of the spinal nerves. It is formed by rootlets that spring out of the side of the cord.

The *Vessels of the Spinal Cord.*

The MIDDLE SPINAL ARTERY begins beneath the cord, in the ring of the atlas. It is here formed on the mesial plane by the fusion of right and left branches, each of which is the posterior branch formed by the bifurcation of the cerebro-spinal artery. The middle spinal artery passes backwards beneath the inferior median fissure of the cord. Its branches are distributed to the cord and its membranes. As it passes backwards giving off its branches, it is reinforced by other arteries entering at the intervertebral foramina. Thus, at each intervertebral foramen in the neck a branch of the vertebral artery enters the spinal canal. In the back similar branches enter from the dorso-spinal division of the intercostal arteries, in the loins the branches emanate from the lumbar arteries, and in the sacral region from the lateral sacral artery. As a rule, the branches entering by adjacent foramina anastomose, and then give off branches to the cord and the vertebral bodies.

The VEINS OF THE SPINAL CORD are tortuous, and form on its surface a plexus from which the blood passes into two large veins that lie one at each side of the superior common ligament. These receive also veins from the vertebral bodies, and they are drained by vessels that issue by the intervertebral foramina to join the vertebral, intercostal, lumbar, or lateral sacral veins.

Directions.—For the examination of the structure of the spinal cord, a few inches of it with the roots of the nerves intact should be procured, and kept in spirit or some other hardening fluid for at least a week. A portion from the spinal cord of any of the domestic animals will serve the purpose ; but, from its larger size, that of the horse is to be preferred.

STRUCTURE of the SPINAL CORD. The meninges having been removed, the student will note the following points regarding the surface of the cord :—It approaches the cylindrical in form, but is slightly flattened above and below. It is traversed in the longitudinal direction by three fissures, and a fourth is sometimes described. The *superior median fissure* is a narrow interval extending into the cord along the middle line of its upper face. It is occupied by neuroglia. The *inferior median fissure* is an actual cleft penetrating the cord along the middle line of its lower face. It is occupied by a process of pia mater. The *supero-lateral fissure* is a faint surface depression extending on the side of the cord, along the line of emergence

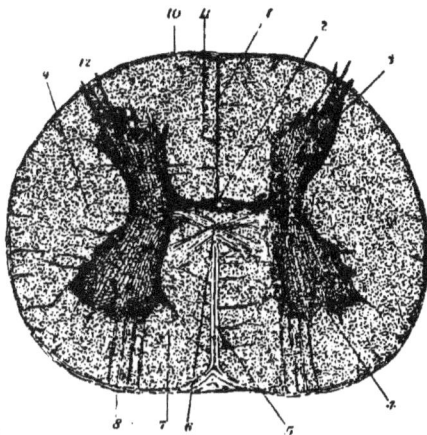

FIG. 17.

TRANSVERSE SECTION OF SPINAL CORD OF CALF (*Klein*).

1. Superior median fissure ; 2. Central canal, in grey (superior) commissure ; 3. Superior horn of grey matter ; 4. Inferior horn of grey matter ; 5. Process of pia mater in inferior median fissure ; 6. White (inferior) commissure ; 7. Inferior column of white matter ; 8. Inferior nerve roots ; 9. Lateral column of white matter ; 10. Pia mater ; 11. Superior column of white matter ; 12. Superior nerve roots.

of the superior roots of the spinal nerves. The *infero-lateral fissure* has no actual existence, but is sometimes described as extending along the line of emergence of the inferior roots.

These fissures will be better seen in a transverse section of the cord. This should be made with a sharp scalpel, so as to leave a clean-cut surface. On examining this surface, the superior median fissure will be seen to extend inwards to near the centre of the cord, while the inferior median fissure, which is wider but not so deep, also extends towards the centre of the cord. The superior and inferior median fissures do not quite meet, being separated by a bridge of tissue connecting the right and left halves of the cord. This bridge of tissue is made up of the *grey* and *white commissures* of the cord. The *grey commissure* stretches across the bottom of the superior median fissure, and in its centre there will be seen a dot-like mark, which is the section of the *central canal* of the spinal cord. This canal extends throughout the whole length of the spinal cord; and where the cord joins the brain, the canal is continued into the medulla oblongata, in which it opens into the 4th ventricle. The *white commissure* forms a thinner stratum than the preceding, beneath which it stretches at the top of the inferior median fissure.

It will be observed that in each half of the cord there are two kinds of nerve tissue, distinguished by a difference in colour.

1. There is the *grey matter*, which lies in the interior, and has a crescentic form. The convex side of each *crescent* is turned inwards, and the right and left crescents are connected by the grey commissure. The extremities of the crescent are termed its *horns*. The *superior horn* is acute, and is prolonged to the supero-lateral fissure by a single bundle of fibres belonging to a superior nerve-root. The *inferior horn* is rounded, and lies some distance beneath the surface of the cord. From it several bundles of fibres pass to form an inferior nerve-root. The grey matter of the cord contains nerve cells, medullated and non-medullated nerve fibres, and delicate nerve fibrillæ. The nerve cells are mostly of the multipolar variety, and are most numerous in the inferior horn. The connective-tissue of the cord, both here and in the white matter, is a delicate substance termed *neuroglia*.

2. The *white matter* in each half of the cord surrounds the crescent, and it is divided into three *columns* by the crescent and the nerve bundles passing from the horns. The *superior column* lies between the superior median fissure and the upper half of the crescent. The *inferior column* is included between the inferior median fissure and the lower half of the crescent. The *lateral column* lies in the concavity of the crescent, its limits being marked at the surface of the cord by the supero-lateral fissure and the line of emergence of the inferior nerve-roots. The white matter of the cord is composed, besides neuroglia, of medullated nerve fibres having for the most part a longitudinal direction.

(For the muscles of this chapter, see the table at page 336.)

CHAPTER IV.

DISSECTION OF THE HEAD AND NECK.

THE UNDER PART OF THE NECK.

As the first stage in the examination of this region, the student should dissect the structures placed below the cervical vertebræ—in other words, the under part of the neck.

Surface-marking.—A well-marked groove extends in the longitudinal direction on the side of this region, beginning at the upper part of the neck, and terminating between the shoulder and the anterior part of the pectoral region. It lodges the jugular vein, and is therefore termed the *jugular channel* or *furrow*. In performing phlebotomy on this vein, pressure is made on the furrow with the fingers, in order to arrest the downward current of blood, and thus distend and make prominent the vessel above the point of pressure. In the lower third of the furrow the vein lies in company with the carotid artery, and it is in this situation that the latter vessel may be most conveniently exposed for ligature or incision. The boundaries of the groove will be learnt after removal of the skin.

Position.—The dissection of this part of the neck should be completed while the dissector of the fore limb is engaged with the pectoral region, the animal being placed on the middle line of its back, and the head being forcibly extended on the atlas in order to put on the stretch the muscles and other structures to be dissected.

Directions.—An incision through the skin should be carried along the middle line, from the cariniform cartilage at the lower part of the neck to the centre of the intermaxillary space. From the latter point a curved incision should be carried outwards a little behind the angle of the jaw, as far as the wing of the atlas. These, in conjunction with the incisions made by the dissector of the pectoral region, will permit the skin to be reflected upwards as far as the middle of the side of the neck. The cutaneous nerves and the cervical panniculus should then be examined.

CUTANEOUS NERVES of the neck. Five stellate groups of nerves will be seen perforating the mastoido-humeralis muscle. These are derived from the inferior primary branches of the cervical spinal nerves from the

2nd to the 6th. The first of these appears behind the wing of the atlas; and, besides twigs to the upper part of the neck, it sends into the intermaxillary space a branch which may be traced to near the symphysis, and auricular branches which will subsequently be followed to the skin of the ear. Some branches from the lowest group turn downwards and backwards over the mastoido-humeralis in front of the shoulder, and spread over the anterior part of the pectoral region (Plate 1). The branches of the other groups are disposed upwards, downwards, and laterally, to supply the skin of the neck.

CERVICAL PANNICULUS (*Platysma myoides* of man). This is the representative in the neck of the muscle which is much more strongly developed in connection with the skin over the trunk and shoulder. It may be said to take origin at the lower part of the neck, where its fibres are fixed to the cariniform cartilage (Plate 27). At this point it is a band of considerable thickness; but as it passes up the neck, it widens and becomes thinner. At the upper part of the neck its fibres do not form a complete layer, but are scattered in an aponeurosis which prolongs the muscle into the intermaxillary space and over the face. Along the middle line it is joined by means of a fibrous raphe to the muscle of the opposite side. The outer edge of the muscle is continued by an aponeurosis over the mastoido-humeralis, splenius, and trapezius muscles. In the lower half of the neck the muscle is intimately adherent to the inferior edge of the mastoido-humeralis, and a careful dissection is necessary to separate them. It covers the jugular furrow, and the sterno-maxillaris, sterno-thyro-hyoideus, and subscapulo-hyoideus muscles. It is supplied by the cervical branch of the 7th cranial nerve, which should be found entering it at the upper part of the jugular furrow, and running on the deep face of the muscle or in its substance where it covers the furrow.

Action.—The cervical panniculus, unlike the panniculus of the trunk, is but slightly adherent to the skin, which, therefore, it can twitch only slightly. Its principal action seems to be to brace the muscles over which it is spread, and by its adhesion to the mastoido-humeralis it may aid in depressing the neck.

Directions.—Beginning at the middle line of the neck, the dissector should carefully remove the foregoing muscle. This will expose the jugular furrow lodging the jugular vein. After that vessel has been examined, a little dissection will serve to separate the muscles in relation to the trachea.

The JUGULAR VEIN (Plate 27) is the large vessel which drains away the blood from the head and the upper part of the neck. It is formed by the junction of the superficial temporal and internal maxillary veins, which unite at the deep face of the parotid gland, below and behind the temporo-maxillary articulation. It passes outwards through the parotid, and then lies in a groove on the surface of the gland; but this part of its course is not to be examined at present. At the lower extremity of the parotid it is joined by a large branch—the submaxillary vein; and it then passes into the jugular furrow, in which it descends to the entrance to the chest. The upper boundary of the furrow, it will now be observed, is formed by the mastoido-humeralis, and the lower by the sterno-maxillaris. In the upper half of this groove the vein rests on the subscapulo-hyoideus muscle, which there separates the vessel from the carotid artery; but in the lower half the vein rests on the side of the trachea, and is in direct contact with the carotid, which lies above and slightly internal to it. The jugular of the left side differs from the right in being related, in the lower part of the groove, to the œsophagus as well as the trachea. At the entrance to the chest the right and left jugulars unite with one another and with the axillary veins, thus forming the initial portion of the anterior vena cava.

The jugular receives, in the part of its course now exposed, the following branches :—

1. The *Submaxillary* or *Facial* vein, which joins the jugular at an acute angle in which lies the inferior extremity of the parotid gland.

2. The *Thyroid* vein, bringing blood from the thyroid body and larynx.

3. Innominate *cutaneous, muscular, œsophageal,* and *tracheal* branches, whose disposition is not constant.

4. The *Cephalic* vein, which enters the jugular near its termination.

[The single jugular of the horse is generally said to be the representative of the external jugular of man; the internal jugular, under that view, being undeveloped in the soliped. This I believe to be a mistake, and for the following reasons. The external jugular of man runs on the surface of the platysma (panniculus), and never beneath it as does the vein of the horse; moreover, it is a vessel of very variable volume, being frequently small, and sometimes absent. On the other hand, the cervical part of the internal jugular of man has a situation exactly corresponding to that of the horse's vein, save that the latter vessel is generally superficially placed to the subscapulo-hyoid (omo-hyoid); and in the horse I have seen the jugular, otherwise normal, pass under that muscle, keeping company with the carotid artery for the whole of its course.]

The STERNO-MAXILLARIS (Plate 27). This muscle corresponds to the inner portion of the sterno-mastoid of man. It *arises* from the cariniform cartilage of the sternum, and is *inserted* by a flat tendon into the angle of the inferior maxilla. In the lower half of the neck the muscle lies below the trachea, and covers the sterno-thyro-hyoideus muscle. In this position the right and left muscles are in contact, but about the middle of the neck they diverge, and cross obliquely upwards and forwards over the trachea and the subscapulo-hyoideus muscle. Its terminal tendon is included between the parotid and submaxillary glands. The upper edge of the muscle forms the lower boundary of the jugular furrow. In its lower part the muscle is thick and rounded, but it becomes more slender and flattened as it is traced upwards.

L

Action.—To depress (flex) the head or give it a lateral inclination, according as the right and left muscles act singly or in concert.

The STERNO-THYRO-HYOIDEUS (Plates 27 and 28). This is a long and slender muscle, extending along the lower face of the trachea, and closely applied along the middle line to its fellow of the opposite side. It takes *origin* from the cariniform cartilage of the sternum. About the middle of the neck its muscular substance is interrupted by a short tendinous portion, rendering the muscle digastric. Above this central tendon the muscle divides into two portions. The outer or *thyroid* band passes obliquely outwards and forwards between the trachea and the sub-scapulo-hyoideus muscle, and becomes *inserted* into the edge of the thyroid cartilage of the larynx. The inner or *hyoid* band is continued directly forwards in company with the corresponding branch of the opposite muscle, and becomes *inserted* into the body of the hyoid bone.

Action —To depress the hyoid bone and larynx.

The SUBSCAPULO-HYOIDEUS (*Omo-hyoid* of man) (Plates 27 and 28). This is a thin, ribbon-shaped muscle having a breadth of three or four inches. It takes *origin* at the inner side of the scapula, from the fascia covering the subscapularis muscle. It then passes downwards and for-wards between the scalenus and rectus capitus anticus major muscles inwardly; and the supraspinatus, anterior deep pectoral, and mastoido-humeralis muscles outwardly. Appearing at the lower edge of the last-named muscle, to which it adheres closely, it passes between the jugular vein and carotid artery ; and crossing over the upper part of the trachea in a direction obliquely forwards and downwards, it applies itself at the outer edge of the hyoid band of the sterno-thyro-hyoideus, and becomes *inserted* along with that muscle into the body of the hyoid bone. In the lower part of the neck the ascending branch of the inferior cervical artery and the prescapular group of lymphatic glands are included between this muscle and the mastoido-humeralis.

Action.—To depress the hyoid bone.

NERVES. At the upper part of the neck a branch from the spinal accessory nerve enters the sterno-maxillaris, and branches from the 1st spinal nerve enter the sterno-thyro-hyoid and subscapulo-hyoid muscles. These, however, will be better dissected at a later stage.

Directions.—The jugular vein should now be ligatured at the upper and lower ends of the jugular furrow, and the intermediate portion of the vessel should be cut away. The excised portion of the vein should be laid open to expose its valves. The part of the subscapulo-hyoideus which passes over the trachea may be cut out after the manner of Plate 28, and the sterno-maxillaris may be similarly treated. This will expose for examination the trachea, the œsophagus, the carotid artery, the pneumogastric and sympathetic nerves, and the recurrent nerve.

VALVES OF VEINS. Three or four valves are placed in the jugular vein.

Each valve is composed of two or three semilunar folds of the inner coat of the vein, the folds having a close resemblance to the semilunar segments of the aortic valve (Fig. 11, page 129). Each flap with the adjacent part of the wall of the vein forms a small pouch with its mouth directed towards the heart. When the blood tends to regurgitate, it distends these pouches until the segments meet across the vessel and thus arrest the backward current. In most veins throughout the body similar valves are found; but they are most numerous in the veins of the limbs. In the small veins each valve may be composed of only a single flap. The following veins have few or no valves :—the pulmonary system of veins, the veins of the portal system, the hepatic veins, the anterior and posterior venæ cavæ, and the veins of the brain.

The TRACHEA, or wind-pipe (Plate 28), begins beneath the altanto-axial articulation, where it is continuous with the larynx. It descends in the middle plane of the neck, beneath the spinal column ; and passing between the first two ribs, it gains the thorax, where it bifurcates to form the bronchi. In the neck the muscles of the region envelop the trachea, and are related to it as follows :—The longus colli is related to its upper aspect, the sterno-thyro-hyoideus extends along its lower face, the sterno-maxillaris crosses its direction obliquely upwards and forwards, the subscapulo-hyoideus crosses it obliquely downwards and forwards, and at the lower part of the neck it contacts on each side with the scalenus. It is also related to the œsophagus, the carotid artery, the jugular vein, and the pneumogastric, sympathetic, and recurrent nerves.

The THYROID BODY or GLAND (Plate 29) is related to the upper part of the trachea on each side, resting on its first four rings. The gland has a rounded form, and a reddish-brown colour ; and it is richly supplied with blood, which it receives from the thyroid and thyro-laryngeal branches of the carotid artery. Sometimes a narrow *isthmus* connects the right and left glands across the lower face of the trachea. The gland has an investing capsule of fibrous connective-tissue, continuous with an internal trabecular framework. Under the microscope the substance of the organ is seen to contain numerous spherical spaces, each lined by a single layer of epithelium, and filled by a viscid *colloid* material.

The ŒSOPHAGUS, or gullet (Plate 28), is a segment of the alimentary canal. It begins above the larynx, where it is continuous with the pharynx. It descends on the upper face of the trachea, and in the first few inches of its course it lies in the middle plane of the neck, being related to the longus colli muscle above. It soon, however, begins to deviate to the left side, so that below the middle of the neck it lies rather on the upper part of the left side of the wind-pipe. Maintaining this relationship, the two tubes enter the thorax in company, the gullet being prolonged through that cavity to pass by the foramen sinistrum of the diaphragm into the abdomen, where it terminates in the stomach. The gullet is related to the muscles of the left side already enumerated as contacting with the trachea, the sterno-thyro-hyoideus excepted. It is also related to the carotid artery, the jugular vein, and the pneumogastric, sympathetic, and recurrent nerves of the left side. The cervical part of the œsophagus has the external appearance of a voluntary muscle, for which it is often mistaken at first sight by the student. The examination of its structure, as well as that of the trachea, must be postponed until the accompanying vessels and nerves have been examined.

The COMMON CAROTID ARTERY (Plate 28). This is the vessel that conveys the blood to the head and upper part of the neck. It begins on the under aspect of the trachea, at the entrance to the thorax, where it results from the bifurcation of a short vessel termed the cephalic trunk—a branch of the arteria innominata. It ascends in the neck in company with the trachea, and terminates above the cricoid cartilage of the larynx by dividing into the external carotid, the internal carotid, and the occipital artery. It thus crosses the trachea very obliquely, being at first on its under surface, then on its lateral aspect, and finally above it. It is in contact with the scalenus, longus colli, rectus capitis anticus major, and subscapulo-hyoideus muscles, the last-mentioned intervening between the artery and the jugular vein in the upper half of the neck. In the lower half of the neck the artery and vein are in direct contact, the carotid being above and slightly internal to the jugular. The common cord of the pneumogastric and sympathetic nerves is on the upper side of the artery, and the inferior laryngeal (recurrent) nerve is below it. At the entrance to the thorax the pre-pectoral group of lymphatic glands is in contact with the artery. The left carotid differs from the right in being related for a considerable part of its course to the œsophagus, which separates it from the trachea. The collateral branches of the carotid are as follows :—

1. Innominate and slender *muscular*, *œsophageal*, and *tracheal* branches.

2. The *Thyroid* artery, which arises a few inches behind the thyroid body, and passes obliquely forwards to enter the gland on its posterior aspect. Sometimes this artery is distributed mainly or entirely to the neighbouring muscles.

3. The *Thyro-laryngeal* artery.—This is the largest of its collateral branches. It arises a little in front of the preceding vessel, and passing to the inner side of the thyroid body, it divides in front of it into thyroid and laryngeal branches. The former turn back to enter the gland in front, while the latter pass to the larynx and pharynx.

The terminal branches of the carotid are not to be followed at present.

The PNEUMOGASTRIC and SYMPATHETIC NERVES in the neck (Plate 28). The pneumogastric, vagus, or 10th cranial nerve has its origin from the medulla oblongata. It leaves the cranium by the posterior part of the foramen lacerum, and inclining downwards and backwards on the guttural pouch, it meets the cervical cord of the sympathetic, with which it becomes in nearly all cases closely united. The common cord resulting from the fusion of the two nerves descends in company with the carotid artery, lying on the upper side of that vessel. At the lower part of the neck the two nerves, in passing into the thorax, again become separate.

The cervical cord of the sympathetic begins at the superior cervical ganglion, which rests on the guttural pouch. After a short course it unites, as just described, with the vagus.

No branches are given off from either the vagus or the sympathetic in the part where they form a common cord.

The INFERIOR LARYNGEAL (RECURRENT) NERVE (Plate 28). This is a branch of the vagus, given off within the thorax. The right nerve has its origin in front of the heart, and is reflected round the dorso-cervical artery. The left nerve is longer than the right, having its point of detachment at the base of the heart, where it is reflected round the root of the posterior aorta. The nerves pass forwards on the trachea, and enter the neck by passing between the first pair of ribs. In the neck each nerve ascends below the carotid artery, the right nerve resting on the trachea, but that of the left side being, for the greater part of its course, on the œsophagus. The nerves will subsequently be followed in their distribution to the larynx. In the neck each recurrent nerve throws off branches to the trachea and œsophagus.

Directions.—At this stage the dissector of the fore limb will be engaged with the dissection of the axilla, and the dissector of the neck should co-operate with him in the examination of the mode of formation of the brachial plexus of nerves, and, thereafter, of the levator anguli scapulæ muscle. For the brachial plexus turn to page 3, and for the levator anguli scapulæ to page 8.

PREPECTORAL LYMPHATIC GLANDS. This is a large group of glands placed beneath and at the side of the great vessels at the entrance to the chest. They are placed on the course of the lymphatic vessels of the head, neck, and fore limb.

Directions.—A segment about six inches in length may now be cut from the trachea, and a similar segment from the œsophagus. These are to be dissected to display the structure of the two tubes.

STRUCTURE OF THE TRACHEA. This comprises (1) a framework of cartilages united by (2) fibro-elastic membrane ; (3) an incomplete layer of non-striped muscular tissue ; (4) a submucous layer ; and (5) a mucous lining.

The *Cartilages* of the trachea number between fifty and sixty, and are of the hyaline variety. Although usually denominated the *rings* of the trachea, they do not form complete circles, but have rather a resemblance to the letter C with its ends overlapping. In consequence of this configuration of the rings, the trachea is not circular on section, but flattened in the vertical direction; and the overlapping of the extremities of the cartilages takes place on the middle of the upper aspect of the tube. The breadth of the rings is not quite uniform, but averages about half an inch. Here and there, however, two adjacent rings may be more or less fused by the obliteration of the uniting fibro-elastic membrane. In the thoracic portion of the tube the extremities of the rings do not meet, and the deficiency is there made up by a number of thin cartilaginous pieces of irregular size and shape, and somewhat imbricated in their arrangement. The rings are thickest and strongest in their central portion, and thinnest at their extremities.

The *Fibro-elastic Membrane.*—This connects the adjacent edges of the cartilages, and at the upper wall of the tube it connects their overlapping extremities. Its extensibility and elasticity permit the length of the trachea to be accommodated to the movements of the neck, and these properties will be made very evident by alternately extending and relaxing a segment of the tube containing five or six rings.

The *Trachealis Muscle.*—This is a layer of non-striped muscular tissue having its fibres directed transversely. It does not extend all round the tube, but is confined to its upper part, where the fibres lie internal to the extremities of the rings or the fibro-elastic membrane. The fibres form a continuous band, being not only placed under each ring, but also in the interval between adjacent rings.

The *Submucous Coat* is composed of areolar connective-tissue with numerous elastic fibres longitudinally disposed. It also contains many small compound racemose glands, whose mucous secretion is discharged by ducts opening on the free surface of the mucous membrane.

The *Mucous Membrane*, which forms a complete internal lining to the tube, possesses a stratified epithelium, the surface layer of cells being ciliated.

STRUCTURE OF THE ŒSOPHAGUS. This comprises (1) a muscular coat, arranged in two layers; (2) a submucous coat; and (3) a mucous lining.

The *Muscular Coat* consists of (*a*) an outer layer of fibres longitudinally disposed, and (*b*) a deeper layer in which the fibres are arranged as transverse or oblique rings. In the cervical part of the tube, and in the thoracic part about as far as the heart, the muscular fibres are for the most part of the striped variety, and the tube has there the external appearance of a voluntary muscle. About the centre of the thorax, however, the character of the fibres gradually changes to the pale, non-striped variety of muscular tissue, and behind that point the tube is therefore pale like the stomach or the intestines.

The *Submucous Coat* is composed of areolar connective-tissue containing the alveoli of numerous mucous glands, whose ducts penetrate the mucous membrane. It forms a very loose bond of connection between the muscular and mucous layers; and when the œsophagus is cut across, the mucous coat appears almost as if it lay independently within the muscular layer.

The *Mucous Membrane* has a thick stratified epithelium; and, except during the act of deglutition, its free surface is thrown into longitudinal folds which meet with one another and obliterate the lumen of the tube. In colour it is whitish, owing to its low vascularity and the thickness of its epithelial covering.

THE UPPER PART OF THE NECK.

Position.—The animal should be suspended in imitation of the natural standing posture, in the manner described at page 8, for the dissection of the outer scapular region.

Directions.—The whole of the neck behind the atlas should be denuded of skin. The cutaneous nerves of the region should then be noticed, and the spinal accessory nerve should be found crossing obliquely backwards and upwards on the surface of the splenius muscle.

CUTANEOUS NERVES. For the most part, the cutaneous nerves of this region are derived from the stellate groups already seen perforating the mastoido-humeralis. These are derived from the inferior primary branches of the cervical spinal nerves from the 2nd to the 6th. Other branches, which are derived from the superior primary branches of the same nerves, emerge near the middle line of the neck above, and are distributed to the integument beneath the mane.

The SPINAL ACCESSORY (11TH CRANIAL) NERVE (Plate 27). This nerve derives its fibres from the medulla oblongata and the cervical part of the spinal cord. It leaves the cranium by the foramen lacerum basis cranii, passes backwards on the guttural pouch, turns upwards over the edge of the wing of the atlas, and passes obliquely backwards and upwards beneath the mastoido-humeralis muscle. Appearing at the upper edge of the last-named muscle, it is continued in the same direction on the surface of the splenius, and disappears beneath the cervical trapezius, in which and the dorsal trapezius it terminates. While the neck is elevated, the trunk of the nerve is thrown into numerous short sinuosities, apparently to obviate stretching of the nerve when the neck is depressed.

Directions.—The cervical portions of the trapezius and rhomboideus muscles should now be examined in co-operation with the dissector of the fore limb; and, thereafter, the mastoido-humeralis is to be dissected.

The TRAPEZIUS. See page 9.

The RHOMBOIDEUS. See page 10.

The MASTOIDO-HUMERALIS, or LEVATOR HUMERI (Plates 27 and 28).

This is a long and powerful muscle, extending between the head and the shoulder, on the side of the spinal column. It takes *origin* from the mastoid process and crest, from the wing of the atlas, and from the transverse processes of the 2nd, 3rd, and 4th cervical vertebræ. The tendon of origin from the mastoid process and crest, which is not to be exposed at present, is thin and aponeurotic; that from the wing of the atlas is common to the splenius and trachelo-mastoideus muscles; while the succeeding slips of origin are fleshy. The muscle passes over the shoulder-joint, and becomes *inserted* into the outer lip of the musculo-spiral groove. As already seen, the lower edge of the muscle forms the upper boundary of the jugular channel, and at the lower part of the neck it is closely united to the sternal band of the panniculus.

Action.—It is an extensor and inward-rotator of the shoulder-joint. When the limb is fixed, it bends the neck laterally.

This muscle represents the greater part of the *sterno-mastoid* of man (the rest being represented by the sterno-maxillaris), combined with the clavicular part of the *deltoid*, this fusion resulting from the absence of a clavicle.

Directions.—If the mastoido-humeralis has not already been cut, it should be divided in front of the shoulder, and turned upwards (Plate 28) to show the prescapular glands and a branch of the inferior cervical artery. The stellate groups of cutaneous nerves may thereafter be traced through the mastoido-humeralis to their source.

The PRESCAPULAR LYMPHATIC GLANDS. These are arranged in the form of a chain between the mastoido-humeralis and subscapulo-hyoid muscles at the lower part of the neck.

The INFERIOR CERVICAL ARTERY is a branch of the axillary, arising at the first rib. It divides into a descending (Plate 1) and an ascending branch, the latter being here seen between the mastoido-humeralis and subscapulo-hyoid muscles, to which and the above-mentioned glands it is distributed.

CERVICAL SPINAL NERVES. There are eight pairs of these, the 1st issuing from the spinal canal by the antero-internal foramen of the atlas, the 2nd by the foramen (converted notch) at the anterior edge of the arch of the axis, and the others in succession by the intervertebral foramina. They have all a common disposition in that each divides at its point of exit into superior and inferior primary branches. Only the inferior primary branches present themselves now for consideration, and of these the 1st is more conveniently taken at a later stage. The remaining six behave as follows:—

The 2nd, 3rd, 4th, 5th, and 6th communicate, each with the preceding and succeeding branches of the series, and divide into three sets of branches, viz., (1) communicating branches to the middle cervical ganglion (see vertebral nerve, page 157); (2) muscular branches to the mastoido-humeralis, longus colli, scalenus, and rectus capitis anticus major muscles, and to the diaphragm (see phrenic nerve, page 6); (3) cutaneous branches which pierce the mastoido-humeralis and are distributed as the stellate groups already seen. Besides these, the 6th nerve sends branches to the levator anguli scapulæ and rhomboideus muscles, and its phrenic branch sends a twig to the brachial plexus.

The branches of the 7th and 8th nerves are expended in the brachial plexus after each has detached a communicating filament to the middle cervical ganglion, that from the 7th joining the vertebral nerve, while that from the 8th passes to the ganglion independently.

Directions.—The dissector of the fore limb will now be in a position to separate the limb from the trunk, which will permit the dissection of the remainder of the neck. The levator anguli scapulæ, as left by the dissector of the fore limb, must now be entirely removed in order to expose the posterior part of the splenius. The mastoido-humeralis may also be cut away as far forwards as the vertebra dentata. The insertions of the splenius, trachelo-mastoideus, and complexus muscles into the head are not to be exposed at present, as that would involve the destruction of the muscles of the ear and other structures not yet dissected.

The SPLENIUS (Plate 27). This is a flat, fleshy muscle of a triangular form, having its fibres directed downwards and forwards. It takes *origin* from the 2nd, 3rd, and 4th dorsal spines, and from the funicular portion of the ligamentum nuchæ. Its origin from the dorsal spines is aponeurotic, and confounded with that of the anterior serratus and complexus muscles. It is *inserted* into the mastoid crest, the wing of the atlas, and the transverse processes of the succeeding four cervical vertebræ. The mastoid insertion is flat and aponeurotic, and is united to the mastoid tendon of the trachelo-mastoideus. The insertion into the atlas is tendinous and riband-like, and is common to the trachelo-mastoideus and mastoido-humeralis. The other insertions are fleshy.

Action.—The right and left muscles acting together elevate the head and neck ; acting singly, they incline the head and neck to the side of the acting muscle.

Directions.—The origin of the splenius should be carefully detached, and the muscle should be raised and turned downwards so as to expose the subjacent structures. Nerves from the superior primary branches of the cervical nerves, and branches from the superior cervical, dorsal, and vertebral arteries, will be found entering its deep face. The trachelo-mastoideus and complexus muscles, now exposed, should be dissected, the branches of nerves and vessels found in connection with them being as far as possible preserved.

The TRACHELO-MASTOIDEUS. This muscle consists of two parallel fleshy portions extending along the spine, under cover of the splenius, and resting on the complexus. The fibres of the muscle *arise* by successive slips from the transverse processes of the first two dorsal vertebræ; and, in common with the complexus, from the articular processes of the last six cervical vertebræ. The upper division of the muscle is *inserted*, by a flat tendon common to the splenius, into the mastoid crest; the lower division terminates in a riband-like tendon, common to the splenius and mastoido-humeralis, and *inserted* into the wing of the atlas.

Action.—Acting singly, to bend the neck laterally; acting with the opposite muscle, to extend the occipito-atlantal articulation (elevate the head).

The COMPLEXUS. This is one of the most powerful muscles of the neck. It covers the lamellar portion of the ligamentum nuchæ, which separates the right and left muscles. It *arises* from the 2nd, 3rd, and 4th dorsal spines, in common with the splenius; from the transverse processes of the first six or seven dorsal vertebræ; and from the articular processes of the cervical vertebræ with the exception of the first. From these different points of origin the fibres converge towards the poll, where they terminate in a tendon *inserted* into the occipital bone.

Action.—Both muscles will extend the occipito-atlantal joint (elevate the head); the muscle of one side will, while elevating the head, turn it slightly to the same side.

NERVES. As already stated, each cervical nerve resolves itself into a superior and an inferior primary branch. The superior primary branches of the last six may now be found distributing nerves to the splenius, trachelo-mastoideus, complexus, and semispinalis colli muscles; and if the complexus be raised from the ligamentum nuchæ, other branches will be found to ascend between the muscle and the ligament to be distributed to the integument near the middle line.

The muscles and other structures which lie above the cervical vertebræ receive their chief blood supply from three vessels, viz., the occipital, dorsal, and superior cervical arteries. The first of these gives branches to the neighbourhood of the poll, but it is not to be sought at present.

The DORSAL ARTERY will be found distributing branches to the upper part of the neck, in front of the withers. The artery has its origin within the thorax, where, on the left side, it is a branch of the axillary artery, and, on the right side, of the arteria innominata. It leaves the chest by the upper part of the second intercostal space; and inclining backwards and upwards, it divides on the longissimus dorsi into a number of branches distributed to the withers, and the neck in front of that region. These will be found ascending between the splenius and complexus muscles.

The SUPERIOR CERVICAL ARTERY, like the preceding, arises within the chest, and from the same source. It leaves the thorax by the upper part of the first intercostal space ; and placing itself on the inner surface of the complexus, it ascends between that muscle and the ligamentum nuchæ, as far as the 2nd or 3rd cervical vertebra, where its terminal branches anastomose with those of the occipito-muscular and dorsal arteries.

VEINS. The dorsal and superior cervical arteries are accompanied by veins of the same names, which, after entering the chest, discharge themselves into the anterior vena cava.

Directions.—The complexus, splenius, and trachelo-mastoideus muscles may now be removed as far as the hinder end of the axis. This will expose the semispinalis colli muscle and the ligamentum nuchæ.

The SEMISPINALIS COLLI MUSCLE (Fig. 18) rests on the laminæ of the cervical vertebræ, and consists of five bundles. The most posterior of these bundles may be described as taking origin from the anterior articular process of the 7th cervical vertebra, and passing forwards and inwards to be inserted into the superior spine of the 6th vertebra. The most anterior bundle passes in the same way between the 3rd vertebra and the axis, while the intermediate bundles have corresponding attachments.

Action.—The right and left muscles, acting in concert, will extend (elevate) the cervical part of the spinal column. The muscle of one side, acting singly, will rotate and incline the spinal column to the opposite side.

The INTERTRANSVERSALES COLLI MUSCLES (Fig. 18). These form a set of six muscular bundles with strong tendinous intersections, and cover the sides of the cervical vertebræ. There is one bundle for each intervertebral articulation except the first. Each muscular bundle consists of an upper and a lower slip; and it may be described as *arising* from the articular process of one vertebra, and passing forwards to be *inserted* into the transverse process of the vertebra in front. The muscles conceal the intervertebral foramina and the vertebral vessels ; and they are perforated by branches of these vessels, and by the superior and inferior primary branches of the spinal nerves of the neck.

Action.—To bend the neck laterally.

The RECTUS CAPITIS ANTICUS MAJOR (Plate 28 and Fig. 18). This muscle begins by a tapering point on the side of the vertebral column at the 5th cervical vertebra ; and passing forwards and inwards, it gains the inferior face of the atlas, in passing to the base of the skull. It *arises* from the transverse processes of the 5th, 4th, and 3rd vertebræ, the slip of origin from the first of these crossing the point of the scalenus. Its *insertion*, which is not to be exposed at present, is into the tubercular processes at the junction of the basilar process with the body of the sphenoid.

Action.—The right and left muscles, acting together, flex the head. When only one muscle acts, it inclines the head to the same side.

The SCALENUS (Plate 3 and Fig. 18). This muscle is situated on the side of the lower half of the neck. It *arises* from the transverse processes

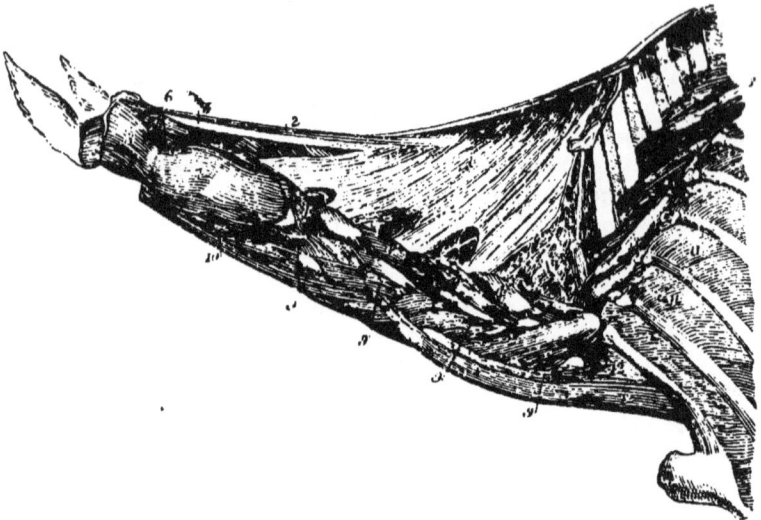

FIG. 18.

LIGAMENTUM NUCHÆ AND DEEP MUSCLES OF THE NECK (*Chauveau*).

1. Lamellar portion of the ligamentum nuchæ ; 2. Funicular portion of the same ; 3. Semispinalis of the back and loins ; 4. Semispinalis colli ; 5. Rectus capitis posticus major ; 6. Rectus capitis posticus minor ; 7. Obliquus capitis inferior ; 8. Obliquus capitis superior ; 9. Intertransversales colli ; 10. Rectus capitis anticus major ; 11. External intercostals ; 12. Upper and lower divisions of the scalenus.

of the last four cervical vertebræ. In front of the 1st rib it is perforated by the roots of the brachial plexus, which there divide it into an upper and a lower portion. The first of these is much the smaller of the two, and it is *inserted* into the outer surface of the 1st rib near its upper end. The lower portion is *inserted* into the anterior border and outer surface of the same rib, the lowest fibres being immediately above the smooth impression left on the anterior border of the bone by the axillary vessels.

Action.—To pull forwards or fix the 1st rib, and thus to aid in inspiration. When the rib becomes the fixed point, the muscles will flex the neck or incline it to the side, according as the right and left muscles act in concert or singly.

The LONGUS COLLI (Plate 28). This muscle clothes the inferior face of the spinal column from the 6th dorsal vertebra to the atlas, the right and left muscles being closely united along the middle line, while at its outer edge each muscle is partially blended with the intertransverse muscles. The dorsal portion of the muscle is seen in the dissection of the thorax (Plates 22 and 25), where its fibres take *origin* from the bodies of the first six dorsal vertebræ, and pass forwards to terminate in a tendon *inserted* into the 6th cervical vertebra. In the neck the fibres of the muscle take *origin* from the transverse processes, and each bundle passes with a forward and inward direction, to be *inserted* into the body of a vertebra anterior to that from which it arises. The most anterior fasciculi terminate in a tendon *inserted* into the tubercle of the atlas.

Action.—To bend the neck downwards.

Directions.—The vertebral vessels and the accompanying nerve should now be exposed by the careful removal of the intertransversales muscles, attention being at the same time directed to the superior and inferior primary branches of the spinal nerves. These nerves emerge in common from the intervertebral foramina, but separately pierce the muscles.

The VERTEBRAL ARTERY is a branch of the axillary artery, given off from that vessel before it leaves the thorax. It enters the neck by passing forwards to the inner side of the 1st rib a little below its upper end. It then ascends along the side of the spinal column, passing first beneath the transverse process of the 7th vertebra, and then in succession through the vertebral foramina of the other bones as far as the axis. Between the last-mentioned bone and the atlas it joins directly the retrograde branch of the occipital artery, but this is not to be exposed at present. It throws off in its course (1) *muscular* and (2) *spinal* branches. The former are very numerous and consist of an upward, a downward, and an outward set. Many of the upward set cross over the vertebræ, and anastomose with branches from the superior cervical artery. The spinal branches are detached from the inner side of the artery; and entering the spinal canal by the intervertebral foramina, they join the middle spinal artery in supplying the spinal cord and its coverings.

The VERTEBRAL VEIN accompanies the artery, and within the chest joins the anterior vena cava.

The VERTEBRAL NERVE runs in close company with the vessels. It is formed by the union of filaments from the inferior primary branches of the cervical nerves from the 2nd to the 7th. In the thorax it joins the inferior cervical ganglion of the sympathetic nerve. It is thus a composite nerve made up of the afferent filaments sent by the before-mentioned spinal nerves to the sympathetic cord.

The LIGAMENTUM NUCHÆ (Fig. 18). This is the largest ligament in the body. It is placed on the middle plane of the neck, above the vertebræ, and it consists of a right and a left division, each of which, again, comprises a *funicular* and a *lamellar* portion. The entire ligament, like most of the other ligaments of the neck, is composed of yellow elastic tissue.

The *funicular* portion has the form of a flattened cord united by its inner edge to the corresponding structure of the opposite side. Posteriorly, behind the summit of the 3rd dorsal spine, this cord is continuous with the supraspinous ligament of the back. Anteriorly the cord is inserted into a special tubercle on the occipital bone. Between these points of attachments the cord extends with a slight upward concavity when the ligament is relaxed, and above it there is developed, in varying amount, a quantity of fatty-elastic tissue supporting the integument from which the mane grows.

The *lamellar* portion is triangular in form, occupying the interval between the funicular portion and the vertebral column. Its fibres have a downward and forward direction, being fixed above to the funicular portion or to the spines of the 2nd and 3rd dorsal vertebræ, and below to the spines of the last six cervical vertebræ. The fibres are stronger and more closely aggregated in proportion as they are anterior, the lamella forming a complete septum in its anterior two-thirds, but having the form of a network in its posterior third. The right and left lamellæ are applied together on the mesial plane, their inner faces being united by areolar connective-tissue.

The ligamentum nuchæ assists in suspending the head; and when the head has been depressed, it aids the muscles in elevating it again. But for its presence, a large additional amount of muscular tissue would have been necessary in the neck. In man, in whom the head is supported by the spinal column, the ligament is very rudimentary, and has lost its elastic texture.

Directions.—The spinal column should now be disarticulated between the 3rd and 4th cervical vertebræ; and the head should be laid aside on a clean table, while the student proceeds to the dissection of the remaining ligaments of the neck. It will suffice to dissect carefully the ligaments of one intervertebral articulation,—say that between the 4th and 5th bones.

The intervertebral joints of the neck posterior to the vertebra dentata are constructed after a common plan, which is also that of the dorsal and lumbar regions. Each vertebra is articulated to the preceding and the succeeding bone (1) by an amphiarthrodial union of the bodies, and (2) by diarthrodial joints between the articular processes.

The atlanto-axial and the occipito-atlantal joints, which are purely diarthrodial articulations, will be dissected at a later stage.

LIGAMENTS AND ARTICULATIONS OF THE NECK POSTERIOR TO
THE DENTATA.

The ligaments may be classified into (1) those connecting the processes and neural arches, and (2) those connecting the adjacent vertebral bodies.

LIGAMENTS of the PROCESSES and NEURAL ARCHES :—

The *Ligamentum Nuchæ.*—This has already been dissected.

The *Interspinous Ligaments.*—These are composed of yellow elastic tissue. Each consists of two narrow parallel bands stretching between adjacent superior spinous processes.

Capsular Ligaments of the articular processes.

Ligamenta subflava.—For these two series of ligaments, see page 135, where the corresponding ligaments of the back and loins are described. In the neck these ligaments differ from those of the other regions in being composed of yellow elastic tissue.

LIGAMENTS of the BODIES :—

The *Superior Common Ligament* lies on the floor of the spinal canal, and terminates in front at the axis. See page 135.

The *Intervertebral Substance.*—See page 136.

MOVEMENTS of the cervical part of the spinal column. These are *flexion, extension, lateral inclination, rotation,* and *circumduction,* the last being a combination of the first three. In *flexion* the vertebræ are carried downwards in a vertical plane, and *extension* is the opposite movement. *Rotation* is the twisting, or turning, of a vertebra round a longitudinal axis passing through its body. In consequence of the thickness of the intervertebral substance, and the feeble development of the transverse and spinous processes, all these movements have here a greater range than in the back or loins ; and within the cervical region the greatest range of movement is permitted in the posterior joints.

THE EXTERNAL EAR (FIG. 19).

The organ of hearing consists of three divisions : the *external,* the *middle,* and the *internal* ear. Only the first of these will now be examined. The middle and the internal ear, which are cavities within the petrous temporal bone, are described at page 267. The external ear comprises the external auditory process of the petrous temporal bone ; three cartilages—conchal, scutiform, and annular ; muscles which move these cartilages ; vessels ; and nerves.

Directions.—An incision through the skin is to be begun a few inches behind the summit of the occipital bone, and carried down the middle line as far as the supraorbital process. It is here to be carried outwards along the supraorbital process, and then backwards along the zygomatic arch. On reaching the articulation of the jaw, the incision should be carried along the edge of the vertical ramus, and inwards to the middle line. All the skin mapped out by this incision is to be removed, the conchal cartilage being also denuded of its outer covering. This will expose not only the parts of the external ear, but also the parotideal

region and the poll, and the dissection of these parts is to be made as
soon as the ear is finished. On one side the muscles and cartilages of
the ear may be dissected ; and then the other side may be denuded of
skin in the same manner as the first, in order to follow the vessels and
nerves.

Muscles of the Ear.—These are divided into *extrinsic* and *intrinsic*.
The former have their origin from extraneous parts, but the latter both
arise from, and are inserted into, the cartilages of the ear. The cartilages of the ear cannot be fully exposed until the muscles have been
examined, but it may be premised that the conchal cartilage is the large
trumpet-like cartilage which mainly gives to the outer ear its form ;
that the annular cartilage is a short tube, or ring, which is telescoped
on to the external auditory process, and is itself embraced by the constricted base of the concha ; and that the scutiform cartilage is a thin
plate which rides on the surface of the temporal muscle, in front of the
base of the concha.

Extrinsic Muscles.

The PAROTIDO-AURICULARIS, or DEPRIMENS AUREM (Plate 29). This
muscle has the form of a broad riband. It *arises* from the outer
surface of the parotid gland ; and passing vertically upwards, it is
inserted into the outer part of the base of the concha, below the
opening.

Action.—To incline the ear downwards and outwards.

The CERVICO-AURICULARES, or RETRAHENTES AUREM. There are
three of these, distinguished as the cervico-auricularis (or retrahens)—
externus, medius, and *internus.* They all *arise* from the poll in the
neighbourhood of the insertion of the ligamentum nuchæ, and they are
here superposed the one to the other. Suppose the ear to be placed
with the opening of the concha looking directly outwards, then the
externus is *inserted* into the middle of the inner face of the concha ; the
medius into the outer side of the concha, beneath the opening, and
under cover of the parotido-auricularis muscle ; and the internus into
the base of the concha, on its posterior aspect, and under cover of the
parotid gland.

Action.—In moderate contraction, these muscles give the opening of
the concha an outward direction ; and when forcibly contracted, they
direct the opening backwards as well as outwards, and incline the ear
towards the poll.

The PARIETO-AURICULARIS EXTERNUS, or ATTOLENS MAXIMUS. This is
a wide, membranous muscle covering the temporalis muscle. It *arises*
from the parietal crest ; and it is *inserted* by its upper fibres into the
scutiform cartilage, and by its lower into the front of the conchal
cartilage. Its upper fibres are, at their origin, continuous across the
middle line with the opposite muscle.

Action.—To prick the ear, that is, to erect it and give its opening a forward direction.

The ZYGOMATICO-AURICULARIS, or ATTOLENS ANTICUS. This muscle is continuous with the preceding by an intermediate aponeurosis, and its own muscular substance is generally divided into two slips by intermediate fascia. It *arises* from the zygomatic process of the squa-

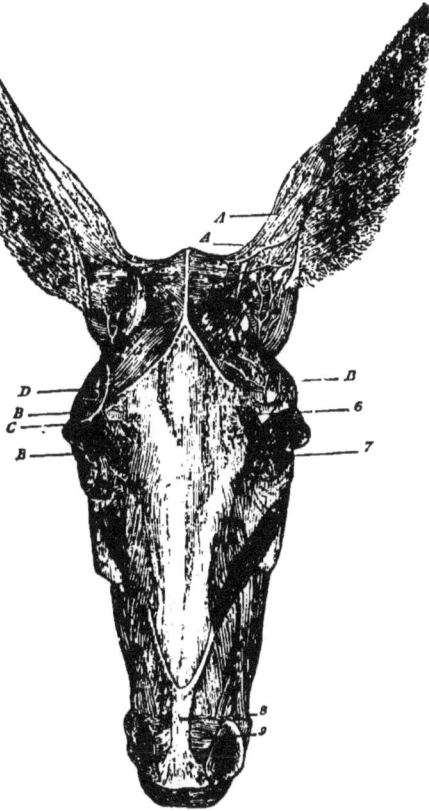

FIG. 19.

AURICULAR MUSCLES AND NERVES OF A MULE (*Chauveau*).

1. Parieto-auricularis externus ; 2. Parieto-auricularis internus ; 3. Scutiform cartilage ; 4. Scuto-auricularis externus ; 5. Temporalis ; 6. Corrugator supercilii ; 7. Orbicularis palpebrarum ; 8. United tendons of the levatores labii superioris proprii ; 9. Dilatator naris transversalis ; A. Auricular branches of 1st cervical nerve ; B. Anterior auricular nerve (of 7th); C. Supraorbital nerve ; D. Auricular branch of the lachrymal nerve.

mous temporal bone ; and it is *inserted* by an inner slip into the scutiform cartilage, and by an outer slip into the outer aspect of the base of the concha.

M

Action.—To prick the ear, like the preceding muscle.

The PARIETO-AURICULARIS INTERNUS, or ATTOLENS POSTICUS. This muscle is to be exposed by the removal of the parieto-auricularis externus, beneath the upper part of which it lies. It *arises* from the upper part of the parietal crest ; and it is *inserted* into the inner side of the concha, beneath the cervico-auricularis externus.

Action.—It is the opponent of the parotido-auricularis, bringing the ear into the erect position.

The MASTOIDO-AURICULARIS. This muscle is to be exposed by cutting the preceding and the cervico-auricular muscles, and forcibly depressing the ear outwards. This will expose, at the base of the ear, a considerable quantity of fat, which is constantly present, and facilitates the movements of the ear. On clearing away this fat, the muscle will be found at the inner side of the base of the ear. It forms a slender fasciculus *arising* from the auditory process, and *inserted* into the base of the concha.

Action.—To telescope the conchal on the annular cartilage.

Intrinsic Muscles.

Besides some scattered fibres on the outer and inner surfaces of the concha, this group includes, the following two distinct muscles passing between the conchal and scutiform cartilages.

The SCUTO-AURICULARIS EXTERNUS consists of two bundles of fibres passing between the outer surface of the scutiform cartilage and the inner side of the concha.

Action.—To assist the parieto-auricularis externus in pricking the ear.

The SCUTO-AURICULARIS INTERNUS. In order to expose this muscle, the scutiform cartilage is to be raised from the surface of the temporal muscle, and turned upwards and outwards. The muscle is stronger than the preceding, and consists of two distinct crossed bundles, which *arise* from the inner surface of the scutiform cartilage, and pass round the inner side of the base of the concha to .get *inserted* into its posterior aspect, above and internal to the insertion of the retrahens internus

Action.—It opposes the preceding muscle, and assists the retrahentes in rotating the concha so as to turn the opening outwards and backwards.

Vessels.

The ear is supplied with blood by the anterior and posterior auricular arteries.

The ANTERIOR AURICULAR ARTERY (Plate 28) is one of the two terminal branches of the superficial temporal artery, which vessel divides under the parotid gland, about an inch below the condyle of the lower jaw. It ascends behind the capsular ligament of the jaw ; and after detaching muscular and cutaneous branches in front of the ear, it enters the temporalis muscle.

The POSTERIOR AURICULAR ARTERY (Plate 28) is a collateral branch of the external carotid, detached while that vessel lies over the great cornu of the hyoid bone, and beneath the parotid gland. It divides on the occipito-styloid muscle into an anterior and a posterior branch. The anterior branch ascends in the parotid, and ramifies on the concha behind the posterior edge of its opening, after giving branches to the base of the ear, and to the interior of the concha. The posterior branch ascends in the parotid gland, and crosses behind the base of the ear, beneath the retrahentes muscle. It then passes under the parieto-auricularis internus, and ascends on the inner surface of the concha as far as its tip.

The blood is drained away from the ear by the anterior and posterior auricular veins.

The ANTERIOR AURICULAR VEIN is a larger vessel than the satellite artery. It joins the subzygomatic vein to form the superficial temporal trunk.

The POSTERIOR AURICULAR VEIN is formed at the base of the ear by two roots which unite at the posterior edge of the parotido-auricularis muscle. It descends at first on the surface of the parotid, and then in its substance, where it joins the jugular vein.

Nerves.

These are derived from the 7th cranial nerve, from the 1st and 2nd cervical nerves, and from the lachrymal nerve of the trifacial.

The POSTERIOR AURICULAR NERVE is detached from the 7th cranial nerve as it issues from the stylo-mastoid foramen. It ascends beneath or in the substance of the parotid gland, in company with the artery of the same name; and passing immediately behind the mastoid process, it gains the back of the ear, and is distributed to the cervico-auriculares muscles.

The MIDDLE AURICULAR NERVE is detached at the same point as the preceding. It ascends over the annular cartilage, behind the peaked process of the concha, which it enters at its base. It is here distributed to the scattered muscular fibres on the interior of the cartilage.

The ANTERIOR AURICULAR NERVE is much larger than either of the preceding nerves. It is given off from the 7th midway between the stylo-mastoid foramen and the edge of the inferior maxilla. It ascends in the parotid, turns over the zygomatic arch, passes downwards beneath the parieto-auricularis muscle, then internal to the root of the supraorbital process of the frontal bone, and terminates below the nasal canthus of the eyelids. It supplies the attolentes muscles as well as the corrugator supercilii and the orbicularis palpebrarum, and its terminal filaments enter the levator labii superioris alæque nasi.

The CERVICAL BRANCH of the 7th nerve. This nerve comes out through the substance of the parotid gland, near or at the same point as the jugular vein, and under cover of the parotido-auricularis muscle.

It supplies that muscle, and is continued down the neck, as already seen (page 144).

AURICULAR BRANCHES of the 1st CERVICAL NERVE. These, which are derived from the superior primary branch of that nerve, appear in the poll between the obliquus capitis superior and the rectus capitis posticus muscles. Crossing the cervico-auriculares muscles, they are distributed at the inner side of the base of the ear.

AURICULAR BRANCHES of the 2nd CERVICAL NERVE. These are derived from the stellate group which the inferior primary branch of that nerve forms on the mastoido-humeralis, behind the wing of the atlas. They reach the ear by crossing over the parotid gland, and are distributed mainly to the skin of the concha on its posterior aspect (when the opening is directed outwards), but some branches reach its inner side.

AURICULAR BRANCH of the LACHRYMAL NERVE. This nerve, which emerges from the orbital sheath, crosses the direction of the anterior auricular branch of the 7th on the zygomatic arch, and is distributed to the skin in front of the ear.

Cartilages of the Ear.

The CONCHAL CARTILAGE. This and the other cartilages are composed of yellow (elastic) fibro-cartilage. Although its name expresses some likeness to a shell, it bears more resemblance to a trumpet. The opening of the trumpet is somewhat elliptical, and can be directed forwards, outwards, or backwards. The margins of the opening meet above and below in acute angles. Beneath the lower angle, or commissure, the cartilage forms a complete tube, which is slightly inflated in form. At its termination it becomes narrow, and slightly embraces the annular cartilage, over the outer side of which it sends a peaked process, whose fibrous extremity is attached to the wall of the guttural pouch.

The ANNULAR CARTILAGE. This has the form of a ring surrounding the edge of the auditory process; and it is itself embraced by the conchal cartilage, the three structures being related to one another like the tubes of a telescope. This connection between the cartilages is maintained by connecting elastic tissue, and by the lining membrane of the ear in passing from the one structure to the other.

The SCUTIFORM CARTILAGE is superposed to the temporal muscle in front of the base of the concha, to which it is connected only by the muscles already described. It is thin, flexible, and irregularly triangular in shape.

THE PAROTIDEAL REGION.

The PAROTID GLAND (Plates 27, 29, and 30). This is the largest of the salivary glands. It derives its name from its proximity to the ear, below the root of which it is placed. From that point it stretches downwards, filling up the space between the wing of the atlas and the edge of the vertical ramus of the inferior maxilla. The outer surface of the gland is flat, and is separated from the skin by the parotido-auricularis muscle, and by the continuation of the cervical panniculus, which here takes the form of an aponeurosis with scattered muscular bundles. The jugular vein lies in a depression on the lower half of this surface, after having become superficial by passing through the substance of the gland. Below and behind the ear the posterior auricular vein is visible for some distance before it passes into the gland to join the jugular. Finally, the outer surface is crossed obliquely upwards and forwards by the auricular branches of the 2nd cervical nerve, and obliquely downwards and backwards by the cervical filament of the 7th cranial nerve, which comes out through the gland at the same point as the jugular, and descends under cover of the parotido-auricularis and panniculus muscles to be continued along the jugular channel of the neck. The anterior edge of the gland is related to the border of the vertical ramus, which it overlaps slightly. This edge is most intimately adherent to the bone and to the masseter muscle, and at it the facial branches of the 7th and 5th cranial nerves, and the transverse facial and maxillo-muscular vessels pass on to the face by emerging between the gland and the bone, or by perforating the edge of the former. The posterior edge of the gland is related to the edge of the wing of the atlas covered by the mastoido-humeralis muscle, and the connection between them is merely by loose areolar tissue. The upper extremity of the gland is notched to embrace the root of the ear, and beneath or through it the auricular nerves and arteries pass to the ear. The inferior extremity of the gland is margined by the submaxillary vein, which joins the jugular beneath the postero-inferior angle of the gland. The deep face of the gland has numerous and important relationships, which will be exposed by the removal of the gland; but its duct must first be examined.

STENSON'S DUCT. This is formed as a single duct by the union of secondary branches at the anterior edge of the gland, a little above its lower extremity. It crosses over the tendon of the sterno-maxillaris muscle, and enters the intermaxillary space, where it will subsequently be followed.

Directions.—The parotid gland should now be removed in order to expose the objects beneath it. Its removal must be effected with great care, so as to leave, as far as possible, the vessels and nerves which lie beneath it, or pass through its substance. This will be best done by following the 7th nerve, and the transverse facial and maxillo-muscular vessels, which emerge at the anterior edge of the gland, and the jugular vein, which passes through its substance. In removing the gland, its vessels and nerves must be cut. Its arteries are derived from the external carotid or its collateral branches; its veins empty themselves into the jugular or auricular veins; its nerves come from the 7th cranial nerve.

The deep face of the parotid gland is related to the following structures:—The mastoid insertion of the mastoido-humeralis, the terminal tendon of the sterno-maxillaris, the stylo-maxillaris, the digastricus (upper belly), the occipito-styloid, the stylo-hyoid, the submaxillary gland, the great cornu of the hyoid bone, the guttural pouch, the external carotid artery (with its terminal, and some of its collateral, branches), the initial part of the jugular vein (formed by the junction of the superficial temporal and internal maxillary veins), and the 7th nerve.

The *Tendons* of the MASTOIDO-HUMERALIS and STERNO-MAXILLARIS. These muscles are described at pages 152 and 145 respectively. The two tendons are connected by a fibrous expansion, which is included between the parotid and submaxillary glands. When they have been examined, the fibrous expansion and the tendon of the sterno-maxillaris may be removed.

The STYLO-MAXILLARIS. This muscle is not distinct from the digastricus (upper belly). It *arises* in common with that muscle from the styloid process of the occipital bone, and it is *inserted* into the angle of the jaw.

Action.—To depress the lower jaw, and assist in opening the mouth.

The DIGASTRICUS (Plate 31). This muscle will be only imperfectly exposed at present. It consists of an upper and a lower muscular belly, with an intermediate tendon. The lower belly will be met in the dissection of the intermaxillary space; and the tendon, in the dissection for the exposure of the mouth and pharynx. The upper belly *arises* from the styloid process of the occipital bone, in front of the origin of the preceding muscle, with which it is confounded. It is succeeded by the intermediate tendon, which plays through a perforation in the tendon of the stylo-hyoid, and is continued by the lower belly. The latter becomes *inserted* into the edge of the inferior maxilla, behind the symphysis.

Action.—To depress the lower jaw and open the mouth.

The OCCIPITO-STYLOID (Plates 31 and 32). This muscle *arises* from the front of the styloid process of the occipital bone; and it is *inserted* into the extremity of the styloid (great) cornu of the hyoid bone, behind its point of articulation with the skull. The deep face of the muscle is lined by the mucous membrane of the guttural pouch; and in the operation for opening the pouch, the muscle is perforated.

Action.—To flex the temporo-hyoideal joint, and carry backwards the hyoid bone and the parts attached to it.

The STYLO-HYOID (Plate 31). Only the *origin* of this muscle, from the heel-like part of the great cornu, will at present be seen. It should be identified, and preserved for examination at a later stage.

The submaxillary gland is to be preserved without disturbance until it can be exposed in its entirety. The guttural pouch will be described to more advantage at a later stage.

The EXTERNAL CAROTID ARTERY (Plate 28). Only the termination of that vessel is here seen. It is one of the terminal branches of the common carotid, which divides above the cricoid cartilage, under cover of the submaxillary gland. The first part of the artery—at present concealed—rests on the guttural pouch, and is covered by the stylo-maxillaris, digastricus, and stylo-hyoid muscles. As now seen, it appears between the last of these muscles and the hinder edge of the great cornu; and crossing obliquely upwards and forwards on the surface of that bone, it terminates by dividing into the superficial temporal and internal maxillary arteries. The vessel detaches three collateral branches, viz., the submaxillary, maxillo-muscular, and posterior auricular arteries, of which the first is concealed by the digastricus and stylo-hyoid muscles.

The MAXILLO-MUSCULAR ARTERY. This branch is given off at the upper edge of the stylo-maxillaris muscle, immediately after the parent trunk emerges between the stylo-hyoid muscle and the great cornu. It forms with the continuation of the main trunk a very obtuse angle. Passing forwards and downwards, it divides into an outer and an inner branch, which embrace between them the edge of the vertical ramus. The outer branch appears on the face at the anterior edge of the parotid gland, and enters the masseter muscle. The inner branch passes to the internal pterygoid muscle.

The POSTERIOR AURICULAR ARTERY. This vessel has its origin a little beyond the preceding, but from the opposite side of the carotid. Its distribution to the ear is given at page 163.

The SUPERFICIAL TEMPORAL ARTERY (Plate 28). Originating by the division of the external carotid on the great cornu, this artery, after a very short course below and behind the condyle of the lower jaw, divides into the transverse facial and anterior auricular arteries.

The TRANSVERSE FACIAL ARTERY turns round the ramus below the condyle, and will be followed in the dissection of the face.

The ANTERIOR AURICULAR ARTERY ascends to the front of the ear, on the capsular ligament of the temporo-maxillary joint. Its distribution is given at page 162.

The INTERNAL MAXILLARY ARTERY. This, much the larger terminal branch of the external carotid, passes within the condyle of the lower jaw, where it will subsequently be followed.

VEINS.—The *jugular* vein is formed within the substance of the parotid gland, close behind the articulation of the jaw, and superficial to the termination of the external carotid, a few lobules of the gland separating the artery and vein. The vessels which unite to form it are

the *superficial temporal* and *internal maxillary* veins.   It passes out
through the substance of the parotid, and then lies in a groove on
its surface, where it receives *maxillo-muscular* and *posterior auricular*
branches.

The 7TH CRANIAL NERVE (*Portio dura*) (Plate 28) emerges from
the aqueduct of Fallopius by the stylo-mastoid foramen of the petrous
temporal bone.   It passes downwards and forwards at the inner face of
the parotid or within its substance ; and turning round the inferior
maxilla, it reaches the face with the transverse facial vessels.   In this
course it is crossed superficially by the posterior auricular artery, and
passes over the angle of division of the superficial temporal artery.   As
it turns round the inferior maxilla, it is joined by the sensory subzygomatic branch from the inferior maxillary division of the 5th cranial
nerve.   In this part of its course the 7th nerve detaches the following
branches :—

1. The *Anterior, Middle,* and *Posterior Auricular Nerves.*—The first two
are given off at the stylo-mastoid foramen, the last is detached midway
between the foramen and the edge of the ramus.   The nerves ascend in
or beneath the parotid gland, and their distribution is given at page 163.

2. Nerves to the occipito-styloid, stylo-hyoid, digastricus (upper
belly), and stylo-maxillaris muscles.   These are given off at the stylo-
mastoid foramen.

3. The *Cervical Branch*, which is given off at nearly the same point as
the anterior auricular, but from the opposite side of the trunk.   It
passes through the parotid, and reaches the surface of the gland (see
page 163).

4. Numerous small and irregular branches to the parotid gland and
guttural pouch.

The SUBZYGOMATIC NERVE (*Auriculo-temporal* of man) is a branch
of the inferior maxillary division of the 5th nerve, given off at
the foramen lacerum basis cranii.   It descends behind the capsular
ligament of the jaw ; and crossing over the termination of the superficial temporal artery, it joins the 7th as it turns round the ramus.   It
sends a branch to accompany the transverse facial vessels (Plate 29).

THE REGION OF THE POLL.

Directions.—It will be convenient at this stage to dissect a group of
muscles (with their nerves and vessels) placed above the occipito-atlantal
and atlanto-axial joints (fig. 18, page 156).   The cervico-auricular muscles
of one side having been cleared away, the mastoid tendon of the
mastoido-humeralis (page 152) will present itself.   Beneath that, again,
is the mastoid tendon common to the splenius and trachelo-mastoideus
(page 153) ; and still deeper, there is the occipital insertion of the
complexus.   Each of these, having been identified, may be cut away ;

and the following muscles are to be isolated, the nerves and vessels being thereafter dissected on the other side.

The OBLIQUUS CAPITIS INFERIOR. This muscle, the most powerful of the group, covers the atlanto-axial joint on each side. It *arises* from the superior spine of the dentata; and it is *inserted* into the wing of the atlas on its upper aspect.

Action.—To rotate the atlas (and head) around the odontoid process of the axis.

The OBLIQUUS CAPITIS SUPERIOR. This muscle covers the occipito-atlantal joint on each side. It *arises* from the free edge of the wing of the atlas ; and it becomes *inserted* into the mastoid crest and styloid process of the occipital bone.

Action.—To extend the head on the atlas.

The RECTUS CAPITIS POSTICUS MAJOR. This muscle is composed of two parallel portions which, although not distinctly separated from one another, were described as distinct muscles by Percivall. It *arises* from the spinous process of the axis. Its most superficial fibres (*complexus minor* of Percivall) join the occipital insertion of the complexus; while its deeper portion (*rectus capitis posticus major* of Percivall) is *inserted* into the back of the occipital bone, beneath the insertion of the complexus.

Action.—The same as the preceding muscle.

The RECTUS CAPITIS POSTICUS MINOR. This is the smallest muscle of the group. It lies beneath the preceding, and covers the occipito-atlantal joint. It *arises* from the upper aspect of the ring of the atlas; and it is *inserted* into the back of the occipital bone, beneath the last-described muscle.

Action.—The same as the two preceding muscles.

The 1st CERVICAL NERVE (*Suboccipital* of man) issues from the spinal canal by the antero-internal foramen of the atlas. At its point of emergence it resolves itself into superior and inferior primary branches. The latter is immediately directed down through the antero-external foramen, and will be followed at a later stage. The superior primary branch appears between the obliquus capitis superior and the rectus capitis posticus muscles, where it gives (1) muscular branches to these muscles, and (2) auricular branches already followed to the skin of the ear.

The 2nd CERVICAL NERVE issues by the foramen at the anterior edge of the arch of the axis, where it is covered by the obliquus capitis inferior. It divides into superior and inferior primary branches, the latter of which has already been referred to (page 144). The superior branch gives twigs to the superior and inferior oblique muscles of the head, and is continued like the succeeding members of the cervical series (page 154).

The OCCIPITAL ARTERY. This vessel will be found ascending through the antero-external foramen of the atlas, and dividing there into *cerebro-spinal* and *occipito-muscular* branches.   The former enters the spinal canal by the antero-internal foramen; the latter divides for the supply of the muscles and other structures of the poll.

The *Retrograde* or *Anastomotic* branch of the occipital artery will be found issuing with a backward course from the posterior foramen of the atlas, and inosculating with the termination of the vertebral artery.

The *Mastoid* branch of the occipital artery will be found beneath the obliquus capitis superior.   It ascends behind the styloid process; and crossing over the mastoid crest, immediately above the mastoid process, it passes under the edge of the squamous temporal bone, and enters the parieto-temporal conduit, in which it anastomoses with the spheno-spinous branch of the internal maxillary artery.

VEINS.—Satellite veins accompany these arteries.

THE INTERMAXILLARY SPACE.

Directions.—Incise the skin along the middle line, from the mental symphysis upwards, and raise it on each side as far as the edges of the rami.

CUTANEOUS NERVES.   The skin of the intermaxillary space is supplied by a nerve derived from the 2nd cervical nerve.   It comes from the first stellate group of cutaneous nerves already seen on the surface of the mastoido-humeralis; and crossing obliquely downwards into the space (the long axis of the head is supposed to be vertical), it extends to near the symphysis of the lower jaw.

The PANNICULUS.   This is here extremely thin.   It hardly forms a continuous layer, but consists of muscular fasciculi scattered in an aponeurosis.

The SUBMAXILLARY LYMPHATIC GLANDS (Plate 27).   This group of glands is placed on the inner side of the horizontal ramus, a little above the point where its edge is crossed by the submaxillary vessels and Stenson's duct.   It rests on the lower belly of the digastricus muscle, being related inwardly to the insertion of the subscapulo-hyoid muscle, and outwardly to the submaxillary artery.   The right and left groups extend towards each other, and nearly meet below the extremity of the glossal (spur) process of the hyoid bone.   These glands are placed on the track of the lymphatic vessels coming from the mouth and nose; and in morbid states of these parts, such as glanders, the glands become inflamed and enlarged from the irritant matters conveyed in the lymphatic vessels.   They should be carefully excised to expose the submaxillary artery and the inferior belly of the digastricus muscle.

The SUBMAXILLARY or FACIAL ARTERY (Plate 27) appears at the upper part of the space, descending between the subscapulo-hyoid and internal pterygoid muscles.   At this point the inferior extremity of

the submaxillary salivary gland lies internal to it.   In passing obliquely
backwards and downwards, it rests on the internal pterygoid muscle, and
is partly covered by the lymphatic glands, under cover of which it detaches its submental branch.   It then comes into company with the vein
of the same name, and with Stenson's duct; and the three vessels turn
round the edge of the ramus to reach the face, the artery being below,
the vein in the middle, and the duct superior.   Where the artery turns
round the ramus, it is very favourably placed for taking the pulse, since
it is a vessel of considerable size, is in an easily accessible position, rests
directly on the bone, and is almost subcutaneous, only the thin panniculus intervening between it and the skin.

The *Submental Artery* crosses downwards beneath the inferior belly of
the digastricus, then along the surface of the mylo-hyoideus, which it
perforates a few inches above the symphysis.   It will be followed to its
termination in the dissection of the mouth.

The SUBMAXILLARY VEIN (Plate 27) is in contact with the artery where
the vessels turn round the ramus; but as it passes backwards, it recedes
slightly from the artery, and follows the posterior border of the submaxillary gland.   It leaves the space above the angle of the jaw, and is
continued along the lower edge of the parotid to join the jugular.   At the
lymphatic glands it receives the *submental vein*, a larger vessel than the
artery of the same name.

STENSON's DUCT (Plate 27), after crossing the sterno-maxillaris tendon,
passes into the space, at the posterior edge of the submaxillary salivary
gland.   It passes downwards on the internal pterygoid muscle, placing
itself in contact with the submaxillary vein, in company with which and
the artery it turns round the bone to reach the face.   It is here the most
superior of the three vessels; and from its being superficially placed
and resting on the bone, it is liable to be opened when a blow is delivered
over this region.

The DIGASTRICUS.   This muscle is named from its having two fleshy
bellies, with an intermediate tendon.   The upper belly *arises* from the
styloid process of the occipital bone, being confounded with the stylo-
maxillaris.   It is succeeded by the intermediate tendon, which plays
through the tendon of the stylo-hyoid muscle; but this and the upper
belly are at present concealed within the jaw.   The lower belly is placed
in the intermaxillary space, where it is partly covered by the lymphatic
glands, and is related anteriorly to the mylo-hyoid muscle.   It is *inserted*
by a flat fasciculated tendon into the edge of the horizontal ramus, a little
distance above the symphysis.

Action.—To depress the lower jaw and open the mouth.

The MYLO-HYOID MUSCLES stretch across the intermaxillary space, and
form a support for the tongue.   Each muscle *arises* from a line on the
inner surface of the horizontal ramus behind its alveolar border; and its

fibres pass transversely inwards, the most superior getting *inserted* into
the body and glossal process of the hyoid bone, and the others into a
median fibrous raphe between the two muscles. The muscle is to be
left intact at present, and its attachments will be better seen in the
dissection of the mouth.

Action.—To raise the body of the tongue towards the roof of the
mouth, and thus assist in mastication and deglutition.

NERVE to the mylo-hyoid and inferior belly of the digastric. This
is a branch of the inferior maxillary division of the 5th cranial nerve.
It descends between the internal pterygoid muscle and the vertical ramus;
and passing above the upper edge of the mylo-hyoid, it runs downwards
on the surface of that muscle, in company with the submental artery.
In part of its course it is covered by the lower belly of the digastricus,
and it sends to that muscle a distinct branch, which enters it on its
outer side.

THE APPENDAGES OF THE EYE.

These are—the eyelids, the membrana nictitans, the caruncula
lachrymalis, the conjunctival membrane, and the lachrymal apparatus.
The lachrymal gland—the most important part of the last mentioned
apparatus—will be dissected with the interior of the orbit; but the
other structures enumerated are to be examined now.

The EYELIDS. The front of the eye is protected by two movable cur-
tains—the upper and lower eyelids; and at the inner side of the eye there
is placed another structure—the *membrana nictitans*, which plays the
part of a third eyelid. The upper lid is larger than the lower, and has
a greater range of movement. Each eyelid presents two surfaces, two
borders, and two extremities. The outer or facial surface is formed by
a continuation of the skin, and is covered by short hairs. Among these
there occur in the lower lid some long tactile bristles. The inner or ocular
surface is lined by the conjunctival membrane, and is moulded on the
front of the eye. If the upper lid be everted there will be found on its
inner surface, near the outer angle, a number of minute openings, into
which bristles should be passed. These are the openings of the
excretory ducts of the lachrymal gland. On the same surface, but near
the opposite angle, and close to the free edge of each lid, there is a
round opening of larger size, but still minute. These are the *puncta
lachrymalia*, the orifices of the lachrymal ducts, by which the lachrymal
secretion is conveyed away from the eye. The free borders of the eyelids
circumscribe the palpebral fissure, which is a mere line when the eye is
closed, but is ovoid or elliptical, with the long axis directed obliquely
upwards and outwards, when the eye is open. The free edge of each lid
is somewhat stiff, this stiffness being due to a slender rod of cartilage
which extends along it. The *meibomian glands* are lodged in depressions
on the ocular surface of this cartilage, and may be seen through the

conjunctiva as close-set yellow lines having a direction at right angles to the edge of the lid. They number about fifty or sixty in the upper lid, but they are fewer and less distinct in the lower. Each gland consists of a main tube with lateral sacculi opening into it on each side, and it discharges its secretion by a dot-like orifice on the edge of the eyelid. The free edge of each lid carries a fringe of stiff hairs—the eyelashes, which tend to prevent the entrance of foreign particles into the eyes. The attached edge of each lid is marked on the ocular side by the angle of reflection of the conjunctiva from the lid to the eyeball, but on the facial side the eyelid passes into the surrounding skin without any defined line. At each extremity the eyelids join to form a commissure, or *canthus*. The outer or *temporal canthus* is acute, but the inner or *nasal canthus* is rounded, and lodges the caruncula lachrymalis.

The CARUNCULA LACHRYMALIS is a small, rounded, and, generally, dark-pigmented nodule placed within the nasal canthus, and about equidistant from the two puncta lachrymalia. It is covered by conjunctiva, and is composed of connective-tissue with some mucous follicles and the bulbs of a few short hairs, which project from it.

The MEMBRANA NICTITANS. This body is placed at the inner canthus, where, ordinarily, it projects to only a slight extent, but it is capable of being thrust more than half way across the front of the eye. It has for its basis a thin and flexible piece of elastic cartilage, which anteriorly is invested by conjunctiva. Posteriorly this cartilage passes to the inner side of the eyeball, where it becomes connected with the cushion of semifluid fat which is found in the posterior part of the orbit. The membrana nictitans has no muscle to move it directly; but when the eyeball is retracted within the orbit, it presses on the semifluid fat behind it, and this, tending to escape at the side of the eyeball, pushes the membrana nictitans before it. In the eye of a subject just dead, this mechanism may readily be demonstrated by pressing the eyeball backwards into the orbit. About the centre of the outer face of the cartilage, there will be found a cluster of reddish-yellow granules—the *Harderian gland*. The gland secretes an unctuous material which is discharged by a number of ducts that perforate the cartilage and open on its ocular surface.

The *Lachrymal Apparatus* comprises—the lachrymal gland with its excretory ducts, the puncta lachrymalia, the lachrymal canals, the lachrymal sac, and the lachrymal duct.

The LACHRYMAL GLAND is placed within the orbit, beneath the supra-orbital process of the frontal bone. The gland itself will be dissected at a later stage.

The excretory ducts of the gland discharge themselves by a number of minute openings on the inner surface of the upper eyelid, close to the temporal canthus. Sometimes a few of the ducts open on the lower lid close to the same canthus.   The watery secretion which issues from them is carried over the front of the eyeball by the movements of the eyelids, and at the nasal canthus it is drained away by the puncta lachrymalia.

The Puncta Lachrymalia.   Each punctum is placed on the inner surface of the lid near its free edge, and distant about $\frac{1}{4}$ of an inch from the caruncula.   The lower punctum is generally larger and more easily found than the upper.   If a flexible bristle be passed into each punctum, it may be directed along the lachrymal canal, into the lachrymal sac. Taking the bristle as a guide, each canal may then be slit open with scalpel or scissors.

The Lachrymal Canals will be found to converge towards the roof of the lachrymal sac, into which they open by distinct orifices.   The upper canal is a little longer than the lower.

The Lachrymal Sac is a small reservoir lodged in the fossa of the same name on the orbital surface of the lachrymal bone.   It receives the lachrymal secretion from the lachrymal canals, and it is directly continued as the lachrymal duct.

The Lachrymal Duct (ductus ad nasum) conveys the lachrymal secretion from the sac to the lower part of the nasal fossa.   In the first part of its course it is lodged in an osseous canal, along which it passes to reach the middle meatus of the nose, where it will subsequently be found.   The lachrymal canals, sac, and duct have a fibrous wall with a mucous lining, the epithelium being stratified and squamous in the canals, but ciliated in the sac and greater part of the duct.

The Conjunctiva.   This is a mucous membrane, consisting of a palpebral part lining the inner aspect of the eyelids, and an ocular portion which is reflected on the front of the eyeball.   The palpebral portion, including that covering the caruncula and membrana nictitans, consists of a stratified epithelium and a papillated layer of vascular subepithelial connective-tissue containing small mucous glands.   The ocular portion where it covers the sclerotic resembles the preceding in structure, but is thinner, nonpapillated, and less vascular; where it passes over the cornea, it consists of the epithelium only, which is generally enumerated as one of the constituent layers of the cornea itself, being termed its anterior epithelium.   Through the puncta lachrymalia, the conjunctival epithelium is continuous with that lining the lachrymal canals ; and at the free margin of the lids it is continuous with the epidermis.

Structure of the Eyelids.—Each lid is composed of the following parts :—a layer of skin outwardly, the palpebral conjunctiva inwardly, a portion of the orbicularis palpebrarum muscle, the palpebral tendon, the tarsal cartilage, vessels, and nerves.

Besides these, there is found in the upper lid the tendon of a special muscle—the levator palpebræ superioris. Of these, the skin and conjunctiva have already been sufficiently noticed.

The ORBICULARIS PALPEBRARUM (Fig. 19, page 161). This muscle will be exposed by removing the layer of skin from the eyelids, and from around the orbital rim for the breadth of an inch. The fibres of the muscle are closely adherent to the skin of the eyelids, and have a circular or elliptical disposition around the palpebral fissure. Below the nasal canthus a number of the fibres are inserted by a slender tendon into the lachrymal tubercle on the bone of the same name, but for the most part the fibres are without bony attachment. Above the orbit the muscle is confounded with some thin fibres—the *corrugator supercilii*—which wrinkle the overlying skin.

Action.—The orbicularis muscle closes the eye by approximating the free edge of the eyelids, and in this action the upper lid has a much wider range of movement than the lower.

The PALPEBRAL TENDON. This is a fibrous layer which will be exposed by removing the orbicularis muscle. At the free edge of the eyelid it is margined by the tarsal cartilage, while by its opposite border it is fixed to the rim of the orbit.

The TARSAL CARTILAGE. This is a slender rod of elastic cartilage imbedded in the free edge of each eyelid. The meibomian glands lie in grooves on its ocular surface. It prevents the margins of the lids from being drawn, or puckered, when the orbicular muscle contracts.

The LEVATOR PALPEBRÆ SUPERIORIS. This muscle is found in the upper eyelid only. The fleshy portion of the muscle is a slender flat fasciculus which takes *origin* at the back of the orbit (Fig. 22, page 209) above the optic foramen, where it will subsequently be exposed along with the other contents of the cavity. Passing along the eyeball and beneath the lachrymal gland, it is succeeded by a thin flat tendon which in the lid is placed beneath the palpebral tendon, and joins the tarsal cartilage.

Action.—To open the eye by raising the upper lid. In this movement the muscle plays over the eyeball like a rope over a pulley. The lower lid, it is to be observed, has no analogous muscle, because, in the first place, it is raised but slightly under the action of the orbicularis, and, secondly, because its own weight and elasticity are sufficient to depress it.

VESSELS. The arteries of the eyelids are derived from the supra-orbital and lachrymal arteries, and from the orbital branch of the superior dental artery.

The *Supra-Orbital Artery* will be found emerging from the orbit by the supra-orbital foramen. It is a branch of the ophthalmic artery.

The *Lachrymal Artery* is also a branch of the ophthalmic, and is distributed mainly to the gland of the same name, but it sends some twigs to the upper eyelid.

The *Orbital Branch of the Superior Dental Artery* creeps over the lower part of the rim of the orbit, and descends to anastomose with the submaxillary artery. It gives some twigs to the lower eyelid.

NERVES. The sensory nerves of the eyelids are derived from the supra-orbital and palpebro-nasal branches of the ophthalmic division of the 5th cranial nerve, and from the orbital branch of the superior maxillary division of the 5th. The supra-orbital nerve emerges by the foramen of the same name, and is distributed to the upper eyelid, and to skin around its point of exit. The palpebro-nasal nerve sends a branch to supply the lower lid and the structures at the inner canthus. The orbital branch of the superior maxillary division of the 5th nerve is distributed in the neighbourhood of the outer canthus. The motor nerve to the orbicularis comes from the 7th, which crosses internal to the nasal canthus. The motor nerve to the levator palpebræ comes from the 3rd, but cannot be reached at present.

THE FACE.

The NOSTRILS, or the INFERIOR or ANTERIOR NARES. Some points in connection with these may conveniently be noticed before the student proceeds to dissect the face. The nostril is the entrance to the nasal chamber or fossa—the first segment of the respiratory passages. It is a large, somewhat oval opening bounded laterally by the *alæ*, or *wings*, of the nostril. The alæ meet above and below, forming the *commissures*, the lower of which is wide and rounded, while the upper is acute. The outer ala is concave in the whole of its extent; but the inner ala, while concave below, forms a convex projection close to the upper commissure. If now the inner wing be manipulated, it will be felt to contain a piece of cartilage which begins above in the convex projection close to the upper commissure, extends downwards and round the inferior commis-

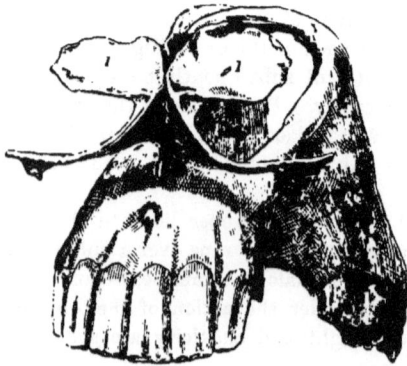

FIG. 20.

1, 1. Comma-like cartilages of the nostril ; 2. Septal cartilage.

sure, and terminates in the lower part of the outer ala. This alar cartilage, when dissected out, displays a close resemblance to a comma, the broad part being placed in the upper part of the inner wing, and the point in the lower part of the outer wing. The cartilages are movably attached by fibrous tissue to the lower extremity of the septal cartilage of the nose, and they give the necessary firmness to the edges of the nostrils, preventing these from falling together in the act of inspiration. The nostrils are covered outwardly by skin continuous with that of the face. This skin, which is thin and adherent to the subjacent textures, is carried round the margins of the alæ, and for a short distance into the nasal chamber. If the finger be introduced below the upper commissure, it will be felt to enter a peculiar diverticulum termed the *false nostril*. This blind pouch extends upwards for about four inches, and it is lined by a continuation of the skin. If the wings of the nostril be separated as widely as possible, the opening of the lachrymal duct will be seen a few inches within the lower commissure. The orifice is circular, appearing as if a small piece of skin had been punched out. Just beyond the orifice the skin is continued by the mucous membrane of the nasal chamber, which is distinguished from the common integument by being non-pigmented, vascular-looking, and destitute of hairs.

Directions.—The skin on the face should be entirely removed. Over the lips and false nostril care is necessary to avoid going deeper than the skin. One side should first be used for the muscles, and then the vessels and nerves should be dissected on the opposite side.

The muscles now to be dissected include a series of thin muscles which terminate in and move the lips and nostrils ; the buccinator, which forms the basis of the cheek ; and the masseter, a powerful muscle of mastication, covering the vertical ramus of the lower jaw. A reference to Plate 29 will enable the dissector to identify these muscles.

The PANNICULUS CARNOSUS does not form a continuous muscle on the face. It consists of disconnected bundles developed in the subcutaneous fascia. It is best developed over the masseter muscle, where it is continuous over the edge of the inferior maxilla with the same structure in the parotideal and intermaxillary regions. On the cheek a few bundles reach the angle of the mouth, in whose retraction they assist.

Below the rim of the orbit there is found a thin layer of pale muscular fibres and intermediate fascia, which may be viewed as a part of the facial panniculus. It wrinkles the overlying skin. This is the *lachrymal* muscle of French authors, and the *inferior palpebral* muscle of Leyh.

The LEVATOR LABII SUPERIORIS ALÆQUE NASI (Plate 29). This muscle has a narrow, thin, and aponeurotic *origin* from the frontal and nasal bones, below and internal to the orbit. It passes over the levator

N

labii superioris proprius ; and widening as it descends, it divides into
an anterior and a posterior branch, between which the lateral dilator of
the nostril emerges. The anterior is the larger branch, and it passes
beneath the last-mentioned muscle to *end in* the outer wing of the
nostril and the adjacent part of the upper lip. The posterior branch is
inserted into the angle of the mouth.

Action.—To dilate the nostril and elevate the upper lip by its
anterior branch ; to raise the angle of the mouth by its posterior
branch.

The LEVATOR LABII SUPERIORIS PROPRIUS (Plates 29 and 30, and fig.
19).   By its superior extremity, which is rounded and fleshy, this
muscle takes *origin* below the orbit, from the malar and superior maxillary bones.   Becoming narrower and thicker, it passes downwards and
forwards beneath the last-described muscle, whose direction it crosses.
Appearing in front of this muscle, it lies on the false nostril ; and here its
muscular portion is succeeded by a narrow tendon, which passes inwards
above the superior commissure of the nostril to gain the upper lip. In
the lip it passes over the transverse dilator of the nose ; and approaching the muscle of the opposite side, it becomes united to it across the
middle line by an intermediate fascia, and is *inserted* into the texture of
the upper lip.

Action.—Acting together, the right and left muscles elevate the
upper lip vertically. When the muscle of either side acts singly, the
lip is raised and inclined towards that side.

The DILATATOR NARIS LATERALIS.   (Doubtfully the homologue of
the *caninus* of man) (Plate 29).   This muscle is triangular in form.
It is narrow and pointed at its upper extremity, where it *arises* from
the superior maxilla. It passes between the two branches of the levator
labii superioris alæque nasi ; and widening as it descends, it passes over
the anterior branch of that muscle, and becomes *inserted* into the outer
wing of the nostril.   Its most posterior fibres blend with the orbicularis
oris.

Action.—To dilate the nostril by pulling the external wing outwards.

The DILATATOR NARIS TRANSVERSALIS (Fig. 19).   This is a four-sided,
fleshy muscle, whose fibres pass transversely across the middle line, and
are *inserted* at either extremity into the broad part of the comma-like
cartilage of the nostril.   It is partly covered by the termination of the
levator labii superioris proprius.

Action.—To dilate the nostrils by approximating the internal wings of
opposite sides.

The DILATATOR NARIS SUPERIOR.   This muscle comprises a few pale
fasciculi which *arise* from the projecting edge of the septum nasi at the
side of the nasal peak, and *terminate* in the wall of the false nostril, and
on the extremity of the ethmoidal (anterior) turbinated bone.

Action.—To dilate the false nostril.

The DILATATOR NARIS INFERIOR. The fibres of this muscle *arise* from the free edge of the premaxillary and superior maxillary bones; and they *terminate* in the wall of the false nostril, and on the cartilaginous prolongation of the maxillary (posterior) turbinated bone.

Action.—The same as the preceding muscle.

The ZYGOMATICUS (Plate 29) is a riband-shaped muscle, often extremely thin. It extends vertically between its point of *origin* from the surface of the masseter, behind the lower portion of the zygomatic ridge, and its *insertion* into the buccinator above the angle of the mouth.

Action.—To raise the angle of the mouth.

The BUCCINATOR (Plate 29). This muscle forms the main mass of the cheek. It is not wholly exposed at present, its upper extremity being covered by the masseter. It comprises a superficial and a deep portion, the former being separately described by Percivall as the *caninus.* The superficial portion shows a longitudinal *raphe* from which the muscular fibres pass forwards and backwards in a penniform manner. In front these fibres become *attached* to the superior maxilla, from the alveolus for the canine tooth to that for the first molar inclusive; and behind, to the inferior maxilla opposite the interdental space between the canine and the first molar. The deep portion of the muscle is longer but narrower than the preceding. It *arises* from the superior maxilla in front of the three upper molar alveoli; from the scabrous imprint on the same bone above the last alveolus; and from the edge of the inferior maxilla above the last alveolus. Inferiorly the fibres of the muscle blend with the orbicularis oris at the angle of the mouth. As will be seen at a later stage, after the removal of the masseter, the superior buccal gland lies on the muscle at its anterior edge, while the inferior buccal gland lies under its posterior edge, and the mucous membrane of the mouth lines its inner surface.

Action.—When unopposed by the orbicularis oris, the muscle will retract (elevate) the angle of the mouth; but otherwise, as in mastication, it compresses the cheek against the teeth and their alveoli, and tends to keep the food between the upper and the lower molars. In man, when the cheeks are distended, as in blowing a wind instrument, the muscle compresses the volume of air and propels it as a stream from the mouth : hence the name, from the L. *buccina*, a trumpet.

The DEPRESSOR LABII INFERIORIS (Plate 29). This muscle is placed along the hinder edge of the buccinator, with which it is confounded at its *origin* from the alveolar edge of the inferior maxilla above the last molar tooth. It *terminates* in the texture of the lower lip.

Action.—To depress (retract) the lower lip.

The ORBICULARIS ORIS (Plate 29). This muscle surrounds the aperture of the mouth like a sphincter. It cannot be described as having either origin or insertion, its fibres forming a continuous ellipse. In both lips the fibres are intimately adherent to the skin, and they are partially blended with the labial insertions of some of the muscles already dissected.

Action.—To approximate the lips, as in the simple act of closing the mouth or in prehension.

Directions.—Evert the upper lip, and wipe its exposed mucous lining clean. Observe that it is studded with numerous short, tubercle-like papillæ. Each of these is perforated by the duct of a labial mucous gland lying beneath the mucous membrane. Now dissect away the mucous membrane so as to expose these glands. At the same time, there will be brought into view the following muscle :—

The DEPRESSOR LABII SUPERIORIS. Under this name, Percivall describes a bundle of muscular fibres that in the human subject is reckoned a part of the buccinator. On each side the muscle *arises* from the premaxillary bone above the corner incisor and the interdental space as far as the canine tooth ; and, on the other hand, its fibres *terminate* in the upper lip, blending with the orbicularis. Branches of the infra-orbital and 7th nerves enter the lip between the outer edge of this muscle and the lowest fibres of the buccinator.

Action.—To assist the orbicularis by depressing the upper lip.

The PALATO-LABIAL ARTERY. While the upper lip is kept everted, dissect backwards on the middle line until this artery is found coming forwards from the roof of the mouth by the incisor foramen. At its point of exit it bifurcates, its branches passing right and left to anastomose with the superior labial artery.

Directions.—Evert the lower lip, and wipe its mucous surface clean. Notice that it is smooth, with few or none of the tubercle-like papillæ found on the upper lip. Dissect away the mucous membrane, which is intimately adherent to the orbicularis muscle. There are few or no labial mucous glands in the lower lip.

The LEVATOR MENTI. This is the name given by Percivall to a muscle of the lower lip resembling the *depressor* already dissected in the upper. Its fibres *arise* on each side from the inferior maxilla beneath the intermediate and corner incisors, and from the interdental space as far as the canine tooth. It runs downwards and backwards to *terminate* in the so-called prominence of the chin, being there intermixed with fibrous tissue, and confounded in front with the orbicularis oris. The inferior labial artery enters the lip between the outer edge of the muscle and the lower fibres of the buccinator.

Action.—To elevate the lower lip.

The MASSETER (Plate 29). This muscle covers the vertical ramus of the lower jaw. It has a flattened, semicircular form, and is thick and powerful. In its anterior half the surface of the muscle is glistening and tendinous, but posteriorly it is fleshy.   Other strong tendinous layers are included within the substance of the muscle, and give attachment to many of its fasciculi.   The muscle *arises* from the zygomatic ridge of the malar and superior maxillary bones, and it becomes *inserted* into the outer surface of the vertical ramus.

Action.—To elevate the lower jaw, and aid in mastication by bringing the lower teeth forcibly into contact with the upper.

The SUBMAXILLARY or FACIAL ARTERY has already been dissected in the intermaxillary space.   It reaches the face by turning round the edge of the inferior maxilla, in company with the vein of the same name and Stenson's duct (Plate 29).   Here the artery is the most inferior of the three vessels, and the duct is the most superior.   The three vessels cross the inferior maxilla at the lower edge of the masseter muscle, and preserve the before-mentioned relationship until they pass on to the buccinator.   Here the vessels cross over the duct, being themselves crossed superficially by the facial nerves; and passing beneath the zygomaticus, below the zygomatic spine, they reach the surface of the superior maxilla, on which the artery divides into an *angular* and a *nasal* branch.   The *angular branch* detaches a twig to anastomose with the orbital branch of the superior dental artery, and is then distributed to the muscles and skin beneath the orbit.   The *nasal branch* reaches the false nostril by passing beneath the levator labii superioris alæque nasi.   The collateral branches which the submaxillary gives off in this part of its course are the superior and inferior labial arteries; and unnamed cutaneous or muscular branches, of which those that pass to the masseter muscle are the largest.

The *Inferior Labial Artery* is the largest branch.   It is given off on the surface of the inferior maxilla.   It passes beneath the depressor labii inferioris, and descends to the lower lip, at the hinder edge of the buccinator.   At the mental foramen it anastomoses with the mental branch of the inferior dental artery, and in the lip it anastomoses on the mesial plane with the opposite vessel.

The *Superior Labial Artery* is detached opposite the zygomatic ridge. It passes beneath the levator labii superioris alæque nasi and the lateral dilator of the nostril, and reaches the upper lip, in which it anastomoses with the palato-labial artery.

The SUBMAXILLARY VEIN (Plate 29) runs in close company with the artery.   Where the vessels appear on the face, and throughout the greater part of their course, the vein is above the artery.   It is formed by an *angular* and a *nasal* branch, the first of which begins in the lower eyelid, while the other originates beneath the skin of the false nostril, and passes over the levator labii superioris alæque nasi to join the first. It receives branches corresponding to those of the artery, and, in addition, the alveolar and buccal veins join it at the edge of the masseter.

In thin-skinned, fine-bred animals the course of the vein and of its angular and nasal branches is conspicuous during life.

STENSON's DUCT (Plate 29), or the parotid duct, crosses the face, being at first between the edge of the masseter and the vein. It then passes forwards and upwards beneath the artery and vein, on the surface of the buccinator ; and finally it perforates that muscle under cover of the zygomaticus, and opens on the inner surface of the cheek, opposite the third upper molar tooth.

The TRANSVERSE FACIAL ARTERY (Plate 29) is a branch of the superficial temporal. Given off beneath the surface of the parotid gland, it turns round the edge of the ramus beneath the condyle, and at the anterior border of the gland it gives off a large masseteric branch. Having gained the face, it descends for a few inches on the surface of the masseter muscle, immediately below the zygomatic arch; and then, about midway between the temporo-maxillary joint and the orbit, it penetrates the masseter, and is distributed to it. Where the vessel is on the surface of the muscle, it is covered only by the thin facial panniculus and the skin, and is conveniently placed for the taking of the pulse. Its *masseteric branch* at once plunges into the substance of the masseter, where it anastomoses with the maxillo-muscular artery, and with the posterior deep temporal by a small branch which traverses the corono-condyloid notch.

The MAXILLO-MUSCULAR ARTERY (Plate 29) is a collateral branch of the external carotid. Beneath the parotid gland it bifurcates to form a *pterygoid* and a *masseteric branch*. The former passes within the ramus to reach the internal pterygoid muscle. The latter emerges between the ramus and the parotid gland, above the insertion of the stylo-maxillaris, and penetrates the masseter.

VEINS.—The *transverse facial* and *maxillo-muscular* veins accompany the arteries of the same name. The former joins the anterior auricular to form the superficial temporal vein, the latter empties itself into the jugular.

The 7TH NERVE on the face (Plate 29). This nerve appears on the face a little below the articulation of the jaw, where it emerges from beneath the parotid gland. Before its emergence it is joined by the subzygomatic branch of the inferior maxillary division of the 5th. The nerve divides into a variable number of branches, which anastomose on the surface of the masseter and form a plexus, termed in man the *pes anserinus* (from its resemblance to the foot of a goose). In this plexus it is not possible to distinguish, among the motor fibres proper to the nerve itself, those sensory fibres derived from the subzygomatic nerve. This plexus is covered by the skin and the thin facial panniculus, both of which receive branches from it. Below the inferior edge of the masseter, branches of the plexus are continued over the submaxillary vessels to supply the muscles of the cheek, lips, and nostrils, as well as the panniculus and skin. The largest of these branches is the most anterior, and it passes beneath the zygomaticus muscle in company with the superior labial vessels. It communicates with the infra-orbital nerves emerging from the infra-orbital foramen, and passes with them to the upper lip. Another branch of considerable size passes beneath the retractor of the lower lip, and runs in company with the inferior · labial artery. It reaches the lower lip, in which it is distributed along with the mental nerves.

The SUBZYGOMATIC BRANCH of the 5th nerve (Plate 31). The major portion of this nerve joins the 7th, and is distributed with it on the face ; but before joining with that nerve, it detaches a branch which passes in company with the transverse facial vessels, and is traceable as far as the orbit.

The INFRA-ORBITAL NERVE (Plate 30). This, which is derived from the superior dental branch of the superior maxillary division of the 5th, and is therefore sensory, emerges from the infra-orbital foramen, under cover of the levator labii superioris proprius muscle. The divisions of the nerve descend beneath the levator labii superioris alæque nasi and the lateral dilator of the nostril, where they communicate with a branch of the 7th nerve, and are distributed in the nostril and upper lip.

Directions.—The masseter muscle on one side is now to be removed ; and in doing this, the dissector should find the branch from the inferior maxillary division of the 5th nerve which enters the muscle through the corono-condyloid notch. The masseter having been removed, the buccinator muscle will now be fully exposed (see page 179), and the dissector is to examine the alveolar vein, the buccal glands, the buccal nerve and vessels, and the interior of the cheek.

The ALVEOLAR VEIN. This vessel rests on the superior maxilla, along the alveoli for the molar teeth. Inferiorly it joins the submaxillary vein, and is here comparatively small. It speedily becomes of large size, and it is continued round the upper extremity of the superior maxilla to reach the orbit, where it will afterwards be followed.

The SUPERIOR BUCCAL GLAND. This consists of a string of lobules resting on the anterior edge of the buccinator. The string is thickest above, and thinnest below, where its lobules are placed at intervals. The ducts of the lobules perforate the buccinator, and will afterwards be seen opening on the cheek.

The BUCCAL ARTERY will be found crossing the upper end of the superior buccal gland. It is distributed to the buccinator muscle and the other textures of the cheek. Its origin from the internal maxillary artery, at the floor of the orbit, will subsequently be dissected.

The BUCCAL VEIN begins at the inferior edge of the masseter, where

it joins the submaxillary vein.   Above the cheek it is continued as the
internal maxillary vein.

Directions.—The buccinator muscle should be cut at its anterior
edge, and turned backwards so as to expose the interior of the cheek.
On the mucous membrane will be seen the following :—

1. A linear series of small papillæ opposite the upper molar teeth.
• Each papilla is perforated at its summit by a duct from one of the
lobules of the superior buccal gland.

2. A large rounded elevation opposite the third superior molar.
Stenson's duct opens on its summit.

3. A linear series of small papillæ opposite the inferior molars.   Each
papilla is perforated by a duct from the inferior buccal gland.

The INFERIOR BUCCAL GLAND will be exposed by incising the mucous
membrane along the last-mentioned series of papillæ.   It is composed
of a string of lobules included between the mucous membrane and the
buccinator.

The BUCCAL NERVE will be found in close relation to the inferior
buccal gland.   It is the sensory nerve to the cheek, and is derived from
the inferior maxillary division of the 5th cranial nerve.

THE PTERYGO-MAXILLARY REGION AND THE REGION OF THE GUTTURAL POUCH.

Directions.—The outer surface of the inferior maxilla having been
laid bare from the condyle to the symphysis by the removal of the
muscles, the dissector is to make two sections with the saw.   The first
is to be made obliquely from a point about an inch above the last
molar tooth to the angle of the jaw ; the second, close above the
symphysis.   In making the sections, the edge of the saw must be kept
parallel to the surface of the bone, in order to avoid injury to the subjacent structures.   The jaw is next to be disarticulated by inserting a
strong scalpel into the joint ; and the scalpel is also to be passed round
the coronoid process, which will be felt embedded in the temporal
muscle, in the temporal fossa.   The vertical ramus is now to be entirely
removed, at the same time leaving in position the parts beneath it.
This is to be done by raising the bone at the angle, and cutting the
muscular fibres inserted into its deep face.   The edge of the knife is to
be kept cutting on the bone, which is at the same time to be forcibly
tilted upwards and forwards until the coronoid process is torn out of
the temporal muscle.   The horizontal ramus is next to be folded downwards and outwards ; and to permit this, it is only necessary to cut and
raise slightly the mucous membrane below the molar teeth.   When these
operations have been effected, the dissection will take the form of Plate
30.   The vertical ramus is to be retained to show the insertion of the
pterygoid and temporal muscles.

The EXTERNAL PTERYGOID MUSCLE (Plate 30) is conical in form, with the apex below. Its fibres *arise* from the outer surface of the sphenoid bone above the entrance to the subsphenoidal canal, and from the entire outer surface of the subsphenoidal process. They pass upwards and backwards to be *inserted* into the inner aspect of the neck, or constriction, below the condyle of the lower jaw.

Action.—When the right and left muscles act simultaneously, the lower incisors are made to protrude in front of the upper. When the muscles of one side act singly, the entire inferior maxilla is thrown to the opposite side ; and by the alternate contraction of the two muscles, the triturating action of the jaws and teeth is produced.

The INTERNAL PTERYGOID (Plate 30) is a much more powerful muscle than the preceding, from which it is separated by the inferior dental nerve and vessels. It occupies a position on the inside of the vertical ramus analogous to that of the masseter on the outside, the ramus being included between the two muscles. Its fibres *arise* from the bony crest formed by the subsphenoidal process and the palatine bone, and they pass backwards to be *inserted* into the depressed inner surface of the vertical ramus.

Action.—The principal action of the muscle is to elevate the lower jaw directly, the two muscles acting together; but acting singly and alternately, they assist in producing lateral movement.

The TEMPORAL MUSCLE occupies the fossa of the same name. Its fibres *arise* from the parietal, frontal, and squamous temporal bones where they bound this fossa, and from the sphenoid bone above the orbital hiatus. They become *inserted* into the coronoid process and adjacent part of the anterior border of the vertical ramus.

Action.—Chiefly to elevate the lower jaw, acting in conjunction with the masseter and internal pterygoid. The temporals also oppose the action by which the external pterygoids protrude the lower incisors in front of the upper.

Directions.—Emerging at the line of apposition of the two pterygoid muscles are the inferior dental vessels and nerve, the mylo-hyoid nerve, the internal maxillary vein, and the lingual branch of the 5th nerve. These are now to be examined.

The INFERIOR DENTAL ARTERY (Plate 30) is a branch of the internal maxillary, detached before that vessel enters the subsphenoidal canal. It passes first between the two pterygoid muscles, and then between the inner muscle and the bone, and enters the inferior dental canal of the inferior maxilla. On examining the inner aspect of the vertical ramus, the truncated end of the vessel will be seen at the upper orifice of the canal. It is continued in the bone beneath the molar, canine, and incisor teeth, supplying these, and detaching at the mental foramen a *mental* branch.

The INFERIOR DENTAL VEIN (Plate 30) has a distribution similar to that of the artery which it accompanies, and it joins the internal maxillary vein.

The INFERIOR DENTAL NERVE (Plate 30) is a branch of the inferior maxillary division of the 5th nerve.   It accompanies the vessels in the bone, supplying the teeth, and detaching sensory *mental* branches to the lower lip.

The MYLO-HYOID NERVE (Plate 30) is a branch of the preceding nerve. It descends between the internal pterygoid muscle and the bone, and reaches the intermaxillary space.   There it has already been dissected on the surface of the mylo-hyoid muscle, to which and the lower belly of the digastricus it is distributed.

The LINGUAL or GUSTATORY NERVE (Plate 30) is a large branch of the inferior maxillary division of the 5th.   Between the two pterygoid muscles it is joined by the *chorda tympani* branch of the 7th nerve, and it is continued between the internal pterygoid and the bone to reach the tongue.   It will be followed in the dissection of that organ.

The INTERNAL MAXILLARY VEIN (Plate 30) lies in front of the preceding nerve.   It is the direct continuation of the buccal vein already dissected in the check.   It will subsequently be followed to its termination, where it forms the jugular by union with the superficial temporal vein.

The LINGUAL VEIN (Plate 30) drains blood away from the tongue. It runs in company with the gustatory nerve; and after receiving branches from the soft palate and the pharynx, it joins the buccal vein between the internal pterygoid muscle and the bone.

PTERYGOID VESSELS.   The pterygoid muscles derive their blood from branches of the internal maxillary or of the inferior dental artery. The inner muscle receives also the inner division of the maxillo-muscular artery.   The pterygoid veins join the internal maxillary vein.

Directions.—The external and internal pterygoid muscles must now be removed.   In cutting away the outer muscle, the thick buccal nerve will be found passing through it near its origin, and giving branches to the muscle.   The nerve is to be preserved to show its origin.   Care must also be taken of the internal maxillary vessels, and of the inferior maxillary nerves, which are included between the muscle and the guttural pouch.   The internal pterygoid is to be cut close to its origin, and taken away without severing the vessels and nerves just dissected. The pterygoid branch of the inferior maxillary division of the 5th nerve will be found entering it, and an inch or two of the nerve should be preserved to show its origin.   Extending along the deep aspect of the origin of the muscle is the tensor palati muscle, and care must be taken not to injure it.   The deep face of the internal pterygoid is related to the pharynx, the guttural pouch, the great cornu of the hyoid bone, the intermediate tendon of the digastricus, the larynx, and the submaxillary gland. Besides these, it is related to Stenson's duct and the submaxillary vessels, which have already been dissected on its surface in the intermaxillary space.

THE GUTTURAL POUCHES (Plates 31 and 32). There are two large cavities situated at the base of the skull, above the pharynx, and between the great (styloid) cornua of the hyoid bone. Anteriorly they extend as far as the upper margin of the posterior nares, and posteriorly as far as the atlanto-axial articulation. Inwardly the mucous lining of the two pouches forms a mesial partition. Outwardly each pouch has numerous relations, the chief of which are as follows:—Behind the great cornu the pouch is covered by the submaxillary gland, and the stylo-maxillaris, digastricus (upper belly), stylo-hyoid, and occipito-styloid muscles, and is crossed by the external carotid, internal carotid, and occipital arteries, and by the 9th, 10th, 11th, 12th, and sympathetic nerves. In front of the great cornu the pouch is covered by the parotid gland and the internal pterygoid muscle, and is crossed by the internal maxillary vessels, the chorda tympani nerve, and the inferior maxillary division of the 5th nerve. The pouch is lined by mucous membrane continuous with that of the Eustachian tube, and by that tube it communicates with the pharynx and the middle ear. Normally the pouch contains air, which it receives from the pharynx through the Eustachian tube. When the mucous lining of the pouch becomes inflamed, pus tends to accumulate in the cavity, since the Eustachian orifice, by which the inflammatory products might escape into the pharynx, is, in the ordinary position of the head, situated towards the upper part of the pouch.

The EUSTACHIAN TUBE (Plate 32). This is a fibro-cartilaginous tube of three or four inches in length, extending downwards from the petrous temporal bone to the pharynx. At its upper extremity the tube communicates with the cavity of the middle ear (in the temporal bone), and at its lower extremity it opens into the pharynx by a slit-like aperture. For nearly the whole of its extent the tube is slit open along its outer side, and is thus in free communication with the guttural pouch. The tube is lined by mucous membrane, and through its agency air is admitted from the pharynx to the guttural pouch and the middle ear.

The INTERNAL MAXILLARY ARTERY (Plates 31 and 32) results from the division of the external carotid on the outer surface of the great cornu of the hyoid bone. It is much larger than the superficial temporal, which is the other terminal branch of the external carotid. In passing to enter the subsphenoidal canal, it describes a double or sigmoid curve,— the first convex downwards, the second upwards. In this course the artery is placed within the articulation of the jaw and the external pterygoid muscle, and rests successively on the guttural pouch and the tensor palati muscle. It is crossed superficially by the inferior dental and lingual nerves, and deeply by the chorda tympani. It detaches the following collateral branches :—

1. The *Inferior Dental Artery*, a large branch arising from the convexity of the first curve. It has already been seen entering the inferior dental canal.

2. The *Tympanic Artery*, the smallest of the branches here given off, is detached at nearly the same point as the preceding, but from the opposite side of the parent vessel. It lies beside the chorda tympani nerve, and penetrates the petrous temporal bone to be distributed in the tympanum, or middle ear.

3. The *Great Meningeal* (spheno-spinous) *Artery*, a vessel of variable volume detached from the upper side of the parent artery beneath the inferior maxillary nerve, and entering the cranial cavity by the foramen lacerum basis cranii.

4. The *Pterygoid Arteries*, two or three, arising from the concavity of the second curve.

5. The *Posterior Deep Temporal Artery*, given off from the upper side of the parent trunk about half an inch before it enters the subsphenoidal canal. It enters the temporal muscle by passing in front of the condyle of the temporal articular surface, and it communicates with the masseteric division of the transverse facial artery by a slender branch which passes through the corono-condyloid notch.

The INTERNAL MAXILLARY VEIN passes between the two pterygoid muscles ; and crossing below the articulation of the jaw, it joins the superficial temporal vein to form the jugular. The junction takes place in the substance of the parotid, a few lobules of the gland being interposed between it and the termination of the external carotid artery.

The INFERIOR MAXILLARY DIVISION of the 5TH NERVE (Plate 31). This is a thick cord containing both sensory and motor filaments which issues from the cranium by the forepart of the foramen lacerum basis cranii. It passes obliquely downwards and backwards on the wall of the guttural pouch, in front of the temporo-hyoideal articulation, and divides, about an inch below its point of exit, into two branches of nearly equal size—the inferior dental and lingual nerves. As it issues from the foramen, it gives off the following branches :—

1. The *Subzygomatic Nerve* is detached from the posterior aspect of the trunk. It turns round behind the articulation of the jaw, and has already been seen to join the 7th nerve, which it accompanies in its distribution on the face. It sends a branch in company with the transverse facial vessels.

2. The *Nerve to the Internal Pterygoid* arises from the antero-inferior aspect of the trunk.

3. The *Nerve to the Masseter and Temporal Muscles.* This is given off from the front of the trunk. It detaches branches to the temporal muscle, and is then continued through the corono-condyloid notch to end in the masseter.

4. The *Buccal Nerve* is a larger branch than any of the foregoing. It arises at the same point as the preceding nerve, and passing through the external pterygoid muscle, to which and the temporal muscle it supplies branches, it is continued as a sensory nerve to the cheek.

THE INFERIOR DENTAL NERVE is the larger of the two branches into which the trunk of the inferior maxillary nerve divides. Under cover of the external pterygoid muscle, it crosses over the internal maxillary artery, in company with the lingual nerve, which lies in front of it. It then passes in between the two pterygoid muscles, where it places itself in front of the inferior dental vessels. It has already been seen issuing from between the muscles to gain the inferior dental canal by passing between the inner muscle and the bone.

The *Mylo-hyoid Nerve* is detached from the posterior edge of the foregoing nerve at its point of formation. It crosses the inferior dental vessels between the two pterygoid muscles, and then descends between the inner muscle and the bone. It has already been traced to the digastric (lower belly) and mylo-hyoid muscles.

The LINGUAL or GUSTATORY NERVE is only slightly smaller than the inferior dental. It lies immediately in front of that nerve as far as the posterior edge of the external pterygoid muscle, where it passes forwards between the internal pterygoid and the bone to reach the tongue. It is a sensory branch; and, while between the two muscles, it is joined by the chorda tympani.

The CHORDA TYMPANI NERVE. This is a branch detached from the 7th nerve in the aqueduct of Fallopius. It passes across the tympanum, or middle ear, and issues from the petrous temporal bone by the styloid foramen. It then descends on the guttural pouch; and crossing beneath the internal maxillary artery and the trunk of the inferior dental nerve, it joins the lingual nerve between the two pterygoid muscles.

The OTIC GANGLION. This ganglion is at best minute, and sometimes absent, or at least not well defined. To examine it well, it is necessary to make a special preparation, exposing the inner aspect of the inferior maxillary nerve at its point of emergence from the cranium. The ganglion is placed on the inner side of the before-mentioned nerve-trunk near the origin of its buccal branch. Its *afferent* branches are: (1) twigs from the buccal branch of the inferior maxillary nerve; (2) the small superficial petrosal nerve from the 7th; (3) twigs from the sympathetic branches that accompany the internal maxillary artery. It supplies *efferent* branches to: (1) the tensor palati muscle; (2)

the tensor tympani; (3) the pterygoid muscles; (4) the Eustachian tube.

The SUBMAXILLARY GLAND (Plates 27 and 31) is, in point of size, the second of the salivary glands. It is elongated, with blunt, rounded extremities; and it is curved, the concavity being directed upwards and forwards. Its outer surface is related to the tendon of the sterno-maxillaris muscle, to the fibrous band connecting that tendon to the mastoid insertion of the mastoido-humeralis, and to the internal pterygoid. The sterno-maxillaris tendon and the above-mentioned fibrous band separate it from the overlying parotid. Its inner surface is related to the guttural pouch, the larynx, and the thyro-hyoid muscle; and it conceals above the larynx the terminal part of the common carotid artery, and the 10th and 11th nerves. Its posterior border is, about its centre, near or in contact with the thyroid gland; and below that point it is margined by the submaxillary vein. Its anterior border is related above to the stylo-maxillaris muscle, and for the rest of its extent it is traversed by Wharton's duct—the excretory duct of the gland. The superior extremity of the gland is loosely maintained beneath the wing of the atlas; the inferior extremity is situated within the intermaxillary space, and is crossed outwardly by the submaxillary artery.

WHARTON'S DUCT is formed by the union of small branches which emanate from the gland structure along its anterior border. It descends along that border, and at the lower extremity of the gland it crosses to the inner side of the submaxillary artery, and is continued beneath the tongue, where it will subsequently be followed in its course towards the barb.

The DIGASTRICUS (Plate 31). The upper belly of this muscle has already been seen to *arise* from the styloid process of the occipital bone, where it is confounded with the stylo-maxillaris muscle. It is succeeded by an intermediate tendon, which plays through a perforation in the tendon of the stylo-hyoid muscle, in front of which it joins the lower belly. The lower belly has already been dissected in the intermaxillary space, where it is *inserted* by tendinous slips into the posterior border of the inferior maxilla, above the symphysis.

Action.—To depress the lower jaw.

The STYLO-HYOID (Plate 31). This muscle *arises* from the extreme upper part of the hinder edge of the great cornu. Its inferior tendon is perforated for the passage of the digastricus, and is *inserted* into the base of the thyroid cornu of the hyoid bone. The external carotid emerges between the belly of the muscle and the great cornu.

Action.—It carries the base of the tongue and the larynx upwards and backwards, by flexing the joint between the great and small cornua, and the joint between the small cornu and the body.

Directions.—Pin the lower extremity of the submaxillary gland and

Wharton's duct in position, and then remove the remainder of the gland together with the stylo-hyoid, the digastricus, and the remains of the stylo-maxillaris. This will expose the posterior part of the guttural pouch, the pharynx, the larynx, the pharyngeal lymphatic glands, the upper part of the external carotid artery (and its terminal branches—the external and internal carotids, and the occipital), the superior cervical ganglion of the sympathetic (and the upper part of the cervical cord), the first parts of the 9th, 10th, 11th, and 12th cranial nerves, and the inferior primary branch of the 1st cervical nerve.

PHARYNGEAL LYMPHATIC GLANDS. These form an elongated cluster situated at the upper part of the side of the pharynx. They are placed on the course of all the lymphatic vessels of the head.

The COMMON CAROTID ARTERY (Plate 32) divides above the cricoid cartilage of the larynx, and under cover of the submaxillary gland or the stylo-maxillaris muscle, into three branches, viz., external carotid, internal carotid, and occipital arteries. The first of these continues the direction of the parent trunk, and is much larger than either of the others, which are of nearly equal size.

The OCCIPITAL ARTERY (Plate 32). The root of this vessel is external to, and slightly in advance of, the root of the internal carotid. It passes upwards and slightly forwards over the anterior straight muscles of the head, and enters the antero-external foramen of the atlas. In the groove which connects this and the antero-internal foramen, the artery divides into occipito-muscular, and cerebro-spinal branches. Before its passage through the foramen the vessel detaches three collateral branches, viz., prevertebral, mastoid, and retrograde or anastomotic arteries.

1. The *Prevertebral Artery.*—This is the first and most slender of the three branches. Passing upwards and forwards, it supplies *muscular* twigs to the anterior straight muscles of the head, and *meningeal* twigs that pass into the cranium by the foramen lacerum basis cranii and the condyloid foramen.

2. The *Mastoid Artery,* a considerable vessel, is detached about one-third of an inch above the preceding. It crosses over the edge of the rectus capitis lateralis, and ascends behind the styloid process of the occipital bone, where it has already been exposed (page 170).

3. The *Retrograde* or *Anastomotic Branch* varies considerably in volume. Arising between the obliquus capitis superior and the rectus capitis lateralis, beneath the wing of the atlas, it passes backwards through the posterior alar foramen, and anastomoses with the termination of the vertebral artery.

The INTERNAL CAROTID ARTERY (Plate 32) is a long vessel which is the main source of supply to the brain. It passes obliquely upwards and forwards, supported by the membrane of the guttural pouch, and

enters the cranium by the foramen lacerum basis cranii. Its mode of
entrance and its distribution to the brain will be examined at a later
stage.

The EXTERNAL CAROTID ARTERY (Plates 31 and 32). This vessel may,
for the purposes of description, be divided into two portions. The first
portion, comprising two-thirds of the artery, is included between the guttural pouch inwardly, and the stylo-maxillaris, digastric, and stylo-hyoid
muscles outwardly. It emerges from beneath the last-named muscle,
and joins the second portion by passing between the muscle and the
posterior edge of the great cornu of the hyoid bone. It is crossed
inwardly by the 9th, and outwardly by the 12th nerve. The second
portion passes obliquely upwards and forwards on the great cornu, at
the upper part of which, and under cover of the parotid gland, it divides
into the superficial temporal and internal maxillary arteries. The
collateral branches of the vessel are: the submaxillary, maxillo-muscular,
and posterior auricular arteries. The last two spring from the second
portion of the artery, and have already been dissected.

The SUBMAXILLARY or FACIAL ARTERY is detached from the
external carotid beneath the digastric muscle. It is a vessel of large
calibre, being nearly equal to the parent vessel beyond its point of
detachment. It descends over the pharynx, being nearly parallel to the
posterior edge of the great cornu, and about an inch behind it, At first
under cover of the digastric and stylo-hyoid muscles, it then turns
round the anterior edge and outer surface of the stylo-hyoid where the
intermediate tendon of the digastric plays through it. Continuing its
course, it crosses Wharton's duct and the lower extremity of the submaxillary gland to the outer side, and appears in the intermaxillary
space between the internal pterygoid and subscapulo-hyoid muscles.
Its further course in the intermaxillary space and on the face has already
been followed. Behind the great cornu the artery is in company with
the 12th nerve. From its origin to the extremity of the submaxillary
gland, it detaches three collateral branches, viz., the pharyngeal, lingual,
and submental arteries.

1. The *Pharyngeal Artery* is a small branch given off at the anterior
edge of the stylo-hyoid or under cover of that muscle. It reaches the
pharynx by passing beneath the great cornu, crossing either outwardly
or inwardly the 9th nerve. It is distributed to the pharynx, giving also
a forward branch to the soft palate.

2. The *Lingual Artery*, whose volume is about equal to that of the
distal part of the parent trunk, has its point of origin over the tip of
the thyroid cornu. It descends on the cerato-hyoid muscle, and reaches
the tongue by passing beneath the great hyo-glossus muscle. It is the
main vessel of supply to the tongue, and will be followed in the dissection of that organ.

3. The *Submental Artery* is detached at the extremity of the submaxillary gland, as the parent artery appears in the intermaxillary space. It has already been seen in the dissection of that space.

The Occipital Vein. This vein descends from beneath the wing of the atlas, in company with the artery of the same name ; and joins the jugular at the posterior edge of the parotid gland, a little above the termination of the submaxillary vein. It is formed by the union of branches corresponding to those of the artery; and, besides these, it receives a branch which comes from the spinal canal by traversing the ring of the atlas, beneath the wing, and another from the subsphenoidal sinus. This last will be exposed at a later stage.

The 9th Cranial Nerve, also called the Glosso-Pharyngeal (Plate 32), issues from the cranium by the posterior part of the foramen lacerum basis cranii. It descends on the guttural pouch, behind the great cornu of the hyoid bone, and under cover of the digastric and stylo-hyoid muscles ; and it here crosses to the inner side of the external carotid artery. Reaching the pharynx, it continues to descend either close behind the posterior edge of the great cornu or under cover of it ; and here it is crossed by the pharyngeal artery, which may pass either over or under it. It next passes within the articulation of the great and small cornua to reach the root of the tongue, where its terminal branches will be seen at a later stage. It gives off the following branches, the first of which will not now be seen :—

1. The *Nerve of Jacobson*, given off from Andersch's ganglion—a minute ganglion placed on the nerve where it issues from the cranium. Jacobson's nerve enters a minute foramen in the petrous temporal bone, and is distributed to the tympanum.

2. *Branches of communication* with the superior cervical ganglion of the sympathetic.

3. *A Branch to the Carotid Plexus.*

4. *A Pharyngeal branch*, which is given off at or near the point where the nerve crosses the external carotid. It is as large as the glossal continuation of the trunk, and it passes on to the wall of the pharynx, behind the stylo-pharyngeus muscle.

5. *A Branch to the Stylo-pharyngeus muscle.*—This may be detached either before or after the preceding branch. It enters the outer side of the muscle.

The 12th Cranial Nerve, called also the Hypoglossal (Plate 32), leaves the cranium by the condyloid foramen. It passes through the angle of separation of the 10th and 11th nerves, and descends on the guttural pouch, crossing to the outer side of the external carotid at or near the origin of the submaxillary artery. It next crosses the pharynx in company with the submaxillary artery, and passes under the angle formed by that vessel and its lingual branch. It is continued to

o

the muscles of the tongue. Where the nerve lies on the guttural pouch, it is covered by the stylo-maxillaris, digastric, and stylo-hyoid muscles, and below that point it is covered by the internal pterygoid muscle.

On the pharynx the hypoglossal is joined by a considerable twig from the inferior primary branch of the 1st spinal nerve. On the guttural pouch it constantly communicates with the superior cervical ganglion of the sympathetic. It has no other branches until it reaches the tongue.

The 10TH CRANIAL NERVE, also termed the VAGUS or PNEUMOGASTRIC (Plate 32), issues from the cranium by the extreme posterior part of the foramen lacerum basis cranii. For about one inch and a half of its course it forms a common cord with the 11th nerve, which issues at the same point. The two nerves then separate (the 12th nerve passing through the angle), and the vagus passes downwards and backwards on the guttural pouch. It passes over the internal carotid artery, and under the occipital; and above the first part of the trachea it meets the cervical cord of the sympathetic, the two nerves then uniting to form a common cord, which applies itself to the upper side of the common carotid artery, and descends with it in the neck. Between the foramen lacerum and the point where the nerve joins the sympathetic, it detaches the following branches:—

1. *Branches of Communication* with the superior cervical ganglion.

2. A *Pharyngeal Branch* is detached near the point where the vagus passes under the occipital artery. It passes to the inner side of the external carotid artery, and reaches the pharynx. There it unites with the sympathetic and the pharyngeal branch of the 9th, forming a plexus from which branches pass to the constrictors and mucous membrane of the pharynx, and to the first part of the œsophagus.

3. The *Superior Laryngeal Nerve* is given off near the termination of the common carotid artery; and crossing beneath the external carotid or the termination of the common carotid, it passes over the pharynx to penetrate the thyroid cartilage at the anterior edge of the thyro-pharyngeus muscle. Within the larynx, as will afterwards be learned, it is distributed to the mucous membrane; and also gives branches to the pharynx, œsophagus, and root of the tongue. Near its origin it detaches an *external laryngeal branch*, which passes to the crico-thyroid and crico-pharyngeus muscles.

The trunk of the vagus is sometimes distinctly gangliform at the point of detachment of its superior laryngeal branch. This is the *ganglion of the trunk* of human anatomy.

The 11TH CRANIAL NERVE, called also the SPINAL ACCESSORY NERVE, (Plate 32), issues from the cranium by the posterior part of the foramen lacerum, in company with the vagus. For the space of about one inch and a half it forms a common cord with that nerve. It then parts company with the vagus, and passes backwards at the edge of the rectus capitis anticus major muscle, where it is crossed superficially by the occipital artery. It then turns round the wing of the atlas at its most prominent point; and passes beneath the mastoido-humeralis muscle, crossing the branches of the 2nd spinal nerve. Before it disappears beneath the mastoido-humeralis, it communicates with the superior cervical ganglion, and gives a branch to the sterno-maxillaris muscle. Its distribution in the neck has already been followed.

The SYMPATHETIC NERVE (Plate 32). The initial part of the *cervical cord* is here seen passing back from the superior cervical ganglion. After a course of a few inches it places itself beside the vagus, and forms a common cord with it.

The SUPERIOR CERVICAL GANGLION is placed on the guttural pouch, above the internal carotid artery. It is about half an inch in length, fusiform in shape, and of a reddish-grey colour. Below it tapers into the cervical cord, and above it tapers into the ascending offsets with the internal carotid artery. Connecting branches unite the ganglion with the 1st spinal nerve, and with the last four cranial nerves. Communications with some of the other cranial nerves are also established through the carotid offsets of the ganglion. The distributory branches from the ganglion pass to the adjoining vessels and the pharynx. Of the former set, two branches accompany the internal carotid into the skull, and form the carotid and cavernous plexuses. Other branches follow the external carotid, and are continued on the branches of that artery.

The 1ST CERVICAL NERVE. The inferior primary branch of this nerve descends through the antero-external foramen of the atlas, and appears in company with the occipital vessels, between the rectus capitis lateralis and the obliquus capitis superior. It passes towards the upper extremity of the trachea, and splits into branches that enter the terminal parts of the subscapulo-hyoid, sterno-hyoid, and sterno-thyroid muscles. Beneath the atlas it gives branches to the anterior and lateral straight muscles of the head; and beyond that point it furnishes a branch which supplies the thyro-hyoid muscle, and gives a twig to join the hypoglossal nerve. It sends a communicating branch to the superior cervical ganglion of the sympathetic.

THE TONGUE.

Directions.—With the cavity of the mouth exposed as in Plate 30, any portions of food found in the cavity should be removed, and the mucous membrane made clean. By moving the tongue about, the following points will be observed.

The inferior portion of the tongue lies free on the floor of mouth, from which it can readily be drawn out. This part is two-sided, being flattened from before to behind, and rounded at its extremity like a spatula. The superior portion, on the other hand, cannot be displaced, and is

thick and of a three-sided form. The entire organ is invested by the
mucous membrane of the mouth, which, in passing on to the tongue,
forms certain folds, or doublings. Thus, if the free portion of the tongue
be raised from its position on the floor of the mouth, there will be seen
on the middle line a double fold of mucous membrane termed the *frœnum
linguæ*. At the extreme upper part of the mouth, again, the mucous membrane, in passing between the root of the tongue and the soft palate,
forms on each side a fold termed the *anterior pillar of the fauces*.

The mucous membrane of the tongue has its surface raised into the
form of papillæ of which there are three varieties :—

1. The *Filiform Papillæ* are the most numerous, and are found all
over the tongue. In shape they are conical, having a tapering summit
either simple or bearing secondary papillæ. They are largest on the
upper half of the dorsum (anterior or upper surface), to which they
give a distinct pile.

2. The *Fungiform Papillæ* are mushroom-shaped, being expanded at
the summit, which bears secondary papillæ. They are scattered along
the dorsum and sides of the tongue, being most numerous in its middle
portion.

3. The *Circumvallate Papillæ* are generally two in number, and are
placed on the dorsum, one on each side of the middle line, about five
inches from the epiglottis. Sometimes there is a third and smaller
papilla, placed on the middle line, about three-quarters of an inch above
the other two. Each is isolated by a circular trench, and is terminated
by a flat summit, which is level with the surrounding surface, and
bears numerous secondary papillæ. They contain the peculiar *gustatory
bodies*, to which fibres of the glosso-pharyngeal nerve are distributed.

The *Sublingual Ridge*.—This is a longitudinal elevation of the mucous
membrane at the floor of the mouth, on each side of the fixed portion of
the tongue. It is caused by the underlying sublingual salivary gland,
whose ducts open on the summits of little tubercles which stud the
ridge.

The *Barbs*.—These are two flattened, leaf-like papillæ situated on the
inferior part of the floor of the mouth, one on each side of the frœnum
linguæ. Wharton's duct—the duct of the submaxillary gland—discharges
itself by a minute opening on the summit of the barb.

STRUCTURE OF THE TONGUE. The tongue possesses a mucous covering,
a collection of mucous glands, a median fibrous cord, muscles, nerves,
and connective-tissue.

Mucous Membrane of the Tongue.—This has already been partly noticed.
Like the rest of the lining membrane of the mouth, it has a stratified
squamous epithelium. It is intimately adherent to the subjacent structures.
It is thickest on the dorsum of the tongue, where it is harsh and wrinkled.
On the sides and posterior surface of the tongue it is thin and smooth.

The *Lingual Fibrous Cord.*—This is a fibrous cord extending along the middle line of the dorsum of the tongue, immediately beneath the mucous membrane. With a sharp scalpel incise the tongue along this line to the depth of half an inch or more, beginning the incision behind the circumvallate papillæ, and terminating it in the free portion of the tongue. This will expose the cord in its entire length. It is a little less in thickness than a goose quill. It begins between the two large circumvallate papillæ, and it terminates towards the junction of the free and the fixed portion of the tongue, being about seven inches in length.

Lingual Glands.—These are aggregated as a thick layer under the mucous membrane at the upper part of the dorsum. The mucus which they secrete coats the bolus of food as it passes through the fauces. They are of the racemose type, and their ducts open on little tubercles of the mucous membrane.

Directions.—The mucous membrane is to be raised from the sublingual gland and the sides of the tongue. It is convenient to describe the gland here, although it is not a part of the tongue.

The SUBLINGUAL GLAND (Plate 30). This, the smallest of the salivary glands, is placed at the floor of the mouth, and at the side of, rather than under, the tongue. It extends from the level of the 5th molar tooth to the symphysis. It is in contact outwardly with the mylo-hyoid muscle. Inwardly it is related to the stylo-glossus, genio-glossus, and genio-hyoideus muscles, to Wharton's duct, and to the lingual nerves. Its posterior border is included between the mylo-hyoid and the genio-hyoid muscles, and is related inferiorly to the submental artery. Its anterior border projects the mucous membrane at the side of the tongue, so as to form the sublingual ridge. The upper extremity is related to the lingual nerve and vein. It is a compound racemose gland, and it discharges its secretion by from fifteen to twenty ducts—the *ducts of Rivinius*—which perforate the little tubercles on the sublingual ridge.

Directions.—The gland is to be carefully excised without injury to Wharton's duct or the adjacent vessels and nerves. It will be observed to receive a nerve from the lingual branch of the 5th, while its vessels are branches of the submental artery and vein.

WHARTON's DUCT (Plates 30 and 31) is the excretory canal of the sub-maxillary gland. Leaving the lower extremity of the gland, where it is crossed outwardly by the submaxillary artery, it passes between the mylo-hyoideus outwardly, and the great hyo-glossus and the stylo-glossus muscles inwardly, its position here being immediately behind the body of the hyoid. A little in advance of the superior extremity of the sublingual gland, it passes to its inner side, and is continued downwards between the gland and the genio-glossus muscle. Finally, it opens on the summit of the flattened papilla, or *barb*, at the side of the frænum linguæ. As the duct passes to the inner side of the sublingual gland, a branch of the lingual nerve turns round it. The duct will be readily recognised and distinguished from a blood-vessel by its slender and uniform calibre, and by its clear contents.

The LINGUAL or GUSTATORY NERVE is a branch of the 5th. It contains here, however, not only its own proper fibres, but also fibres derived from the 7th through the *chorda tympani*, which joins it in the first part of its course. At the root of the tongue it passes between the mylohyoideus and the stylo-glossus and great hyo-glossus muscles, until it passes to the inner side of the sublingual gland. There it turns forwards between the muscles, and is continued in a flexuous manner, giving off branches that are distributed to the mucous membrane of the tongue in its lower two-thirds. At the base of the tongue it detaches a few filaments to the mucous membrane there, to Wharton's duct and the submaxillary gland, and a larger branch for the sublingual gland and the adjacent mucous membrane.

The HYPOGLOSSAL (12th) NERVE (Plate 30) will be found at the base of the tongue, in front of Wharton's duct, where it is included between the mylo-hyoid and great hyo-glossus muscles. Passing to the inner side of the sublingual gland, it comes into relation with the lingual nerve, and divides. The branches of the hypoglossal are motor to the muscles of the tongue.

The SUBMENTAL ARTERY. This vessel, a branch of the submaxillary, has already been seen in the dissection of the intermaxillary space. It leaves the space by passing forwards through the mylo-hyoid muscle, and extends along the posterior border of the sublingual gland, into which, and the muscles, it throws branches. It extends beyond the lower extremity of the gland, and terminates in small branches to the mucous membrane.

The SUBMENTAL VEIN. This is relatively larger than the artery, which it accompanies.

The LINGUAL VEIN (Plate 30). This vessel will be found at the upper part of the tongue, in company with the gustatory nerve. It receives branches from the soft palate and pharynx, and joins the buccal vein.

The lingual artery (which runs its course separate from the vein of the same name) and the 9th nerve cannot be followed until some of the muscles have been dissected. Plate 31 will serve as a guide in the isolation of these muscles.

The MYLO-HYOID MUSCLE has already been seen in the dissection of the intermaxillary space, and it is now seen on its opposite aspect. It is described at page 171.

The STYLO-GLOSSUS. (*Hyo-glossus longus* of Percivall). (Plate 31). This is a long, riband-shaped muscle, *arising* by a thin aponeurotic tendon from the outer surface of the great cornu of the hyoid bone near its lower extremity. It extends along the side of the tongue to near its tip, where its fibres are confounded on the middle line with those of the opposite muscle.

Action.—To retract the tongue, and at the same time to incline it laterally if only one muscle acts.

The GREAT HYO-GLOSSUS (*Hyo-glossus brevis* of Percivall) (Plate 31). To expose this muscle fully, the stylo-glossus should be cut near its origin and reflected. The great hyo-glossus *arises* from the lateral aspect of the glossal process, body, and thyroid cornu of the hyoid bone. Its fibres extend obliquely forwards and downwards across the side of the fixed portion of the tongue, and turn inwards on reaching the dorsum.

Action.—To retract and depress the tongue.

Directions.—Incise the origin of the foregoing muscle, and raise it forwards after the manner of Plate 32. This will expose more fully the other muscles of the tongue, and also the lingual artery and the 9th nerve.

The MIDDLE HYO-GLOSSUS MUSCLE* (Plate 32) *arises* from the front of the articulation between the great and the small cornu of the hyoid bone. In passing downwards beneath the great hyo-glossus it crosses over the lingual artery. Its fibres are confounded in front with those of the palato-glossus.

Action.—It is a feeble retractor of the tongue.

The PALATO-GLOSSUS (Plate 32). This is a small, thin muscle *arising* from the edge of the soft palate. It passes downwards beneath the great hyo-glossus and over the lingual artery, and reaches the root of the tongue.

Action.—To narrow the fauces.

Directions.—The next two muscles lie internal to the small cornu. To expose them, the middle hyo-glossus should be cut, and the intercornual joint pulled outwards.

The HYOIDEUS TRANSVERSUS is peculiar in that it is an unpaired muscle without a median raphe. It extends transversely across the middle line, being *attached* at its extremities to the small cornua in the whole of their extent. When relaxed it passes between its points of attachment with a curve whose concavity is directed upwards and forwards.

Action.—To raise the root of the tongue.

The SMALL HYO-GLOSSUS. This is a muscle of small size *arising* from the lower extremity of the small cornu and from the body of the hyoid bone. It passes forwards over the preceding muscle and terminates in the root of the tongue.

* This, apparently, is the muscle first described by Brühl, in 1850, as the *middle descending stylo-glossus*.

Action.—To aid in retracting the tongue.

The GENIO-GLOSSUS, or GENIO-HYO-GLOSSUS (Plates 31 and 32). This muscle is fan-shaped. Its *origin* is from a depression on the inner surface of the horizontal ramus of the lower jaw, near the symphysis. From this point, and from a tendon at the posterior edge of the muscle, its fibres radiate into the tip, centre, and base of the tongue, beneath those of the stylo-glossus and great hyo-glossus muscles.

Action.—The upper fibres protrude the tongue by pulling downwards its base, the lower fibres retract the free portion of the tongue, and the intermediate fibres (or the entire muscle) depress the tongue as a whole towards the floor of the mouth.

If the dissector will raise the posterior tendinous edge of the muscle, he will find that he has now reached the middle plane of the tongue, which is here occupied by a quantity of connective-tissue and fat between the right and left genio-glossus muscles.

The GENIO-HYOIDEUS (Plates 31 and 32). This is a muscle of the hyoid bone, rather than of the tongue. It is elongated and fusiform in shape, and is placed beneath the tongue, near the middle line. It *arises* from the inner surface of the horizontal ramus, close to the symphysis; and passing upwards along the inner edge of the mylo-hyoid, it becomes *inserted* into the glossal or spur process of the hyoid bone.

Action.—To pull forwards the hyoid bone.

The LINGUAL ARTERY (Plate 32). This is a large branch of the sub-maxillary artery, and has already been seen at its origin. It passes under cover of the great hyo-glossus, crossing the small cornu of the hyoid bone. In passing to the root of the tongue, it crosses obliquely forwards and downwards beneath the middle hyo-glossus and the palato-glossus. In the body of the tongue it lies beneath the great hyo-glossus, and in the free portion of the tongue it is internal to the stylo-glossus. In its course it becomes reduced in size by detaching lateral branches, and it terminates at the tip of the tongue by turning inwards and anastomosing on the middle line with the vessel of the opposite side. Like the lingual nerve, the artery is flexuous in the inferior part of the tongue, that it may be adapted, without stretching, to the varying length of the organ.

The GLOSSO-PHARYNGEAL (9th) NERVE (Plate 32). The lingual continuation of this nerve will be found crossing the inner side of the articulation between the great and small cornua of the hyoid bone to reach the base of the tongue. Here it divides into branches for the mucous membrane on the superior third of the tongue.

THE HARD PALATE (FIG. 21).

The surface of the hard palate is covered by a dense mucous membrane having a stratified squamous epithelium. It is traversed in its entire length by a median raphe, and is crossed from side to side by from eighteen to twenty curved ridges. The concavity of the ridges is directed upwards (towards the root of the tongue). The ridges are sharpest and the interspaces are narrowest at the upper part of the palate. A layer of connective-tissue with numerous veins connects the mucous membrane to the periosteum of the bones forming the basis of the palate.

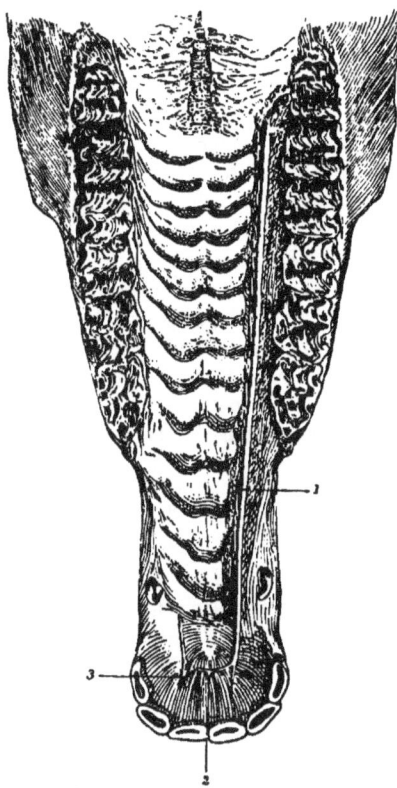

FIG. 21.

HARD PALATE : 1. Palato-labial artery of right side ; 2. Inosculation of right and left arteries, forming a single labial artery which passes forwards through incisor foramen ; 3. Bar of cartilage under which palato-labial artery runs.

The PALATO-LABIAL ARTERY. This vessel passes along the side of the hard palate, resting in a groove on the bone, close to the alveoli. An incision should be made down to the artery in this position, and it should be followed backwards and forwards. It is accompanied by a satellite vein and nerve. The vessel is the continuation of the internal maxillary artery. Beginning at the maxillary hiatus, it reaches the upper extremity of the palate by traversing the palatine canal. It descends at the side of the palate, and at the level of the corner incisor it curves inwards towards the incisor foramen, where, on the middle line, it unites with the vessel of the opposite side. The single labial vessel resulting from this union passes forwards through the incisor foramen to reach the upper lip. In passing in to join its fellow at the incisor foramen, the palato-labial artery runs under a small flexible bar of cartilage, which is fixed to the bone by its upper extremity, while its lower extremity is free. Where the artery curves inwards, it detaches a branch that passes downwards to be distributed in the palate below the level of the incisor foramen.

PALATINE VEINS. Over the whole extent of the hard palate there exists a rich network of veins in the submucous connective-tissue. This network is drained by a large vein which accompanies the palato-labial artery as far as the lower orifice of the palatine canal. There it parts company with the artery, and passes along the staphyline groove, with the staphyline artery and nerve. It joins the alveolar vein. The variable thickness of the palate depends principally on the amount of blood in these veins, this being greatest in the young animal.

The PALATINE NERVE is a branch of the superior maxillary division of the 5th. It emerges from the palatine canal along with the artery, around which its branches interlace. It is the sensory nerve to the hard palate.

THE SOFT PALATE, or VELUM PENDULUM PALATI.

This is an oblique valvular curtain placed on the limit of the oral and pharyngeal cavities. The oral surface of the curtain looks downwards and backwards, and is covered by mucous membrane continuous with that of the hard palate. The pharyngeal surface has the opposite direction, and its mucous covering is continuous with that of the nasal chambers. The anterior edge is fixed at the posterior margin of the hard palate. The lateral edges are attached on the limits of the mouth and pharynx. The posterior edge is free, and extends across the root of the tongue, in front of the epiglottis. The mucous membrane in passing between the soft palate and the root of the tongue is raised on each side into a fold, and two similar but less prominent ridges of mucous membrane extend from the soft palate to the sides of the pharynx. These are termed respectively the *anterior* and the *posterior pillars* of the soft palate. Between them is a space which marks the situation of the *tonsil* in most mammals, and into which numerous mucous glands open. The soft palate of the horse is remarkable for its large size—a fact which explains the difficulty with which the horse can expire or eject regurgitated matters through the mouth.

The ISTHMUS OF THE FAUCES. This is the aperture of communication between the mouth and the pharynx. It is bounded in front and laterally by the free edge of the soft palate, and by its anterior pillars. Behind it is bounded by the extreme upper part of the dorsum of the tongue. In the horse, owing to the length of the soft palate, this aperture is closed except during the passage of solids or liquids in deglutition.

STRUCTURE OF THE SOFT PALATE. This comprises two layers of mucous membrane, and, included between these, a layer of mucous glands, a fibrous aponeurosis, muscles, vessels, and nerves.

Mucous Membrane.—The membrane covering the oral aspect of the curtain is directly continuous with that of the hard palate, and through the anterior pillars it is also continuous with the mucous membrane of the tongue. It has a thick tesselated epithelium, and shows numerous small papillæ perforated by the ducts of the subjacent glands. The mucous membrane of the pharyngeal surface of the curtain is continuous with that of the nasal chambers, and will be exposed in the dissection of the pharynx.

Staphyline * *Mucous Glands.*—These form a thick granular layer which will be exposed by removing the mucous membrane from the oral surface of the soft palate. The ducts of the glands open on the oral surface of the curtain, and the bolus of food thus gets a mucous coating as it passes through the isthmus.

Fibrous Aponeurosis.—If the glandular layer be removed, this will be exposed in the anterior part of the curtain. It is fixed in front to the margin of the hard palate; and behind it is continuous with the palato-pharyngeus, to which it serves as a tendon of origin.

Muscles of the Soft Palate.

The PALATO-GLOSSUS MUSCLE (Plate 32). This muscle has been dissected with the tongue (page 199).

The PALATO-PHARYNGEUS (Plate 32). This muscle will be found beneath the glandular layer in the posterior half of the curtain. At the middle line it is continuous with the opposite muscle, and in front it is continuous with the fibrous aponeurosis. At the side of the soft palate it is continued to the wall of the pharynx; and passing beneath the hyo-pharyngeus muscle, it becomes *inserted* into the edge of the thyroid cartilage. In the latter part of its course it need not be exposed at present.

Action.—To tense the velum and carry its free edge upwards towards the pharynx.

The TENSOR PALATI (Plates 31 and 32). This muscle and the next will be found parallel to the Eustachian tube, and on its outer side. It *arises*, in common with the levator palati, from the styloid process of the petrous temporal bone, and from the Eustachian tube. Its terminal

---

* Strictly speaking, this adjective applies to structures pertaining to the uvula, but it may conveniently be used to distinguish parts belonging to the soft palate of the horse, in which the uvula is not developed.

tendon is reflected inwards on the *hamular* or pulley-like process of the
pterygoid bone, a synovial bursa intervening, and expands on the posterior surface of the fibrous aponeurosis of the palate.

Action.—To tense the anterior half of the palate.

The LEVATOR PALATI (Plate 32). This muscle is placed between the
preceding and the Eustachian tube. It *arises* with the tensor palati
from the styloid process and the Eustachian tube. Reaching the upper
wall of the pharynx, it passes beneath the pterygo-pharyngeus to gain
the pharyngeal surface of the palate, where it expands.

Action.—To raise the velum towards the roof of the pharynx, and
thus to shut off the communication between the pharynx and the nasal
chambers.

The AZYGOS UVULÆ, which in man is situated in the uvula, was
named from the belief that it was a single muscle. It consists, however, of right and left halves applied together on the middle line of the
soft palate. In order to expose them, it will be necessary to remove
from the oral surface of the velum a portion of the palato-pharyngeus
muscles on each side of the middle line. The *origin* of the muscle
is from the fibrous aponeurosis of the palate, and it terminates at the
free edge of the curtain.

Action.—To raise the free edge of the velum.

The STAPHYLINE ARTERY is a slender vessel arising from the internal
maxillary artery, above the maxillary hiatus. It reaches the velum by
passing in the staphyline groove, along with the nerve of the same name
and the palatine vein.

The PHARYNGEAL ARTERY (Plate 32), a branch of the submaxillary,
crosses the pharynx, and gives its terminal branches to the soft palate.

The VEINS of the soft palate enter either the lingual vein or the
palatine vein, which lies in the staphyline groove.

Nerves.—The staphyline nerve will be found with the artery, in the
groove of the same name. It comes from the superior maxillary division
of the 5th nerve. It carries not only sensory fibres to the mucous
membrane of the palate, but also motor fibres, which come from the 7th
nerve through the spheno-palatine ganglion, and are distributed to the
levator and probably also the azygos uvulæ.

The nerve to the tensor comes from the 5th, through the otic ganglion.

THE PHARYNX, THE HYOID BONE, AND THE BASE OF THE SKULL.

Directions.—Take Plate 32 as a guide, and remove the greater portion
of the large cornu of the hyoid bone, making the upper section with the
bone-forceps just below the point at which the external carotid artery
crosses the hinder edge of the cornu, and the lower a little above the
articulation of the two cornua. Preserve the 9th nerve at the posterior

edge of the great cornu, and the pharyngeal artery passing beneath it.

The PALATO-PHARYNGEUS has already been in part dissected with the muscles of the soft palate. See page 203.

The PTERYGO-PHARYNGEUS. This muscle *arises* from the pterygoid process, from which point its fibres diverge to the upper and lateral aspect of the forepart of the pharynx. At its lower edge its fibres are parallel to, and with difficulty separated from, those of the palato-pharyngeus. Its outer surface is constantly covered by a layer of fatty elastic tissue.

The HYO-PHARYNGEUS, *origin*—thyroid or heel process of the hyoid bone.

The THYRO-PHARYNGEUS, *origin*—thyroid cartilage.

The CRICO-PHARYNGEUS, *origin*—cricoid cartilage.

These three muscles succeed one another in the order named, the first-mentioned being the most anterior. They pass upwards over the side of the pharynx, and terminate on the middle line of its roof.

The STYLO-PHARYNGEUS. This is a thin, strap-like muscle *arising* from the inner surface of the great cornu of the hyoid bone near its upper extremity. It descends to the pharynx, and expands on it at the outer edge of the pterygo-pharyngeus.

The SMALL STYLO-PHARYNGEUS. This muscle is not constant, though frequently present. It is a delicate, worm-like muscle *arising* from the inner surface of the great cornu, about an inch above its lower extremity. It passes upwards on the pharynx at the anterior edge of the hyo-pharyngeus.

The ARYTENO-PHARYNGEUS. This muscle will not be seen at present. It is a small slip *arising* from the arytenoid cartilage, and passing to the pharynx at its junction with the œsophagus.

Action of the Pharyngeal Muscles.—The stylo-pharyngeus dilates the anterior part of the pharynx for the reception of the bolus. All the other muscles are constrictors, grasping in succession the bolus, and carrying it on to the œsophagus.

The CERATO-HYOID (*Hyoideus parvus* of Percivall) (Plate 33). This muscle is most conveniently dissected at this stage, and it is therefore here described although not belonging to the pharynx. It is a small flat muscle occupying the angle between the small and the thyroid cornu of the hyoid bone. It *arises* from the posterior edge of the small cornu, and from the great cornu immediately above the inter-cornual articulation. It is *inserted* into the upper edge of the thyroid cornu. The lingual artery, in crossing down to the base of the tongue, passes over the muscle.

Action.—To elevate the thyroid cornu and with it the larynx.

The PHARYNGEAL ARTERY. This is a small branch of the submaxillary which passes beneath the great cornu, crossing either outwardly or

inwardly the 9th nerve.   It gives branches to the pharynx, and is continued to the soft palate.

The 9TH or GLOSSO-PHARYNGEAL NERVE descends on the guttural
pouch, behind the great cornu ; and crossing to the inner side of the
external carotid artery, it gives off its *pharyngeal branch.*   This turns
round behind the stylo-pharyngeus muscle, and ramifies on the pharynx,
meeting there the pharyngeal branches of the 10th and sympathetic
nerves, and forming with them the *pharyngeal plexus.*

Nerve to the Stylo-pharyngeus.—This is a special branch of the glosso-
pharyngeal, given off either from the trunk of the nerve or from its
glossal continuation.   It enters the muscle at the middle of its outer
face.

The *Glossal Continuation* of the 9th nerve passes over the hyo-pharyngeus and palato-pharyngeus to reach the base of the tongue ; and in
its course it detaches fibres which are probably motor to the palato-
pharyngeus, hyo-pharyngeus, cerato-hyoid, and hyoideus transversus.   •

For the pharyngeal branches of the 10th and sympathetic nerves see
pages 194 and 195.

Directions.—The pharynx is to be opened by an incision along its
lateral aspect, when, by hooking back the edges of the incision, a view
of its interior will be obtained.

The PHARYNGEAL CAVITY (Fig. 24, page 217) is irregularly tubular in
form, and presents seven openings.   These are :—

1. The *Isthmus of the Fauces,* already described (page 202).

2. The *Superior Nares* (2), situated vertically over the isthmus, from
which they are separated by the soft palate.   They are wide, gaping
orifices, with rigid bony margins.

3. The *Lower Openings of the Eustachian Tubes* (2), which are situated
on the sides of the pharynx, behind and in line with the superior nares.
The opening has the form of a vertical slit, the outer edge of which
contains the cartilaginous extremity of the tube.

4. The *Upper Aperture of the Larynx.*—This is placed on the floor of
the pharynx, and has the form of a pitcher-mouth.   During deglutition
the epiglottis folds over it like a lid, but at other times it is a large
patent orifice.

5. The *Œsophageal Orifice.*—At the posterior end of the pharynx, its
cavity is continued into the lumen of the gullet.   Except during the
passage of solids or fluids, however, this orifice is not open.

Mucous Membrane.—This lines the cavity completely, and is continuous through the before-mentioned orifices with the lining membrane
of the mouth, nasal chambers, Eustachian tubes and guttural pouches,
larynx, and œsophagus.   It has a stratified squamous epithelium, except
in its upper and anterior part, where it is ciliated.   It has many
mucous glands.

Directions.—Remove the larynx by cutting across the remaining great cornu of the hyoid bone, and the root of the tongue in front of the glossal (spur) process. Cut away the remains of the pharyngeal muscles ; and after examining the articulations of the hyoid bone, set the larynx (and the attached hyoid) aside for future examination. (The larynx is described at page 224).

TEMPORO-HYOIDEAL ARTICULATION. This joint will be found intact at the base of the skull. It is a typical *amphiarthrodial* joint, the toe-like extremity of the great cornu being joined to the hyoid process of the petrous temporal bone by a *short rod of cartilage* (about half an inch in length). The interposed cartilage is sufficiently long and flexible to permit a considerable range of movement.· By these joints the hyoid is suspended to the base of the skull, and swings backwards and forwards in movements comparable to *flexion* and *extension.* These movements are concerned in the actions of mastication and deglutition.

The INTERCORNUAL ARTICULATION. This is another *amphiarthrodial* joint, the opposed extremities of the great and the small cornu being united by *intermediate cartilage.* In this cartilage there is frequently to be found a small pea-like nucleus of bone which is the representative of a third cornu—the *epihyal* of comparative anatomy.

The BASI-CORNUAL ARTICULATION. The lower extremity of the small cornu is articulated to the body of the hyoid in a small diarthrodial joint provided with a *capsular ligament* and a *synovial membrane.* The small cornu and the thyroid cornu form at this joint an angle, and the movements are *extension* and *flexion.*

Directions.—The dissector will now be able to trace the anterior root of the occipital vein to its origin from the subsphenoidal sinus, and thereafter the anterior and lateral straight muscles of the head are to be dissected.

The SUBSPHENOIDAL SINUS (or confluent). This is a venous sinus placed at the inner edge of the foramen lacerum basis cranii. It extends downwards (the head being vertical) for half an inch below the carotid notch, the sphenoid bone being there depressed for the sinus. At this extremity it terminates blindly. Traced upwards, it lies along the inner edge of the foramen lacerum (the basilar process being depressed beneath it), and it is directly continued as one of the roots of the occipital vein. A little above its lower extremity the sinus is penetrated by the internal carotid. Slit open the sinus, and observe that it communicates by an oval foramen with the cavernous sinus of the dura mater. This oval foramen has its broad end circumscribed by the carotid notch of the sphenoid, and for the remainder of its extent it is bounded by the fibrous tissue that fills up the greater part of the foramen lacerum. Through the forepart of this oval foramen the carotid artery passes from the subsphenoidal to the cavernous sinus, the artery here being remarkable in that it is actually within a vein.

At the inferior end of the sinus the vidian nerve will be found descending in the vidian groove to enter the canal of the same name (page 214).

The RECTUS CAPITIS ANTICUS MAJOR. This muscle was, for the most part of its extent, exposed in the dissection of the neck, but its tendon of insertion remains to be examined now (page 155).

The RECTUS CAPITIS ANTICUS MINOR. This small muscle *arises* from the lower aspect of the ring of the atlas ; and passing over the occipito-atlantal joint, at the side of the preceding muscle, it becomes *inserted* into the basilar process of the occipital and the body of the sphenoid at their point of articulation.

Action.—To flex the occipito-atlantal joint (to nod the head).

The RECTUS CAPITIS LATERALIS. This is another small muscle placed beneath the occipito-atlantal joint. It *arises* from the atlas, external to the origin of the preceding muscle ; and passes to be *inserted* into the styloid process of the occipital bone.

Action.—The same as the preceding muscle.

THE ORBIT.

Directions.—With the saw, cut through the supraorbital process external to the foramen, and through the zygomatic arch at either extremity. The palpebral ligament of the upper lid is to be detached from the orbital rim, and the tri-radiate piece of bone marked out by these sections is to be taken away.

The orbital cavity in the skeleton is not separated from the temporal fossa, but in the living animal a fibrous membrane, continuous with the periosteum of the bones circumscribing the cavity, completes the orbit on its outer side, and separates it from the temporal fossa. The orbit is thus lined and completed by a fibrous membrane of a conical form, termed the *ocular sheath.* This membrane is composed of fibrous connective-tissue with some unstriped muscular fibre.

Contents of the Orbital Cavity.—These are : the eyeball with its muscles, the lachrymal gland, the levator palpebræ, the membrana nictitans, vessels, nerves, and a quantity of fat. For the proper dissection of the eyeball, fresh specimens are required, and it is therefore separately described at page 257. The lachrymal gland and the muscles may be dissected on one side, the other being reserved for the display of the vessels and nerves.

The LACHRYMAL GLAND is the organ which secretes the watery fluid that moistens the front of the eye. It is lodged above the eye, beneath the supraorbital process, which is slightly depressed where it covers the gland. In structure it is of the racemose type, resembling the salivary glands, but being looser in texture. Its secretion is discharged by a number of ducts which open on the ocular surface of the upper eyelid, close to the temporal canthus. The gland is to be removed.

The LEVATOR PALPEBRÆ SUPERIORIS is described at page 175.

The MEMBRANA NICTITANS and its connection with the adipose tissue of the cavity are described at page 173.

FIG. 22.

MUSCLES OF THE EYEBALL.

1, 1. Superior oblique; 2. Fibrous loop for the same; 3. Superior rectus; 4. Internal rectus; 5. External rectus; 6. Inferior rectus; 7, 7, 7. Fasciculi of the retractor; 8. Cut origin of the levator of the upper eyelid; 9. Nerve to inferior oblique.

Muscles of the Eyeball.—These are seven in number, viz., four recti, one retractor, and two oblique. (A third oblique muscle was described by the late Professor Strangeways, of the Dick Veterinary College, but its presence is, to say the least, not constant.)

Directions.—The muscles are to be defined by detaching the eyelids and conjunctiva from the front of the eye, and removing the levator palpebræ and the loose fat which forms a packing material between the muscles. Special care must be taken not to injure the fibrous arch for the tendon of the superior oblique at the inner side of the orbit.

The RECTI. There are four of these, distinguished as the *superior rectus*, the *inferior rectus*, the *external rectus*, and the *internal rectus.* They are placed one above, one below, and one on either side of, the eye. They have all a flat riband-like form, and are terminated anteriorly by aponeurotic tendons. They all take *origin* around the optic foramen, and each becomes *inserted* into the forepart of the sclerotic.

The RETRACTOR OCULI is placed within the recti, and around the

P

optic nerve. In form it is funnel-shaped, forming a kind of sheath to the optic nerve ; but frequently it is divided into four distinct fasciculi,

FIG. 23.

MUSCLES OF THE EYEBALL.

1. Inferior oblique ; 2. Inferior rectus ; 3. External rectus ; 4. Internal rectus ; 5. Superior rectus ; 6, 6. Retractor ; 7. Nerve to inferior oblique.

one lying beneath each of the recti. Its *origin* is from the margin of the optic foramen, and it is *inserted* into the sclerotic.

The SUPERIOR OBLIQUE, or TROCHLEARIS, has a remarkable disposition. It *arises* at the back of the orbit, and passes inwards to a fibrous arch, or pulley, through which it plays at the inner wall of the orbit, below the root of the supraorbital process. Having passed through this arch, the muscle is directed outwards above the eyeball ; and it is continued by a tendon which passes beneath the superior rectus to be *inserted* into the sclerotic on its outer side, between the insertions of the superior and external recti.

The INFERIOR OBLIQUE. This muscle has its origin from the lachrymal fossa at the floor of the orbit. It passes outwards below the eyeball, and becomes inserted into the sclerotic between the insertions of the inferior and external recti.

Action of the Muscles of the Eyeball.—The superior and the inferior rectus rotate the eye around a horizontal transverse axis, the former rolling it upwards, the latter downwards. The external and the internal rectus rotate the eye around a vertical axis, the first rolling it outwards, the second inwards. The oblique muscles rotate the eye around an antero-posterior horizontal axis, the superior muscle elevating the nasal angle of the pupil, while the inferior muscle depresses it. The retractor oculi pulls the eyeball directly backwards into its cavity, and is thus, by pressure exerted on the orbital fat, instrumental in protruding the membrana nictitans, as explained at page 173.

The OPHTHALMIC ARTERY is a branch of the internal maxillary, from which it is detached within the subsphenoidal canal. Emerging from that canal at the back of the orbit, it is directed inwards to enter the internal orbital foramen ; and in this course it passes between the superior rectus and the retractor oculi, where the latter surrounds the optic nerve. By the internal orbital foramen it reaches the forepart of the cranial cavity. It is further described at page 238. In the orbital part of its course it gives off the following branches :—

1. The *Supraorbital Artery* ascends on the inner wall of the orbit to pass through the foramen of the same name.

2. The *Lachrymal Artery*, distributed to the gland and the upper eyelid.

3. *Muscular Branches.*

4. *Ciliary Branches* to the eyeball.

5. The *Central Artery of the Retina*, which places itself in the axis of the optic nerve, and enters the eyeball.

The ORBITAL BRANCH of the superior dental artery. This is a long and slender branch detached from the parent vessel before it enters the superior dental canal. It creeps over the floor of the orbit to reach the face, where it anastomoses with the submaxillary artery (Plate 29).

VEINS. The structures within the orbit are drained by vessels which unite to form the ophthalmic vein. This, after uniting with the alveolar vein, passes into the cranial cavity by the foramen lacerum orbitale.

The OPHTHALMIC NERVE is one of the three primary divisions of the 5th nerve. It is a sensory nerve, and divides into the following three branches, which issue in company from the foramen lacerum orbitale :—

1. The *Supraorbital Nerve*, accompanying the artery of the same name.

2. The *Lachrymal Nerve*, to the gland, and giving off a branch which traverses the ocular sheath to reach the skin over the temporal fossa.

3. The *Palpebro-nasal Nerve* divides into a nasal branch which accompanies the ophthalmic artery through the internal orbital foramen, and a palpebral branch to the lower eyelid and inner canthus. The palpebro-nasal nerve also furnishes the sensory filaments to the ciliary ganglion, through which it supplies sensory fibres to the eyeball.

The ORBITAL BRANCH of the superior maxillary division of the 5th nerve issues with the parent nerve from the foramen rotundum. It passes at the outer side of the ocular sheath to gain the temporal canthus of the eyelids, where it is distributed. It is sensory.

The 3RD CRANIAL NERVE, or MOTOR OCULI, issues from the foramen lacerum orbitale, and supplies the following muscles:—the superior, internal, and inferior recti; the corresponding fasciculi of the retractor oculi; the inferior oblique; and the levator palpebræ superioris. It also gives the motor root to the ciliary ganglion, and thus supplies the ciliary muscle and the circular fibres of the iris.

The 4TH CRANIAL NERVE (called also the *trochlear* or *pathetic nerve*) issues by the pathetic foramen, and is wholly distributed to the superior oblique muscle.

The 6TH CRANIAL NERVE, or ABDUCENS, issues by the foramen lacerum orbitale, and is distributed to the external rectus and the subjacent fasciculus of the retractor oculi.

The CILIARY GANGLION, called also the ophthalmic or lenticular ganglion. This minute ganglion should be sought near the origin of the nerve to the inferior oblique muscle. Find that nerve entering the muscle, and trace it back to its origin from the motor oculi. From its minute size, the ganglion is likely to have been disturbed in the previous dissection; and in order to display it satisfactorily, a special dissection is necessary. The nerves which pass to and from the ganglion may be arranged as follows :—

Afferent Branches.—(1) A motor root from the 3rd nerve, (2) a sensory root from the palpebro-nasal nerve, (3) a sympathetic root from the cavernous plexus, joining the ganglion independently or (more commonly) with the sensory root.

Efferent Branches.—These are the ciliary nerves. They pierce the sclerotic and are distributed to the eyeball, and will be again referred to in the dissection of that organ.

The optic foramen, through which the optic nerve issues, is one of a group of foramina termed the orbital hiatus, or the orbital group of foramina. When the head is vertical, there lies below this, at the posterior and inner part of the orbit, another group of foramina—the maxillary group or hiatus. The internal maxillary artery and the superior maxillary division of the 5th nerve pass between these two groups, and in that course detach several important branches. Their dissection is conveniently undertaken after that of the muscles, vessels, and nerves of the eye.

The INTERNAL MAXILLARY ARTERY issues from the lower orifice of the subsphenoidal or pterygoid canal, and descends to the maxillary hiatus, where it is directly continued as the palato-labial artery. While within the canal it detaches the ophthalmic and anterior deep temporal arteries. After its emergence it gives off the buccal, superior dental, staphyline, and spheno-palatine arteries.

The OPHTHALMIC ARTERY issues from the subsphenoidal canal along with the parent artery. It has already been followed.

The ANTERIOR DEEP TEMPORAL ARTERY passes forwards out of the subsphenoidal canal by an un-named foramen above the edge of the orbital hiatus. It is expended in the temporal muscle and the overlying skin.

The BUCCAL ARTERY is detached shortly after the parent vessel emerges from the bone. It has already been followed in its distribution to the cheek.

The SUPERIOR DENTAL ARTERY is a large branch which enters the superior dental canal. The vessel is continued above the roots of the molar, incisor, and canine teeth. Before passing into the canal, it gives off an *orbital branch* which passes across the floor of the orbit to reach the face. Within the canal it emits an *infraorbital branch* which reaches the face by the infraorbital foramen.

The STAPHYLINE ARTERY is a slender branch given off from the posterior aspect of the internal maxillary, close to the maxillary hiatus. It courses along the staphyline groove to reach the soft palate.

The SPHENO-PALATINE (NASAL) ARTERY. This vessel is of considerable size, and passes at once through the spheno-palatine foramen to be distributed in the nasal chamber.

The PALATO-LABIAL ARTERY is the continuation of the internal maxillary. It passes along the palatine canal to reach the hard palate.

VEINS. At this point the veins have a disposition slightly different from the arteries. The *alveolar vein*—a large vessel lying on the superior maxilla in front of the molar teeth—turns round the bone and reaches the maxillary hiatus. Here it receives *superior dental, palatine,* and *spheno-palatine* branches. It then perforates the ocular sheath, within which it joins the *ophthalmic* vein. The *opthalmic vein* passes into the cranial cavity by the foramen lacerum orbitale, and joins the cavernous sinus.

The superior dental and spheno-palatine veins emerge by the same foramina as the corresponding arteries. The palatine vein, however, does not issue from the palatine canal, but turns round the bone in the staphyline groove.

The SUPERIOR MAXILLARY DIVISION of the 5TH NERVE. This sensory division of the trifacial emerges from the cranium by the foramen rotundum, as a large round cord. In company with the internal maxillary artery, it descends to the maxillary hiatus, where it enters the superior dental canal. Within the canal it gives *dental branches* to the roots of the molar, canine, and incisor teeth, and then issues on the face at the infraorbital foramen. In its passage between the orbital and the maxillary hiatus, it gives off the following branches:—

1. An *Orbital Branch.*—See page 211.

2. The *Palatine Nerve* accompanies the palato-labial artery into the palatine canal, and is distributed to the hard palate.

3. The *Staphyline Nerve* passes by the groove of the same name to the soft palate.

4. The *Spheno-palatine Nerve* enters the foramen of the same name, and is distributed to the nasal mucous membrane.

The SPHENO-PALATINE (MECKEL'S) GANGLION. This is a small, greyish, elongated and fusiform enlargement, generally adherent to the spheno-palatine nerve. Slender branches radiate from it, and are divided into *afferent* and *efferent* filaments.

Afferent Filaments.—1. The *Vidian Nerve*, which enters its posterior extremity. This is a composite nerve formed by the union of the large superficial petrosal branch of the 7th with a sympathetic filament. Traced upwards, it enters a minute foramen—the lower orifice of the vidian canal. At the upper orifice of the canal it enters the sub-sphenoidal confluent, and passes into the cavernous sinus by the foramen lacerum basis cranii. There it separates into its petrosal and sympathetic branches.

2. Short branches passing from the spheno-palatine nerve to the posterior part of the ganglion.

The vidian nerve is supposed to combine the motor and sympathetic roots of the ganglion; the spheno-palatine branches represent its sensory root.

Efferent Branches.—Some of these pass to the ocular sheath, to the ophthalmic vessels, and to the muscles and other accessory parts of the eye. Others join the spheno-palatine, palatine, superior dental, and staphyline nerves. The latter it is believed derives from this source the motor filaments which it conveys to the levator palati muscle.

THE OCCIPITO-ATLANTAL ARTICULATION.

This joint possesses two synovial sacs and an enveloping capsule, with accessory fasciculi above and at each side which are sometimes described as distinct ligaments—the cruciform and styloid.

The OCCIPITO-ATLANTAL LIGAMENT is membranous, and closes the interval between the occiput and atlas. It is attached to the occiput at the upper and lower edges of the foramen magnum, and to the outer side of the condyles. Its posterior edge is fixed to the anterior border of the atlas. The most superior fibres pass obliquely, the right and left fibres intercrossing. This is the so-called *cruciform ligament.* On each side a thickened cord-like portion passes to be inserted into the styloid process of the occipital bone, and these constitute the *styloid ligaments.*

SYNOVIAL SACS. Each of these belongs to an occipital condyle and its receiving cavity on the front of the atlas. On the inner side each is related to the dura mater and the occipital continuation of the odontoid ligament, and elsewhere they are supported by the occipito-atlantal ligament.

Movements.—It is at this joint that the *nodding movements of the head* are executed.

THE ATLANTO-AXIAL ARTICULATION.

This joint possesses four ligaments and a synovial capsule.

The INFERIOR ATLANTO-AXIAL LIGAMENT is riband-shaped, and stretches below the joint, from the forepart of the inferior ridge of the axis to the tubercle of the atlas.

The SUPERIOR ATLANTO-AXIAL LIGAMENT is exactly like the *interspinous ligament* of the succeeding joints of the neck. It consists of two parallel bands of yellow elastic tissue connecting the bones above the joint.

The ATLANTO-AXIAL INTERANNULAR LIGAMENT is membranous, and connects the neural arch of the atlas with that of the axis. It represents the *ligamentum subflavum* of succeeding joints.

The ODONTOID LIGAMENT is placed at the floor of the spinal canal in this region. To expose it, it is necessary to remove the upper part of the ring of the atlas. It is strong, flattened, and triangular. It is narrow behind, where it is fixed to the depressed upper surface of the odontoid process. It is widest in front, where it is fixed to the floor of the atlas. A thin continuation of the ligament is carried forwards on each side to be attached to the edge of the foramen magnum.

SYNOVIAL SAC. This is supported by the odontoid ligament above, by the inferior atlanto-axial ligament below, and by the interannular ligament laterally.

MOVEMENTS. It is at this joint that the *movements of the head from side to side* are executed. In these movements the axis remains fixed, while the atlas rotates around the odontoid process, carrying with it the head.

TEMPORO-MAXILLARY ARTICULATION.

This is the joint formed between the articular surface of the squamous temporal and the condyle of the inferior maxilla. An interarticular fibro-cartilage is interposed between the osseous surfaces, and the joint possesses a capsular ligament and two synovial sacs.

CAPSULAR LIGAMENT. This envelopes the joint, being attached around the temporal articular surface above, and around the condyle of the lower jaw below. Its inner surface is adherent to the interarticular cartilage. In front and inwardly the capsule is thin and membranous, but behind and on the outer side it shows thickenings which are sometimes described as distinct *posterior* and *external ligaments*. The first of these stretches from the post-glenoid process to the inferior maxilla below and behind the condyle. The second is attached above to the lower edge of the zygomatic arch, from which it extends downwards and backwards to be fixed to the inferior maxilla below and external to the condyle.

The INTERARTICULAR FIBRO-CARTILAGE should be exposed by removing the capsular ligament on the outer side. The cartilage extends completely across the joint, which it divides into an upper and a lower cavity. Its upper surface is a cast of the temporal articular surface, while its lower is moulded on the condyle of the jaw.

SYNOVIAL SACS. The upper of these belongs to the articulation between the fibro-cartilage and the temporal articular surface; the lower to the articulation between the fibro-cartilage and the condyle.

MOVEMENTS. These are — *depression, elevation, protraction, retraction,* and *lateral movement* of the inferior maxilla.

When the jaw is depressed, as in opening the mouth, the condyles of both jaws are carried forwards, taking with them the fibro-cartilages, until they lie under the condyle of the temporal articular surface ; and at the same time, the maxillary condyles move in the depression on the under surface of the interarticular fibro-cartilage, rotating around a transverse axis. When the lower jaw is *protracted,* the movement consists principally in antero-posterior gliding between the temporal articular surface and the interarticular cartilage ; and when the same movement is executed alternately on opposite sides, a *lateral,* grinding action is produced.

THE CAVITY OF THE NOSE (FIGS. 24 AND 25).

Directions.—Make, with the saw, an antero-posterior vertical section of the head, a little to one or other side of the mesial plane, taking the mesial sutures on the front of the head as a guide.

The cavity of the nose is the first segment of the air-passages, and is thus a part of the respiratory apparatus. It is also in part devoted to the sense of smell, the olfactory nerve being distributed over a part of its boundary walls. It is a large tubular passage tunnelled through the skull in front of the mouth (the head being vertical). A mesial partition —the *septum nasi*—divides the passage longitudinally into the right and left *nasal fossæ*. Each nasal fossa may be described as having anterior, posterior, and lateral walls, and a superior and an inferior extremity.

The *Anterior Wall,* sometimes termed the *Roof,* is narrow and formed by the frontal and nasal bones.

The *Posterior Wall,* sometimes termed the *Floor,* is considerably more extensive than the roof. It is formed by the palatine, superior maxillary, and premaxillary bones, but in much greater proportion by the second of these.

The *Outer Wall* is formed by the nasal and superior maxillary bones, and is occupied by the anterior and posterior turbinated bones, which project into the cavity and separate the *meatuses* of the nose from one another. Thus, the *anterior meatus* is the narrow interval between the anterior (ethmoidal) turbinated bone and the roof of the cavity (the nasal bone); the *middle meatus* is another and larger interval between the two turbinated bones; while the *posterior meatus*, the largest of these

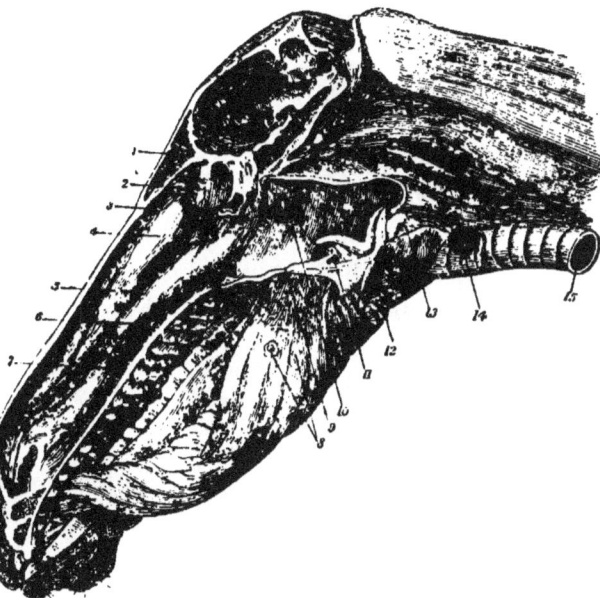

FIG. 24.

LONGITUDINAL SECTION OF THE HEAD, SHOWING THE CAVITIES OF THE MOUTH, NOSE, AND PHARYNX (*Leyh*).

1. Frontal sinus; 2. Lateral mass of ethmoid bone; 3. Anterior meatus of nasal chamber; 4. Anterior turbinated bone; 5. Middle meatus; 6. Posterior turbinated bone; 7. Posterior meatus; 8. Circumvallate papillæ of the tongue; 9. Section of soft palate; 10. Opening of right Eustachian tube on side of pharynx; 11. Isthmus of the fauces; 12. Upper aperture of the larynx; 13. Communication between pharynx and œsophagus; 14. Thyroid body; 15. Trachea.

intervals, is included between the posterior (maxillary) turbinated bone and the floor of the cavity.

The *Inner Wall* (the septum nasi) is partly bony, and partly cartilaginous. In its upper part it is formed by the bony *perpendicular plate* of the ethmoid, and at its posterior edge it is formed by the vomer bone; but for the greater part of its extent, the partition is composed of a plate of cartilage—the *septal cartilage*. This septal cartilage is continuous above with the perpendicular plate, which is merely an ossified portion of it; behind it is received into the cleft of the vomer, and expands on the premaxillary suture; in front it expands on the inner aspect of the internasal suture; and inferiorly the alar cartilages are movably connected to it.

The *Inferior Extremities* of the nasal fossæ are termed the *inferior nares* or, in common language, the nostrils. They have already been described (page 176).

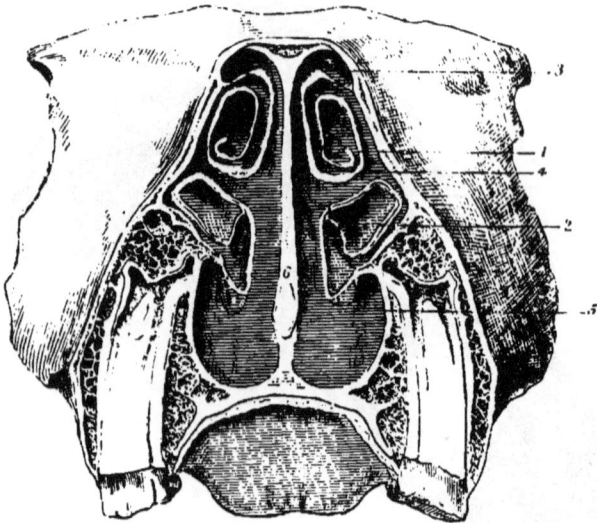

FIG. 25.

TRANSVERSE SECTION THROUGH THE NASAL CHAMBERS.

1. Anterior turbinated bone; 2. Posterior turbinated bone; 3. Anterior meatus; 4. Middle meatus; 5. Posterior meatus; 6. Septum nasi.

The *Superior Extremities* are separated from the cranial cavity by the cribriform plate of the ethmoid bone, and are occupied by the lateral masses of the same bone. Below and behind these are the *superior nares*—the large patent orifices by which the nasal fossæ communicate with the pharynx, the right and left openings being separated by the vomer bone.

The following openings into the nasal fossa should be found :—

1. The *Opening of the Lachrymal Duct* (*ductus ad nasum*).—Look for this on the floor of the nasal fossa, a few inches within the nostril. It is easily seen in the living animal, and has already been referred to in connection with the nostril (page 176). It is a small opening (about the same diameter as a goose quill) with a circular outline, having an appearance as if a small circle of skin had been punched out. The opening, it is to be observed, is on the skin, and not the mucous membrane, taking the presence of hair as distinguishing the former from the latter. The duct passes upwards beneath the mucous lining of the middle meatus until it enters the osseous tube that conducts it to join the lachrymal

sac at the floor of the orbit. The lower portion of the tube has a stratified epithelial lining, but in its upper part the epithelium is ciliated.

2. The *Opening of Stenson's Canal.*—Look for this opening on the floor of the nasal fossa, over the incisor or naso-palatine cleft. Pass a flexible probe into it. It will be found to pass obliquely into the cartilaginous substance that closes this opening. It there joins another canal—the *organ of Jacobson*, which passes upwards at the side of the hinder edge of the septal cartilage, terminating blindly after a course of four or five inches. The organ of Jacobson has a wall of hyaline cartilage, with a mucous lining, and numerous mucous or serous glands. Its epithelial lining is in part a stratified epithelium, and in part it resembles the olfactory epithelium to be presently described ; and to the latter portion some fibres of the olfactory nerve are traceable.

3. The *Opening of Communication with the Sinuses of the Head.*—This is placed towards the upper extremity of the middle meatus. Ordinarily it has the form of a curved slit not visible from the nasal fossa ; but if a flexible probe be insinuated between the two turbinated bones at this point, it may be guided on into the frontal or the maxillary sinus.

The NASAL MUCOUS MEMBRANE (Pituitary or Schneiderean Membrane). As already seen in the examination of the nostrils, the skin is carried round the edges of these, and for a short distance into the nasal fossa. Along an abrupt line it loses its pigment and hair, and is continued by the mucous membrane. This mucous membrane, it will be observed, differs in its upper and its lower portions. Thus, in its lower three-fourths the membrane has a rosy, vascular tint, while in its upper fourth it is distinguished to the naked eye by being of a pale, somewhat yellowish colour. The first of these may be termed the *respiratory portion* of the membrane, as distinguished from the second, or *olfactory portion*. The former has a stratified, columnar, ciliated epithelium similar to that of the air passages in general, and in its submucous tissue are numerous small racemose serous or mucous glands. The olfactory mucous membrane, on the other hand, has its free surface formed by a layer of columnar cells for the most part non-ciliated ; and between the bases of these are peculiar spindle-shaped *olfactory cells.* The olfactory cells are connected by their deep ends with the olfactory nerve fibres, while their opposite extremities are insinuated between the columnar cells, and terminate on the surface of the membrane in a few stiff, hair-like processes. In the submucous tissue are numerous tubular glands—the *glands of Bowman*—which open on the free surface of the membrane. The nasal mucous membrane is continuous with that of Stenson's canal and the organ of Jacobson, the pharynx, and the sinuses of the head.

The OLFACTORY (1st CRANIAL) NERVE. The delicate oval swelling termed the *olfactory bulb*, which is lodged in the fossa of the same name at the forepart of the cranial cavity, gives off from its surface the olfactory nerve fibres. These pass in bundles through the foramina of the cribriform plate and enter the nasal fossa, where they are distributed as a network in the olfactory mucous membrane. As the fibres leave the cranium, they carry with them prolongations from the membranes of the brain ; and they are remarkable among cerebro-spinal nerves in being destitute of the white substance of Scwhann.

SPHENO-PALATINE NERVE. This nerve, already seen at the back of the orbit as a branch of the superior maxillary division of the 5th (page 214), enters the nasal fossa by the spheno-palatine foramen, and divides into an outer and an inner branch for the nasal mucous membrane, on which it confers common sensibility.

The NASAL BRANCH of the OPHTHALMIC NERVE (page 238) is another nerve of common sensation. Entering the upper extremity of the nasal fossa, through a foramen in the cribriform plate, it ramifies in the mucous membrane on both sides of the fossa.

VESSELS. The mucous membrane of the nasal fossa is richly supplied with blood by the spheno-palatine artery and the nasal branch of the ophthalmic artery (pages 213 and 238), satellites of the two preceding nerves. The veins form beneath the mucous membrane a rich plexus which is drained principally by the spheno-palatine vein.

NAME OF MUSCLE.	ORIGIN.	INSERTION.	SOURCE OF NERVE.
Sterno-maxillaris	Sternum, cariniform cartilage	Inferior maxilla, angle	11th (spinal accessory) nerve.
Sterno-thyro-hyoideus	Sternum, cariniform cartilage	Thyroid cartilage; and hyoid bone, body	1st cervical nerve.
Subscapulo-hyoideus	Subscapular fascia	Hyoid bone, body	
Trapezius } see p. 54			
Rhomboideus			
Mastoido-humeralis	Dorsal vertebræ, 2nd, 3rd, and 4th spines; and funicular part of ligamentum nuchæ		
Splenius	Dorsal vertebræ, 1st and 2nd, transverse processes; and cervical vertebræ, last six, articular processes	Mastoid crest, wing of atlas, and transverse processes of 2nd to 5th cervical vertebræ	Cervical and dorsal nerves.
Trachelo-mastoideus	Dorsal vertebræ, 2nd, 3rd, and 4th spines, and transverse processes of first six or seven; and cervical vertebræ, last six articular processes	Mastoid crest, and wing of atlas.	
Complexus	Cervical vertebræ, 3rd to 7th, articular processes	Occipital bone	
Semispinalis colli	Vertebræ, first dorsal and last five cervical, articular processes	Cervical vertebræ, 2nd to 6th spines	
Intertransversales colli	Cervical vertebræ, 5th, 4th, and 3rd, transverse processes	Vertebræ, last six cervical, transverse processes	Cervical nerves.
Rectus capitis anticus major	Cervical vertebræ, last four, transverse processes	Occipital, basilar process; and sphenoid, body	
Scalenus	Dorsal vertebræ, last six bodies; and cervical vertebræ except atlas, transverse processes	1st rib (two insertions)	
Longus colli		Cervical vertebræ except last, bodies	
Obliquus capitis inferior	Axis, spine	Atlas, wing	
Obliquus capitis superior	Atlas, wing	Occipital, mastoid crest and styloid process	
Rectus capitis posticus major	Axis, spine	Occipital	
Rectus capitis posticus minor	Atlas, ring	Occipital	
Parotido-auricularis	Parotid gland	Conchal cartilage, base	7th nerve.
Cervico-auricularis externus	Ligamentum nuchæ	Conchal cartilage, inner side	
Cervico-auricularis medius		Conchal cartilage, outer aspect of base	
Cervico-auricularis internus		Conchal cartilage, posterior aspect of base	
Parieto-auricularis externus	Parietal bone, crest	Scutiform and conchal cartilages	

NAME OF MUSCLE.	ORIGIN.	INSERTION.	SOURCE OF NERVE.
Zygomatico-auricularis	Squamous temporal, zygomatic crest	Scutiform and conchal cartilages	
Parieto-auricularis internus	Parietal bone, crest	Conchal cartilage	
Mastoido-auricularis	Auditory process	Conchal cartilage	
Scuto-auricularis externus	Scutiform cartilage, outer surface	Conchal cartilage	
Scuto-auricularis internus	Scutiform cartilage, inner surface	Conchal cartilage	
Stylo-maxillaris	Occipital, styloid process	Inferior maxilla, angle	
Digastricus	Occipital, styloid process	Inferior maxilla, above symphysis	
Occipito-styloid	Occipital, styloid process	Hyoid, styloid cornu	
Stylo-hyoid	Hyoid, great (styloid) cornu	Hyoid, thyroid cornu	
Orbicularis palpebrarum	Lachrymal bone, tubercle	Surrounds palpebral fissure	
Levator labii superioris alaeque nasi	Frontal and nasal bones	Outer wing of nostril, upper lip, and angle of mouth (two divisions)	
Levator labii superioris proprius	Malar and superior maxillary bones	Upper lip	7th nerve.
Dilatator naris lateralis	Superior maxilla	Outer wing of nostril	
Dilatator naris transversalis	Cartilage of the nostril	Same cartilage of opposite side	
Dilatator naris superior	Septum nasi	False nostril and ethmoidal turbinated bone	
Dilatator naris inferior	Premaxillary and superior maxillary bones	False nostril and maxillary turbinated bone	
Zygomaticus	Surface of masseter muscle	Buccinator muscle, above angle of mouth	
Buccinator { Superficial portion / Deep portion }	Median raphe (of the muscle)	Superior maxilla and inferior maxilla	
Depressor labii inferioris	Superior maxilla and inferior maxilla	Angle of mouth	
Orbicularis oris	Inferior maxilla	Lower lip	
Depressor labii superioris	Surrounds the mouth	Upper lip	
Levator menti	Premaxilla	Prominence of the chin	
Masseter	Malar and superior maxillary bones, zygomatic ridge	Inferior maxilla, inner surface of vertical ramus	
External pterygoid	Sphenoid bone	Inferior maxilla, neck	
Internal pterygoid	Sphenoid and palatine bones	Inferior maxilla, inner surface of vertical ramus	Inferior maxillary division of 5th nerve.
Temporalis	Parietal, frontal, squamous temporal, and sphenoid bones	Inferior maxilla, coronoid process and anterior border	
Mylo-hyoid	Inferior maxilla, horizontal ramus	Hyoid, body and glossal process; and median raphe	

Muscle	Origin	Insertion	Nerve
Stylo-glossus	Hyoid, great (styloid) cornu		12th nerve.
Great hyo-glossus	Glossal process, body, and thyroid cornu of hyoid		
Middle hyo-glossus	Hyoid, intercornual joint		
Small hyo-glossus	Hyoid, small cornu and body.		
Palato-glossus	Soft palate	End in tongue	Spheno-palatine ganglion (7th nerve). 12th nerve.
Genio-glossus	Inferior maxilla, near symphysis		12th nerve.
Cerato-hyoid	Hyoid, great and small cornu	Hyoid, thyroid cornu	9th nerve (?)
Hyoideus transversus	Hyoid, small cornu	Same cornu of opposite side	12th nerve.
Genio-hyoideus	Inferior maxilla, near symphysis	Hyoid, glossal process	
Tensor palati	Petrous temporal, styloid process; and Eustachian tube	Soft palate	Otic ganglion (5th nerve).
Levator palati	Petrous temporal, styloid process; and Eustachian tube	Soft palate	Spheno-palatine ganglion (7th nerve).
Azygos uvulæ	Aponeurosis of soft palate	Free edge of soft palate	Pharyngeal plexus.
Palato-pharyngeus	Soft palate		
Ptergo-pharyngeus	Pterygoid bone		
Hyo-pharyngeus	Hyoid, thyroid cornu		
Thyro-pharyngeus	Thyroid cartilage	End in wall of pharynx	Sup. laryngeal nerve. 9th nerve.
Crico-pharyngeus	Cricoid cartilage		
Stylo-pharyngeus	Hyoid, styloid (great) cornu.		Pharyngeal plexus.
Small stylo-pharyngeus	Hyoid, styloid (great) cornu.		
Aryteno-pharyngeus	Arytenoid cartilage		
Levator palpebræ superioris	Sphenoid, above optic foramen	Upper eyelid	3rd nerve.
Rectus oculi superior			3rd nerve.
Rectus oculi inferior			3rd nerve.
Rectus oculi externus			6th nerve.
Rectus oculi internus	Sphenoid, near optic foramen		3rd nerve.
Retractor oculi		Sclerotic coat	3rd and 6th nerves.
Obliquus oculi superior			4th nerve.
Obliquus oculi inferior	Lachrymal bone, fossa		3rd nerve.

CHAPTER V.

DISSECTION OF THE LARYNX.

THE larynx is a short tube forming the upper part of the windpipe. It is, however, not merely a part of the respiratory apparatus, but is also the organ of voice. It possesses a framework of cartilages, which are movably articulated together, and connected by ligaments or membranes. These cartilages are moved by muscles, some of which pass between the different cartilages and constitute an *intrinsic* group, while others pass between the cartilages and extraneous parts, and constitute an *extrinsic* group.

Directions.—Provided the dissection has to be carried out on one larynx, the study of the muscles must precede that of the cartilages. When another larynx can be procured, it is more advantageous to reverse this order, removing the muscles from the first larynx in order to study the cartilages and their mode of union, and then using the other for the examination of the muscles and remaining structures.

Even when the first method has to be followed, it is advisable, before proceeding to dissect the muscles, to read the description of the cartilages, which is therefore here put first.

CARTILAGES OF THE LARYNX.

These are five in number, viz., the cricoid, thyroid, and epiglottis, which are single ; and the pair of arytenoid cartilages. In man there are two additional pairs—two cornicula laryngis and two cuneiform cartilages. In the horse the first of these are amalgamated with the tips of the arytenoids, while the cuneiform cartilages are small, shot-like bodies included in the aryteno-epiglottic fold of mucous membrane.

In the natural position of the animal at rest, the long axis of the larynx is oblique upwards and forwards. For convenience of description, however, we may assume it to be vertically placed, as indeed it is when the head and neck are extended (elevated) to the fullest degree. In this position the cartilages are related to one another as follows :—The cricoid is the lowest, and is connected to the first ring of the trachea. The thyroid is placed above this, and bounds the tube of the larynx in front and at the sides. The arytenoids surmount the cricoid behind, and the epiglottis is superposed to the thyroid in front of the upper aperture of the tube.

The CRICOID CARTILAGE has the form of a finger ring, from which it receives its name. The depth of the ring is greatest behind, where it presents a portion comparable to the bezel, or part of a ring in which the stone is set. The inner surface of the ring is smooth, and lined by the laryngeal mucous membrane. The outer surface of the bezel is divided by a vertical median ridge which increases the surface of origin of the posterior crico-arytenoid muscle. Towards the outer limit of this surface there will be seen on each side a little cavity which is smooth for articulation with the thyroid cartilage. The inferior border is notched in the middle line of the bezel, and is connected by ligament to the first ring of the trachea. The upper border has a wide notch in front; and posteriorly, over the bezel, it shows a pair of smooth convex facets for articulation with the arytenoid cartilages. In texture the cricoid is composed of hyaline cartilage.

The THYROID CARTILAGE receives its name from covering the front and sides of the larynx like a shield. It consists of a median thickened portion, or body; and two lateral plates—the alæ, or wings. The body is known in human anatomy as "Adam's apple"—the *pomum Adami*. The epiglottis is superposed to it, the two cartilages being united by elastic fibres. On each side it is continuous with the wings. Each ala is a rhomboidal plate of cartilage. The outer surface is slightly convex, and the inner is correspondingly concave. The upper edge of the cartilage is attached to the thyroid cornu of the hyoid bone by the thyro-hyoid membrane. The lower edge slightly overhangs the cricoid, and receives the insertion of the crico-thyroid muscle. The other two edges are directed obliquely, one backwards and upwards, the other forwards and downwards. The first of these receives the insertion of the palato-pharyngeus muscle; the other, in receding from the corresponding edge of the opposite wing, leaves beneath the body a triangular gap which is occupied by the crico-thyroid membrane. Of the four angles of each plate, three demand mention. The supero-anterior angle is acute, and joins the ala to the body of the thyroid. The supero-posterior angle is obtuse, and carries a small bar of cartilage—the *superior cornu* of man—which is articulated to the extremity of the thyroid cornu of the hyoid bone. Close to the base of this process the plate is perforated by a foramen for the passage of the superior laryngeal nerve. The postero-inferior angle is acute like the first, to which it is diagonally opposite. It is drawn out a little, forming a projection—the *inferior cornu* of man—which is terminated by a convex facet for articulation with the cricoid. The thyroid is composed of hyaline cartilage.

The ARYTENOID CARTILAGES. These stand at the upper aperture of the larynx like the mouth of a pitcher, and from this resemblance they are named. They are irregular in shape, but each bears some resemblance to a three-sided pyramid. The inner surface of the pyramid is covered

Q

by laryngeal mucous membrane; the outer surface receives the insertion
of the thyro-arytenoid muscle; the posterior surface is covered by the
arytenoid muscle. The base of the cartilage possesses within its area a
smooth, depressed facet for articulation with the cricoid. Two of the angles
of the base require particular notice, viz., the anterior angle, which is
pointed, and projects horizontally forwards to receive the insertion of the
vocal cord; and the postero-external angle, which is thick and rounded,
and receives the insertion of the crico-arytenoid muscles. The apex is
directed upwards, and is prolonged by a slender piece of yellow fibro-
cartilage representing the *cornicula laryngis* of man. This curves backwards and inwards; and with the corresponding process of the opposite
side forms, behind the upper aperture of the larynx, the pitcher-
like lip. Except in the apical prolongation, which is composed of
yellow or elastic fibro-cartilage, the texture of the arytenoid is hyaline
cartilage.

The EPIGLOTTIS is shaped like an ovate, pointed leaf. Its anterior
surface is concave in the vertical direction, and convex from side to side.
Near the base it receives on the middle line the insertion of the hyo-
epiglottideus muscle. The posterior surface has the converse configuration, and presents numerous pits in which are lodged mucous glands.
The borders of the cartilage are convex, and they are free above, but
below they are enveloped by the aryteno-epiglottic folds of mucous membrane. The apex is pointed, and curved forwards in the upright
position of the cartilage. The base of the cartilage is expanded, and
rests on the body of the thyroid. From each side of it an irregular bar
of cartilage projects horizontally backwards. The epiglottis is composed
of yellow fibro-cartilage.

ARTICULATIONS, LIGAMENTS, AND MEMBRANES OF THE LARYNX.

MODE OF UNION WITH THE HYOID BONE. The larynx is suspended to the
base of the skull through the intervention of the hyoid bone, the tip of
the thyroid cornu (heel process) of that bone being connected by ligamentous fibres (without a synovial membrane) to the so-called *superior
cornu* at the supero-posterior angle of the thyroid ala. The connection
between the hyoid bone and the larynx is further maintained by the
thyro-hyoid membrane, which is attached, on the one hand, to the body and
thyroid cornua of the hyoid, and, on the other, to the body and upper
edge of each wing of the thyroid cartilage.

MODE OF UNION WITH THE TRACHEA. The lower edge of the cricoid
cartilage is connected to the first ring of the trachea by a fibro-elastic
membrane—the *crico-tracheal ligament*.

UNION OF THE CRICOID AND THYROID CARTILAGES. The postero-inferior
angle, or inferior cornu, of each thyroid ala is articulated to the concave
facet on the bezel of the cricoid in a diarthrodial joint, provided with a

small *capsular ligament*, and lined by a synovial sac. The two cartilages are further united by the *crico-thyroid membrane*. This is a fibro-elastic structure consisting of a central and two lateral portions. The central portion is triangular and fills up the space between the adjacent edges of the right and left thyroid alæ. It is attached by its sides to these edges, while by its base it is inserted into the upper border of the cricoid. Each lateral portion lies under cover of the laryngeal mucous membrane, and is shaped somewhat like a quadrant, having an inferior convex edge fixed to the margin of the cricoid in company with the central portion, an anterior edge confounded with the central portion, and an upper straight edge which is thin and free on the side of the larynx. This upper edge is the *true vocal cord ;* and since its fibres are attached in front to the angle of union of the thyroid alæ, and posteriorly to the projecting anterior angle of the base of the arytenoid, it is also termed the *thyro-arytenoid ligament.* Vocal sounds are produced by the vibration of the vocal cords.

Movements.—The movements between the cricoid and thyroid cartilages take place around an imaginary horizontal axis passing through the right and left crico-thyroid joints, and in these movements either cartilage may be supposed to remain fixed while the other revolves around the axis. It should be observed that these movements vary the distance between the angle of junction of the thyroid alæ and the base of the arytenoids, and thus vary the tension of the true vocal cords, which stretch between these points.

UNION OF THE CRICOID AND ARYTENOID CARTILAGES. Each arytenoid cartilage is articulated by the concave facet on its base to one of the convex facets on the upper edge of the cricoid bezel. It is a diarthrodial joint, possessing a *capsular ligament* and a synovial sac.

Movements.—The arytenoid cartilage swings like a door, around a vertical axis passing through the crico-arytenoid joint. When the cartilage is swung outwards, the true vocal cord, which is attached to the anterior angle of its base, is separated from the cord of the opposite side, and the glottis is widened. The glottis is narrowed by the opposite movement.

UNION OF THE THYROID AND EPIGLOTTIS. These cartilages are united by *elastic fibres* passing between them, and forming a kind of amphiarthrosis.

Movements.—Except during the act of deglutition, the epiglottis stands erect in front of the upper aperture of the larynx. During that act the cartilage is bent downwards and backwards so as to cover the aperture like a lid. This movement, however, is executed not exclusively at the joint between the two cartilages, but partly by a bending of the whole cartilage. At the close of the act of deglutition the epiglottis assumes the erect position, owing to its own elastic texture and the

elastic fibres connecting it to the thyroid ; but in the horse this action
is assisted by the hyo-epiglottideus muscle.

THE MUSCLES OF THE LARYNX.

Extrinsic Group.—This includes the sterno-thyroid, the thyro-hyoid,
and the hyo-epiglottideus. The last of these is a single muscle ; the
other two are double.

The STERNO-THYROID MUSCLE (Fig. 26). See page 146.

The THYRO-HYOID MUSCLE (Figs. 26 and 27). This is a dark-coloured,
fleshy muscle taking *origin* from the thyroid cornu (heel process) of the
hyoid bone, and *inserted* into an oblique line on the outer surface of the
thyroid wing.

Action.—Acting alone, the thyro-hyoid muscles would elevate the
larynx between the thyroid cornua of the hyoid bone ; but when they
act in concert with the sterno-thyroid, the thyroid cartilage will be
steadied, and will serve as the fixed point for the crico-thyroid and thyro-
arytenoid muscles.

The HYO-EPIGLOTTIDEUS MUSCLE (Fig. 27) takes *origin* from the
upper face of the body of the
hyoid bone ; and passing backwards in the middle line, it is
inserted into the anterior surface
of the epiglottis at its lower part.
Its fibres are mixed with a quantity of fatty-elastic tissue.

Action.—To assist the natural
elasticity of the epiglottis in
restoring the cartilage to the
erect position at the close of the
act of deglutition.

Intrinsic Group.—This includes
four pairs of muscles, viz., the
crico-thyroid, the thyro-arytenoid,
the posterior crico-arytenoid, and
the lateral crico-arytenoid ; and
a single muscle—the arytenoideus.

FIG. 26.

LARYNX, SIDE VIEW.

1. Glossal Process of Hyoid ; 2. Small Cornu ;
3. Great Cornu ; 4. Arytenoid Cartilage ; 5. Thyro-
Hyoideus ; 6. Insertion of Sterno-Thyroid ; 7. Crico-
Thyroideus ; 8. Crico-arytenoideus Posticus ; 9. 1st
Ring of Trachea ; 10. Thyroid Body.

The CRICO-THYROID MUSCLE (Fig. 26) *arises* from the side of the
cricoid cartilage ; and its fibres, passing obliquely upwards and backwards, are *inserted* into the lower edge of the thyroid wing.

Action.—This muscle acts on the crico-thyroid joint, increasing the
tension of the vocal cord by increasing the distance between the forepart of the thyroid and the base of the arytenoid cartilage. In this

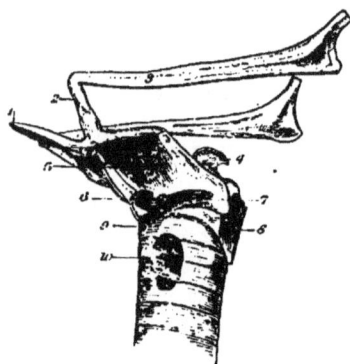

action either the cricoid or thyroid attachment may be the fixed point of the fibres.

Directions.—The thyro-arytenoid and lateral crico-arytenoid muscles lie under cover of the thyroid wing, which must therefore be removed on one side. This is to be done by removing the thyro-hyoid and crico-thyroid muscles, disarticulating the crico-thyroid joint, and incising the ala a little behind the body of the thyroid, after the manner of Fig. 27.

The THYRO-ARYTENOID MUSCLE (Fig. 27) consists of two parallel bundles, between which the mucous membrane of the ventricle of the larynx protrudes as a pouch. Its fibres *arise* from the inner surface of the thyroid wing near its junction with the body, and from the crico-thyroid membrane. The lower fibres are *inserted* into the outer surface of the arytenoid cartilage, while its higher fibres join those of the arytenoideus muscle.

Action.—The muscle is antagonistic to the crico-thyroid, diminishing the tension of the vocal cord by acting on the crico-thyroid joint.

The POSTERIOR CRICO-ARY-TENOID MUSCLE (Figs. 26 and 28). This is the most powerful of the intrinsic muscles. Its muscular tissue is dark red, and mixed with tendinous tissue. Its fibres take *origin* from the outer surface of the cricoid bezel, and are *inserted* into the prominent tubercle on the external angle of the arytenoid cartilage.

Action.—To swing outwards the arytenoid cartilage, and thus to separate the vocal cords and dilate the glottis.

The LATERAL CRICO-ARY-TENOID MUSCLE (Fig. 27). This muscle is placed below the thyro-arytenoid, under concealment of the thyroid

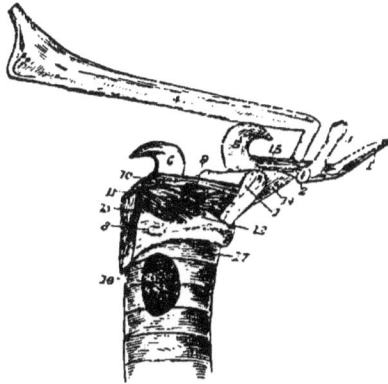

FIG. 27.

LARYNX, SIDE VIEW (THYROID ALA REMOVED).

1. Glossal Process of Hyoid ; 2. Cut Base of Thyroid Cornu ; 3. Small Cornu ; 4. Great Cornu ; 5. Epiglottis ; 6. Arytenoid Cartilage ; 7. Cut Wing of Thyroid Cartilage ; 8. Facet on Cricoid for Articulation with Thyroid Cartilage ; 9. Pouch of Mucous Membrane from Ventricle of Larynx ; 10. and 11. Upper and Lower Bundles of Thyro-Arytenoideus ; 12. Crico-Arytenoideus Lateralis; 13. Crico-Arytenoideus Posticus ; 14. Thyro-Hyoideus ; 15. Hyo-Epiglottideus ; 16. Thyroid Body ; 17. 1st Ring of Trachea.

wing. Its fibres *arise* from the upper border of the side of the cricoid cartilage ; and passing backwards and upwards, they become *inserted* into the same tubercle on the base of the arytenoid as the posterior muscle, and into the outer surface of the arytenoid in front of that tubercle.

Action.—The muscle acts on the crico-arytenoid joint in a manner antagonistic to the preceding muscle, approximating the vocal cords and narrowing the glottis by swinging the arytenoid cartilage inwards.

The ARYTENOIDEUS MUSCLE (Fig. 28). This may be regarded either as a single muscle, or as a double muscle whose right and left fibres meet at a median raphe. Its fibres are *inserted* on each side into the posterior surface of the arytenoid cartilage, and superiorly it is joined by the higher fibres of the thyro-arytenoid muscle.

Action.—To approximate the right and left arytenoid cartilages, and thus narrow the glottis.

NERVES OF THE LARYNX.

Two nerves are distributed to the larynx—the superior and inferior laryngeal nerves. The latter is also known as the recurrent nerve, and both are branches of the vagus, or 10th cranial nerve.

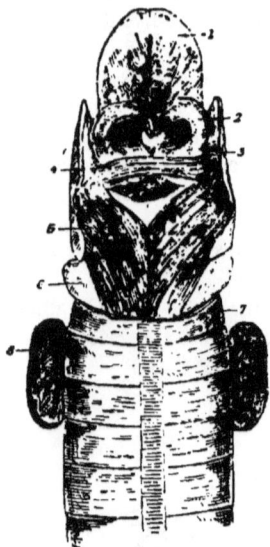

FIG. 28.

LARYNX, BACK VIEW.

1. Epiglottis; 2. Arytenoid Cartilage; 3. Thyroid Cartilage; 4. Arytenoideus; 5. Crico-Arytenoideus Posticus; 6. Cricoid Cartilage; 7. 1st Ring of Trachea; 8. Thyroid Body.

The SUPERIOR LARYNGEAL NERVE has its origin described at page 194. It gives motor filaments to the crico-thyroid and crico-pharyngeus muscles; and then penetrating the thyroid wing by the foramen near its supero-posterior angle, the nerve splits into sensory branches distributed to the mucous membrane of the larynx, giving also twigs to the lining of the pharynx and œsophagus.

The INFERIOR LARYNGEAL (RECURRENT) NERVE has its origin and course described at page 149. It is the motor nerve to all the intrinsic muscles except the crico-thyroid, and it also gives some sensory twigs to the laryngeal mucous membrane.

INTERIOR OF THE LARYNX.

Directions.—A vertical incision should be made along the middle line of the larynx behind, severing the arytenoideus muscle and the bezel of the cricoid cartilage. By separating the lips of this incision, a view of the interior of the larynx from behind will be obtained, and this is to be supplemented by looking into the tube from its upper and lower apertures.

The SUPERIOR APERTURE of the larynx is a large orifice placed at the floor of the pharynx. It is bounded in front by the epiglottis, behind by the pitcher-like lip of the arytenoid cartilages and the fold of mucous membrane uniting them, and laterally by the aryteno-epiglottic fold of mucous membrane. During degluti- tion the epiglottis is folded over the aperture, which it closes like a lid.

The LOWER APERTURE is circum- scribed by the inferior edge of the cricoid cartilage, and is directly con- tinued by the lumen of the trachea.

The GLOTTIS, or RIMA GLOTTIDIS. This is a third aperture, placed about the middle of the tube of the larynx, which it divides into an upper and a lower compartment. In its anterior two-thirds this opening lies between the right and left vocal cords, and in its posterior third it lies between the bases of the arytenoid cartilages. The size of the aperture is varied by the movements executed in the crico- arytenoid joints, as already seen ; and its form varies with its size. It can be completely closed by the apposition of its margins in the mesial plane.

FIG. 29.

When it is only slightly opened, it is a slit-like antero-posterior aperture, widest at the centre ; when moderately open, as in easy respiration, it has the form of an elongated isosceles triangle with the base behind ; when dilated to the fullest extent, it is lozenge- shaped.

The VENTRICLES, or SINUSES, of the larynx. Each of these is a recess, or cavity, placed on the side of the larynx. The entrance to it lies above the vocal cord, whose free straight edge, covered by mucous membrane, forms the lower margin. The upper margin is formed by a concave fold of mucous membrane, containing in man a few fibres designated the *false vocal cord.* The cavity of the ventricle descends to the outer side of the true vocal cord, and a pouch of the mucous lining of the cavity passes out between the upper and lower divisions of the thyro-arytenoid muscle.

The SUB-EPIGLOTTIC SINUS is a depression beneath the base of the epiglottis, and provided with a lunated fold of mucous membrane.

The SUB-ARYTENOID SINUS is a depression beneath the crico-arytenoid joints.

MUCOUS MEMBRANE OF THE LARYNX.   This, which is continuous with
the lining of the pharynx and trachea, is of a pale colour.   It forms
the aryteno-epiglottic folds, and lines the ventricle of the larynx.   It is
provided with numerous mucous glands.   Its free surface is covered by
an epithelium, which is ciliated except over the vocal cords and around
the superior aperture, in which positions it is stratified and squamous.

CHAPTER VI.

DISSECTION OF THE BRAIN, OR ENCEPHALON.

Directions.—The removal of the brain of the horse from its containing cavity is a somewhat difficult operation, in consequence of the thickness of the cranial bones. Supposing the head of an animal recently killed to have been procured for the special purpose, the first steps are the disarticulation of the jaw on both sides, and the removal of the inferior maxilla. Next denude the cranial bones of the muscles and other soft structures, and with the saw remove on each side the zygomatic arch, the supraorbital process of the frontal, and the styloid process of the occipital. Estimating the thickness of the last-named bone at the poll, as much as possible of it may be sawn off without actually encroaching on the cranial cavity. Armed with a chisel, mallet, and strong bone-forceps, the student must now remove as much of the cranial wall as will enable him to extract the brain ; and he may do this by removing either the roof or the floor of the cavity. The first method is the speedier, but the latter has the advantage of permitting the roots of the cranial nerves, the pituitary body, and the cranial vessels to be better preserved. The dura mater is to be left as far as possible intact, but its attachments along the interfrontal and interparietal sutures, and to the oblique ridge between the cerebral and cerebellar divisions of the cranial cavity, must be cut with the scalpel. When the forepart of the cavity is reached, the handle of the scalpel is to be used to scoop the olfactory bulbs out of the fossæ in which they lie.

The brain having been removed in its membranes, it should be laid with its base upwards on a broad strip of calico, and lowered into a vessel of methylated spirit or a ten per cent. solution of nitric acid in water. After a week's immersion, it will be ready for examination.

MEMBRANES, OR MENINGES, OF THE BRAIN.

The brain, like the spinal cord, is surrounded by three envelopes : the dura mater, the arachnoid, and the pia mater.

The DURA MATER is the external of these envelopes. It is a strong fibrous membrane, similar in structure to the spinal dura mater, with which it is continuous at the foramen magnum. It differs, however, from the same envelope of the spinal cord, in that it is closely adherent to the inner surface of the cranial bones, and forms for them an internal periosteum. All over its outer surface it is connected by slender fibrous processes and vessels to the bones; but it is particularly adherent to these along the lines of the sutures, and at the margins of foramina. The meningeal vessels ramify on the outer surface of the membrane, and leave their impressions on the inner surface of the cranial bones. Sometimes the outer surface of the dura mater, on each side of the middle line above, shows numbers of granular processes—the *Pacchionian bodies*, which are developed from the subjacent arachnoid. Occasionally they are large enough to cause the partial absorption of the bones over them. The inner surface of the dura mater is smooth, in virtue of an endothelial layer representing the parietal layer of the arachnoid. This inner surface is closely applied to the brain contained within the other two membranes; and along certain lines it detaches processes which pass inwards, and form partial partitions between the different divisions of the encephalon. These processes are: the falx cerebri and the tentorium cerebelli.

The *Falx Cerebri* is a vertical, mesial, sickle-shaped process which dips in between the two hemispheres of the cerebrum. The convex upper edge of the process is attached to the cristagalli process, and to the interfrontal and interparietal sutures. The concave lower edge is thin and lace-like, and rests free on the corpus callosum. The short posterior edge, or base, is straight, and is attached to the intracranial projection of the interparietal bone.

The *Tentorium Cerebelli* is a vaulted partition extending transversely between the cerebrum and the cerebellum. In outline it is crescentic, having a superior convex, and an inferior concave, border. The former is attached on the middle line to the intracranial projection of the interparietal bone, and on each side of that its attachment descends obliquely forwards and downwards along the crest formed by the parietal and petrous temporal bones. The concave edge is free, and arches over the crura cerebri. The anterior surface of the membrane is convex, and the posterior ends of the cerebral hemispheres rest on it. The posterior surface is concave, and is in contact with the cerebellum.

The *Sinuses of the Dura Mater.*—These are venous passages formed by the splitting of the dura mater. They are as follows:—

The SUPERIOR LONGITUDINAL SINUS is of considerable size, and is found in the falx cerebri at its attached or convex edge. Beginning at the crista galli process, it becomes larger as it passes backwards, and it terminates at the intracranial projection of the interparietal bone.

The INFERIOR LONGITUDINAL SINUS is small and inconstant. It extends along the free or concave edge of the falx; and after receiving the veins of Galen, it is continued backwards in the tentorium cerebelli, terminating at the same point as the preceding sinus.

Where the two foregoing sinuses meet, they form the whirlpool of Herophilus (*torcular Herophili*), from which the blood is drained away by the transverse sinuses.

The TRANSVERSE SINUSES pass right and left at the periphery of the tentorium cerebelli, and enter the parieto-temporal conduit. In that canal each is continued as the parieto-temporal confluent, from which the blood is drained away by the roots of the temporal veins.

The CAVERNOUS SINUSES. Each of these is placed in the dura mater at the side of the sella turcica of the sphenoid bone. Anteriorly each receives the ophthalmic vein, and posteriorly the right and left sinuses become continuous behind the pituitary gland. The venous arch which they thus form discharges its blood through the foramen lacerum basis cranii into the sub-sphenoidal confluent. The internal carotid artery traverses the cavernous sinus, and forms while in it a sigmoid curve.

The PETROSAL SINUSES are small, and pass in the tentorium cerebelli on each side, between the transverse and cavernous sinuses.

The OCCIPITAL SINUSES. These are placed in or external to the dura mater lining the cerebellar division of the cranial cavity. They are continuous through the foramen magnum with the spinal sinuses, and their contained blood is drained away by a large vein that passes through the condyloid foramen to join the occipital vein.

The MENINGEAL ARTERIES. These are derived from the *meningeal branch of the ophthalmic artery*, which enters the forepart of the cavity at the internal orbital foramen; and from the *great meningeal* or *spheno-spinous* branch of the internal maxillary. The *spheno-spinous artery* enters by the foramen lacerum basis cranii, and, after detaching meningeal branches, enters the parieto-temporal conduit to anastomose with the mastoid artery. Some meningeal twigs are also furnished by the prevertebral branch of the occipital artery (page 191).

The MENINGEAL NERVES. Filaments from the 4th, 5th, 9th, and 10th cranial nerves, and from the sympathetic, are said to have been traced to the dura mater.

The ARACHNOID. This, like the same membrane of the spinal cord, is a delicate transparent membrane. In structure and disposition it is comparable to a serous membrane. Its *parietal* layer is represented by the endothelial lining of the dura mater; its *visceral* layer invests the brain and pia mater; and the parietal and visceral portions together enclose a space, which is the *arachnoid cavity*, or *subdural space*. The free surface of the membrane bounding this space is smooth and moist like a serous membrane. Between the visceral arachnoid and the pia mater another space is left, which is termed the *subarachnoid space*. This space is most evident over the intervals between the cerebral convolutions, and over surface depressions at the base of the brain, for at these points the arachnoid does not dip down to line the hollows, but bridges them over. The space is continuous with the same space in the spinal meninges, and contains the limpid *cerebro-spinal fluid.*

The PIA MATER. This is the vascular membrane of the brain. It consists of delicate areolar tissue and bloodvessels. It invests the brain closely, following all its surface irregularities. Behind the cerebral hemispheres it sends towards the interior of the cerebrum a wide process —the *velum interpositum ;* and where the cerebellum is superposed to the medulla oblongata, it forms on each side a thickened granular cord—the *choroid plexus of the 4th ventricle.* These will be exposed at a later stage.

ARTERIES OF THE BRAIN (FIG. 30).

Three vessels are concerned in supplying blood to the encephalon, viz., the basilar, internal carotid, and ophthalmic arteries.

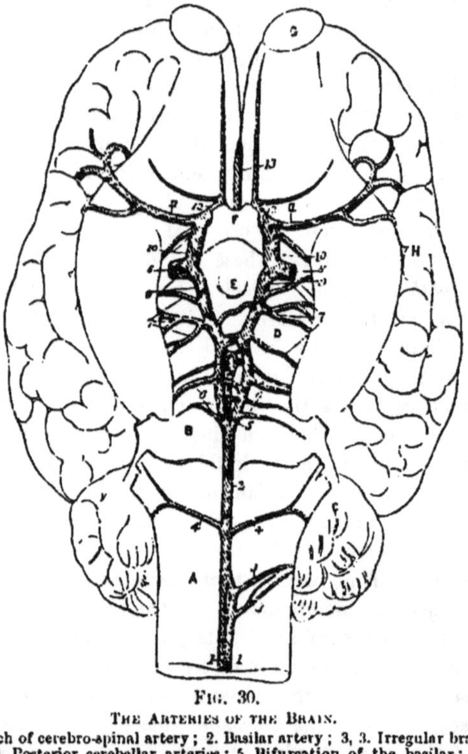

FIG. 30.

THE ARTERIES OF THE BRAIN.

1. Anterior branch of cerebro-spinal artery ; 2. Basilar artery ; 3, 3. Irregular branches to medulla and cerebellum ; 4. Posterior cerebellar arteries ; 5. Bifurcation of the basilar ; 6. Anterior cerebellar arteries ; 7. Posterior cerebral arteries (more numerous and smaller than usual) ; 8. Internal carotid ; 9. Posterior communicating branch ; 10. Anterior branch of internal carotid, which divides to form 11 and 12—the middle and anterior cerebral arteries ; 13. Single vessel formed by the union of 11 and 12, disappearing into great longitudinal fissure ; A. Medulla oblongata ; B. Pons Varolii ; C. Cerebellum ; D. Crus cerebri ; E. Corpus albicans ; F. Optic commissure ; G. Olfactory bulb ; H. Cerebral hemisphere.

The BASILAR ARTERY is formed on the middle line of the lower face of the medulla oblongata, by the union of two vessels. These are the anterior divisions of the right and left cerebro-spinal arteries, whose posterior divisions unite in the same manner to form the middle spinal artery. The basilar artery passes forwards in the median groove of the medulla, and crosses the pons, in front of which it bifurcates to form the posterior cerebral arteries. In its course the basilar artery detaches on each side, besides numerous vessels to the medulla and pons, the posterior cerebellar arteries.

The *Posterior Cerebellar Arteries* are two in number, a right and left. They are detached at different levels from the basilar, behind the pons ; and they turn round the medulla to reach the cerebellum.

The *Posterior Cerebral Arteries* diverge from each other in the inter-peduncular space ; and after being connected together by a short transverse branch of considerable volume, and by numerous smaller reticulate twigs, they are joined by the posterior communicating branch of the internal carotid. Each then turns outwards over the crus cerebri to gain the choroid plexus and the posterior part of the cerebral hemisphere. Behind the point at which the vessels are connected by the short transverse branch, they give off the anterior cerebellar arteries. Sometimes, as in Fig. 30, the posterior cerebral artery, instead of turning outwards as a single vessel, detaches from its outer side two or three branches which wind round the crus.

The *Anterior Cerebellar Arteries* are variable in number and disposition, and may arise as branches of the basilar artery. Generally there are two or three on each side, and they turn backwards and outwards over the crus cerebri to gain the front of the cerebellum.

The INTERNAL CAROTID ARTERY. This vessel begins above the cricoid cartilage of the larynx, as one of the terminal branches of the common carotid. It passes upwards and forwards to the foramen lacerum basis cranii, being sustained in a fold of the guttural pouch, and accompanied by some nervous branches from the superior cervical ganglion of the sympathetic. Piercing the sub-sphenoidal sinus, it passes through the foramen into the cavernous sinus, within which it forms a sigmoid curve. It then leaves the sinus, and gaining the deep face of the dura mater, it divides at the margin of the sella turcica of the sphenoid bone into an anterior and a posterior branch. The latter, termed the *posterior communicating artery*, is reflected backwards to join the posterior cerebral artery. The anterior branch passes forwards, and at the outer side of the optic commissure divides into the middle and anterior cerebral arteries.

The *Middle Cerebral Artery* passes outwards across the hemisphere, in the fissure of Sylvius.

The *Anterior Cerebral Artery* unites in the mesial plane, above the optic commissure, with the corresponding vessel of the opposite side. The single vessel thus formed receives the meningeal branch of the ophthalmic artery, and turns round the anterior end of the corpus callosum to gain the great longitudinal fissure. Here it separates into a right and a left branch, each of which passes backwards along the flat face of the hemisphere.

By the anastomosis of the two anterior cerebral arteries in front, and the junction of the posterior communicating artery on each side with the posterior cerebral, which results from the bifurcation of the basilar artery, a vascular circle is established around the pituitary body. This is termed the *Circle of Willis*, and its object is to keep up a free blood supply to the cerebrum, even should there be an obstruction in one of the main vessels forming the circle. Moreover, the internal carotid arteries of opposite sides are, before they divide, connected by a large transverse branch which further contributes to the freedom of the circulation.

The OPHTHALMIC ARTERY is a collateral branch of the internal maxillary. It enters the cranial cavity from the orbit by the internal orbital foramen, along with the nasal branch of the ophthalmic nerve, and divides into meningeal and nasal branches.

The *Meningeal Branches* of opposite sides give off branches to the dura mater, and then unite to form a single trunk which joins the middle cerebral arteries.

The *Nasal Branch* passes through the cribriform plate to gain the nasal chamber.

The SYMPATHETIC NERVE. Two branches from the superior cervical ganglion accompany the internal carotid artery, and anastomose around it to form the *carotid plexus*. Within the cavernous sinus they form another plexus—the *cavernous plexus*. From these plexuses filaments pass to join the 3rd, 4th, 6th, and ophthalmic cranial nerves. A twig also joins the large superficial petrosal nerve from the 7th, to form the vidian nerve; another passes to the lenticular ganglion, either separately or with the ophthalmic nerve; and some filaments pass to the Gasserian ganglion.

The *Brain*, or *Encephalon*, consists of four principal parts, viz., the medulla oblongata, the pons Varolii, the cerebellum, and the cerebrum. The medulla is the division which is in direct continuity behind with the spinal cord. The pons projects as a thick transverse bar, or ridge, in front of the medulla. The cerebellum is superposed to both medulla and pons. The cerebrum lies in front of the other three segments, and is larger than these taken together. The weight of the whole brain in an average-sized horse is about twenty-three ounces.

THE MEDULLA OBLONGATA, OR BULB (PLATES 35 AND 36).

The medulla oblongata is continuous at the foramen magnum with the spinal cord, of which it appears to be the expanded anterior termination.

It rests by its inferior face on the basilar process of the occipital bone ; and its superior face, which is concealed by the cerebellum, is depressed and forms the floor of the 4th ventricle. Its anterior extremity is limited by the pons Varolii, and is its widest part. The middle line of the medulla above and below is traversed by lines which continue forwards the superior and inferior median fissures of the cord.

The medulla is composed of both white and grey nerve matter. The former occurs at the exposed surface of the medulla, and its nerve fibres are for the most part longitudinal in direction, and are collected into tracts, or bundles. Thus, lying at each side of the inferior median fissure of the organ, there is a tract termed the *inferior pyramid.* To the outer side of this again, and isolated from it by a faint longitudinal groove, is a tract occupying the position of the *olivary fasciculus* and *olivary body* of human anatomy. More externally placed than the last, and forming a thick cord at each side of the medulla, is the *restiform body;* while above the restiform body, and nearer the superior median fissure, is a more slender column of fibres termed the *superior pyramid.* The line of separation between the two last-mentioned tracts is very faint, and in the horse there is seldom or never any surface line of demarcation between the restiform body and the olivary fasciculus.

Where the medulla joins the cord, the inferior pyramids become narrow, and the inferior median fissure shallow or nearly obliterated ; and at that point there is a visible crossing of fibres from one side to the other, constituting the *decussation of the pyramids.* Towards the posterior part of the medulla its lateral aspect is crossed by superficial curved fibres—the *arciform fibres,* and immediately behind the pons Varolii there is a band of transverse fibres termed the *trapezium.* Within the medulla some fibres pass across the median plane and connect its right and left halves.

The grey matter of the medulla oblongata occurs in considerable amount at the floor of the 4th ventricle, where it will subsequently be exposed.

COURSE OF THE LONGITUDINAL FIBRES OF THE MEDULLA OBLONGATA. The inferior pyramid is in part composed of fibres from the inferior column of the same side of the cord, but principally of fibres crossing from the opposite side of the cord at the decussation. These decussating fibres are furnished mainly by the lateral column, but partly also from the superior column. The fibres of the inferior pyramid are continued through the pons to the cerebrum.

The olivary fasciculus of fibres is derived from the inferior column of the cord on the same side, and it is continued through the pons to the cerebrum.

The restiform body derives its fibres from all three columns of the cord on the same side, but in greatest proportions from the superior column. It enters the cerebellum, of which it forms the posterior peduncle. The superior pyramid derives its fibres from the innermost part of the superior column of the cord on the same side, and its fibres are continued through the pons to the cerebrum.

The medulla oblongata shows the superficial origin of the last seven cranial nerves.

THE PONS VAROLII (PLATE 33).

The pons Varolii rests on the basilar process, in front of the medulla oblongata. In front of it the crura cerebri appear. Its inferior face is convex in both directions, and has a faint median furrow. The superior face forms the anterior part of the floor of the 4th ventricle. Its extremities are curved upwards to enter the cerebellum, of which they form the middle peduncles. The pons consists of white and grey nerve matter. The nerve fibres of the white matter are arranged in two sets—a transverse and a longitudinal. The transverse fibres consist of the surface fibres of the pons, and of deeper fibres separated from these by the longitudinal set. It is these transverse fibres that curve upwards at either extremity of the pons to enter the cerebellum as its middle peduncle, and they accordingly play the part of a commissure to the right and left halves of the cerebellum. The longitudinal fibres are the forward continuation of the longitudinal fibres of the medulla oblongata, *minus* the restiform bodies. In front of the pons these longitudinal fibres are continued as the crura cerebri. The grey matter of the pons occurs within its substance, and at the floor of the 4th ventricle. To a group of pigmented nerve cells in the latter position, the term *locus cœruleus* is applied.

The pons shows the superficial origin of the 5th cranial nerve, by two distinct roots springing from its lateral aspect.

THE CEREBELLUM (PLATES 34 AND 35).

The cerebellum is superposed to the medulla and pons, and lies under the supra-occipital division of the occipital bone. The tentorium cerebelli arches downwards in front of it, and isolates it from the posterior extremities of the cerebral hemispheres. It is traversed in the antero-posterior direction by two shallow grooves, which divide it into a middle and two lateral lobes. The *middle lobe* is the smallest, and is known as the *vermiform lobe*. When followed forwards, the vermiform lobe is seen to be reflected round the anterior aspect of the cerebellum to gain its lower surface at the roof of the 4th ventricle ; and it terminates by a blunt end about the middle of this surface. When followed posteriorly, the vermiform lobe behaves in the same way, terminating at the roof of the 4th ventricle by a blunt end opposed to the first. These reflected portions have a distinct resemblance to two caterpillars, and they may be distinguished as the *anterior and posterior vermiform processes*. The anterior vermiform process is adherent to the valve of Vieussens. Each *lateral lobe* is joined on its inferior aspect by three bundles of nerve fibres, which are termed the *peduncles*. The *posterior peduncle* is the termination of the restiform body, the *middle peduncle* is the reflected extremity of the pons, and the *anterior peduncle* passes forwards beneath the corpora quadrigemina.

Besides the grooves which divide the cerebellum into its lobes, numerous smaller fissures occur over its surface, and divide the lobes into *folia*, or *leaflets*. The arrangement of these leaflets will be made much more evident by making an antero-posterior vertical section, at or near the mesial plane of the organ. The peduncles are to be cut as they enter the lower face of the lateral lobe, and the anterior vermiform process is to be carefully separated from the valve of Vieussens with the scalpel. This will enable one half or a little more of the cerebellum to be removed after the manner of Plate 35.

The cerebellum contains both grey and white matter. The white matter forms a large mass in the interior, and from this mass large plates are given off towards the surface. From these primary plates proceed more numerous smaller secondary plates, and these again detach small terminal plates which end in the surface folia. In consequence of this disposition of the white matter, it presents on vertical section a strikingly arborescent appearance, to which the term *arbor vitæ* is applied. The nerve fibres of the white matter are for the most part directly continuous with the peduncles; but some are proper to the organ, and connect different points of the grey matter.

The grey matter of the cerebellum is spread over its surface, and also forms two independent masses within the central mass of white matter. These latter have the form of a corrugated capsule, and each is placed a little to one side of the mesial plane, and is known as the *corpus dentatum* of the cerebellum. The surface layer of grey matter invests the core of white matter within each leaflet, and also extends across the bottom of the fissures between adjacent leaflets. It consists of two strata: an outer *grey layer*, and an inner *rust-coloured layer*.

The FOURTH VENTRICLE (Plates 35 and 36). This is a space between the cerebellum above, and the medulla and pons below. Its boundaries are as follows:—Its floor is formed by the medulla and pons; its roof by the valve of Vieussens, the under suface of the vermiform lobe, and the reflection of pia mater from the medulla to the cerebellum; laterally it is bounded in its anterior third by the anterior peduncle of the cerebellum, and in its posterior two-thirds by the restiform body. The widest part of the space is at the point where the peduncles enter the cerebellum, and it contracts towards both extremities. At the posterior

n

extremity there is a minute hole, which is the entrance to the short tube that continues the central canal of the cord into the posterior end of the medulla. The pointed posterior end of the space is the *calamus scriptorius* of human anatomy, so named from its resemblance to a writing pen. The anterior end of the space lies under the valve of Vieussens, and leads into the *aqueduct of Sylvius*, which is a canal tunnelled beneath the corpora quadrigemina, and opening anteriorly into the 3rd ventricle. The floor of the cavity is traversed by a longitudinal mesial furrow, and it shows the grey matter of the medulla and pons. The cavity is lined by a ciliated epithelium, and it communicates by one or more minute apertures in its floor with the sub-arachnoid space. On each side of the cavity, between the cerebellum and the restiform body, there is a thickened piece of pia mater—the *choroid plexus of the 4th ventricle*.

The VALVE OF VIEUSSENS is a delicate, translucent fold, placed at the anterior part of the roof of the 4th ventricle. The lateral edges of the valve are fixed to the anterior cerebellar peduncles, its anterior edge is attached behind the testes, and its posterior edge stretches across the anterior vermiform process. The upper face of the valve is adherent to the anterior vermiform process, and its lower face is free and forms the anterior part of the roof of the 4th ventricle. The 4th nerve arises in the valve, close behind the testes, the right and left nerves appearing continuous with one another across the middle line.

THE CEREBRUM.

Under the term cerebrum are included all the parts of the encephalon except the medulla, pons, and cerebellum. It forms a mass larger than these taken together, although the amount by which it exceeds them is much less in the horse than in man. The inferior aspect of the cerebral mass is termed its base, and the student should begin by examining the objects to be seen there (Plate 33).

The CRURA CEREBRI are two thick, round, white cords, which appear in front of the pons. At this point they are close together; but as they proceed forwards, they diverge and form the posterior boundaries of a lozenge-shaped area—the *interpeduncular space*, which is completed in front by the optic tracts and commissure. Anteriorly each crus disappears into the cerebral hemisphere, but its point of termination is concealed by the optic tract. The crus is composed of a superficial and a deep layer of nerve fibres with an intermediate thin stratum of grey matter. The superficial layer of fibres is known as the *crusta*, and the deep is termed the *tegmentum*. The fibres of both layers are continuous posteriorly with the longitudinal fibres of the pons; and they are transmitted in front to the optic thalami, corpora striata, and grey matter of the hemisphere. The corpora quadrigemina, which are superposed to the crura, also receive some fibres. The grey matter of the crus contains nerve cells with dark pigment, and is therefore termed the *locus niger.* ⸍ The crura cerebri show the superficial origin of the 3rd pair of nerves.

The Optic Tracts. These are two white cords of nerve fibres which turn round the crura cerebri, and pass forwards and inwards to meet in the middle line and form by their fusion the *optic commissure* or *chiasma.* This commissure rests on the sphenoid bone, in front of the pituitary fossa ; and in front it gives off the diverging optic or 2nd nerves. The optic tracts form the anterior boundary of the interpeduncular space.

The Pons Tarini is the grey matter in the posterior angle of the interpeduncular space. It is also known as the *locus perforatus posticus,* from its being penetrated by numerous vessels.

The Corpus Albicans is a pea-like, white nodule placed on the middle line, about the centre of the interpeduncular space. As will subsequently be learned, the body is formed by the reflection of the anterior pillars of the fornix.

The Tuber Cinereum is a layer of grey matter between the corpus albicans and the optic commissure. It is perforated in its centre, and connected to the upper surface of the pituitary gland by a hollow tube of grey matter—the *infundibulum.*

The Pituitary Body is a reddish-yellow, disc-shaped body, having a diameter about equal to that of a sixpence. It is thickest in its centre and thinnest at its rim. Its lower face rests on the sella turcica of the sphenoid bone ; and its upper face receives the insertion of the infundibulum, and covers the tuber cinereum, and, in part, the corpus albicans and optic commissure. Within its structure it comprises cells resembling those of the blood-vascular or ductless glands, and others that resemble nerve cells. In the foetus it is proportionally larger, and contains a cavity which communicates with the 3rd ventricle through the infundibulum.

The pons Tarini, corpus albicans, and tuber cinereum form the floor of the *3rd ventricle,* a cavity which the dissector will hereafter expose by working from the upper aspect of the cerebrum.

The Lamina Cinerea, or Lamina Terminalis, is a thin, delicate layer of grey matter which is placed above and in front of the optic commissure. It is the anterior boundary of the 3rd ventricle.

The Locus Perforatus Anticus is a spot of grey matter at each side of the optic commissure, penetrated by numerous vessels for the corpus striatum, which lies above the spot.

The Fissure of Sylvius, is a faint and ill-defined groove which begins at the locus perforatus anticus, and extends outwards across the hemisphere.

The Great Longitudinal Fissure. In front of the optic chiasma the cerebral mass is seen to be mesially divided by the great longitudinal fissure. This fissure, as will be better seen when the brain is viewed from above, is a great vertical mesial cleft extending the whole length of the cerebrum, which it partially divides into right and left halves, or hemispheres.

The OLFACTORY BULBS.   The olfactory bulb is the white body situated at the anterior end of the hemisphere.   It occupies the olfactory fossa at the forepart of the cranial cavity; and unless special care is taken in the removal of the brain, the bulb is apt to be separated from the hemisphere and left in that fossa.   From the free surface of the bulb the delicate filaments of the olfactory (1st cranial) nerve pass through the cribriform plate of the ethmoid bone, and enter the nasal chamber.   The bulb is hollow, having a central cavity that is in communication with the anterior cornu of the lateral ventricle.

The OLFACTORY PEDUNCLE is a short, thick, white cord immediately behind the bulb, and in direct continuity with it.   The hemisphere is slightly depressed over the peduncle, the depression being termed the olfactory fissure.   The peduncle divides posteriorly into the olfactory tracts.

The OLFACTORY TRACTS.   These are two white diverging bands—an inner and an outer—that continue the olfactory peduncle backwards. The *internal tract* (inner olfactory root) is short, and passes backwards and inwards to the edge of the great longitudinal fissure.   The *external tract* (outer olfactory root) is a much longer band which curves outwards and backwards across the fissure of Sylvius, and then encircles outwardly the uncinate and hippocampal convolutions, to reach the tentorial aspect of the hemisphere, on which it is lost.   At the fissure of Sylvius the tract seems to lose some of its fibres in front of the uncinate convolution, and behind that point it becomes grey on its surface.

Behind the angle of divergence of the olfactory tracts is a smooth and slightly convex area—the *quadrilateral space* of Paul Broca.   The surface layer of this space consists of grey matter constituting the *middle* or *grey olfactory root;* and, according to Broca, it covers white fibres that connect the olfactory bulb to the crus cerebri and to the anterior cerebral commissure (anterior white commissure of 3rd ventricle).

If now the olfactory peduncle be raised from the olfactory fissure, it will be seen to be connected to the frontal lobe of the hemisphere by a lamina termed by Broca the *superior olfactory root.*   This lamina is grey on its surface and white beneath, and if it be ruptured the cavity of the bulb and the communication between that cavity and the anterior cornu of the lateral ventricle will be brought into view.

External to the outer olfactory tract, each hemisphere shows numerous winding worm-like ridges, termed *convolutions,* and internal to the posterior half of the same tract there is seen a thick ridge—the *hippo-*

campal convolution—which terminates behind the fissure of Sylvius in a nipple-like eminence—the *uncinate convolution* (mastoid lobule, or mammillary eminence). These will presently be more particularly described.

Directions.—The student must now reverse the position of the brain, laying it with its base downwards, while he proceeds to examine its upper aspect.

The GREAT LONGITUDINAL FISSURE is now seen in its entirety. It extends from the anterior to the posterior end of the cerebrum, and appears to completely separate the right and left hemispheres. In the natural state the fissure is occupied by the falx cerebri. Gently separate the contiguous margins of the hemispheres, so as to widen out the fissure. Except towards the hinder end of the fissure, this proceeding requires no dissection, but at that point the hemispheres are united on the middle. It is, however, a mere adhesion through the medium of pia mater. Separate the hemispheres here by traction, or by cutting carefully in the mesial plane. There will now be exposed (Fig. 31) a white body—the *corpus callosum*—which connects the hemispheres at the bottom of the great longitudinal fissure. At the same time there will be brought into view the opposed inner surfaces of the hemispheres.

The CEREBRAL CONVOLUTIONS. In his examination of the base of the brain, and more clearly now, the student will have observed that the surface of the hemisphere is not smooth, but traversed by numerous winding worm-like elevations. These are termed the cerebral *convolutions* or *gyri;* and the intermediate grooves or fissures are technically termed *sulci.*

At first sight it might be supposed, as indeed was believed until a comparatively recent date, that the disposition of these convolutions is quite irregular and hap-hazard. Observation has shown, however, that such is far from being the case, and that the convolutions have a nearly, if not altogether constant, arrangement. In the human subject, indeed, the surface of the hemisphere has been accurately mapped, and each convolution named. In the brain of the horse the plan of these convolutions appears to be as uniform as in man; and although, perhaps, the convolutions are not absolutely identical in any two brains, or even in the two hemispheres of the same brain, still the irregularities are so slight as to permit one to describe with considerable minuteness what might be termed a common plan. The mapping of the surface of the hemispheres derives its chief interest and utility from the discovery that definite areas are associated with particular functions, in such a way that when these areas are destroyed or injured there follows total loss or disturbance of these functions, and that in some cases the exercise of particular functions can be brought about by applying stimuli to particular spots of the cerebral cortex.

In the brain of man the hemisphere is primarily subdivided into five lobes, viz., frontal, parietal, occipital, temporo-sphenoidal, and central, the last being also known as the Island of Reil, or the Insula. The lines of separation between these lobes are certain well-marked fissures, distinguished from the sulci in general by their greater depth and constancy. In each lobe, again, the secondary sulci form the lines of separation between a definite number of convolutions.

In the third edition of Professor Chauveau's admirable work (*Traité d'Anatomie comparée des Animaux domestiques*) an attempt is made to describe the cerebral convolutions of the horse after the plan followed in human anatomy, and to establish an almost complete correspondence of these parts in the two brains. It appears to me, after very careful consideration, that except in a few points, an identity between convolutions in the two brains is not clearly indicated on anatomical grounds alone. That most of the convolutions of the human brain have corresponding convolutions in the brain of the horse is more than probable ; and experimental, pathological, or developmental evidence may yet place this correspondence beyond doubt. In the meantime, however, and provisionally, I think it preferable to describe the cerebral convolutions of the horse according to what appears the most natural plan.

The surface of each hemisphere (excluding from present consideration its inner aspect) is divided into three lobes or areas, viz., an anterior lobe, a postero-superior lobe, and a postero-inferior lobe. This subdivision is effected by certain fissures (Plate 34), as follows :—

1. The CRUCIAL FISSURE. This is a short fissure which begins near the middle of each hemisphere where it margins the great longitudinal fissure. Passing outwards, it joins the great oblique fissure. The crucial fissure separates the anterior from the postero-superior lobe. In the right hemisphere of Plate 34 these lobes are connected across the fissure by a small *annectent* or bridging convolution.

2. The GREAT OBLIQUE FISSURE. This is the most pronounced fissure of the hemisphere. Beginning near the middle of the upper surface of the hemisphere, where it is continuous with the crucial fissure, it is directed obliquely outwards, downwards, and backwards, to reach the tentorial aspect of the hemisphere (Plate 33). It separates the postero-superior from the postero-inferior lobe.

3. The LATERAL FISSURE. This fissure begins on the upper surface of the hemisphere, at the point of junction of the crucial and great oblique fissures. It curves round the side of the hemisphere, with a slightly forward inclination ; and it separates the anterior from the postero-inferior lobe.

4. The FISSURE OF SYLVIUS.* This begins at the base of the brain (Plate 33), at the side of the optic commissure. It passes outwards as a faint and ill-defined depression in front of the uncinate convolution ; and crossing the outer olfactory tract, it divides into four branches, which, however, are mere sulci. One of these is directed backwards between the outer olfactory tract and the postero-inferior lobe ; another passes forwards between the inner olfactory tract and the anterior lobe ; a third ascends into the postero-inferior lobe ; and the fourth is directed forwards into the anterior lobe. In the first part of its course the fourth branch separates the adjacent convolutions of the anterior and postero-inferior lobes, being itself separated by a bridging convolution between these lobes from the lower extremity of the lateral fissure.

To the outer side of the outer olfactory tract, at the point from which these branches of the Sylvian fissure radiate, there is a minute nodular convolution that is partially or entirely concealed from view until the adjacent convolutions are slightly separated. This seems to foreshadow the convolutions of the *insula* of man.

The ANTERIOR LOBE presents four convolutions :—

1. The *First Anterior Convolution* (Plate 33, 1. A) is seen on the under surface of the lobe. It lodges the olfactory peduncle in the olfactory fissure ; and when the peduncle is in position it shows an inner and an outer part, the former occupying the position of the *gyrus rectus* of human anatomy.

2. The *Second Anterior Convolution* begins on the under surface of the lobe (Plate 33,

_____

* The first part of the fissure of Sylvius, as far as the outer olfactory tract, is sometimes and more correctly called the *valley of Sylvius*, and Broca restricts the term *fissure of Sylvius* to the third of the above-described branches.

2. A), external to the preceding. It runs forwards and upwards round the extremity of the hemisphere, and abuts on the antero-marginal convolution (Plate 34).

3. The *Third Anterior Convolution* begins at the under surface of the lobe (Plate 33, 3. A) to the outer side of and behind the preceding. It curves upwards and forwards across the hemisphere, and reaches its upper aspect (Plate 34). Here it is reflected backwards and inwards; and turning upon itself, it descends to near the point from which it started, being connected at its termination by a bridging convolution to the first convolution of the postero-inferior lobe.

4. The *Antero-marginal Convolution* (Plate 34, A. M.) lies at the forepart of the great longitudinal fissure, and is visible on both the upper and inner aspects of the lobe. Beginning at the crucial fissure (being sometimes connected to the postero-marginal convolution of the postero-superior lobe), it passes forwards at the edge of the hemisphere, at the anterior end of which the first and second anterior convolutions abut upon it.

The POSTERO-SUPERIOR LOBE comprises three convolutions :—

1. The *Postero-marginal Convolution* (Plate 34, Γ. M.) extends along the margin of the lobe, appearing on both its upper and inner surfaces, and lying in series with the antero-marginal convolution of the anterior lobe. Beginning at the crucial fissure, it passes backwards at the edge of the hemisphere, and reaches its tentorial aspect.

2. The *First Oblique Convolution* (Plate 34, 1. O) lies external to the preceding. Beginning in front, near the margin of the hemisphere, it passes obliquely backwards and outwards, and curves round the extremity of the lobe to reach its tentorial surface.

3. The *Second Oblique Convolution* (Plate 34, 2. O) passes with an oblique direction between the preceding convolution and the great oblique fissure, and reaches the tentorial surface of the lobe (Plate 33), from which it seems to be in part continued by the outer olfactory tract.

The POSTERO-INFERIOR LOBE. The sulci of this lobe are numerous and small, and it is difficult to divide it naturally into convolutions. For convenience of description, however, two convolutions may be recognised in it :—

1. The *First Postero-inferior Convolution* (Plates 33 and 34, 1. P. I.) is four-sided, and contains within itself several short sulci. It lies behind the third convolution of the anterior lobe, to which it is connected by a bridging convolution.

2. The *Second Postero-inferior Convolution* (Plates 33 and 34, 2. P. I.) lies at the posterior part of the lobe, above the outer olfactory tract; and its posterior extremity appears on the tentorial surface of the hemisphere. Like the preceding, it possesses numerous minor sulci within itself.

Directions.—There still remains for examination the inner surface of each hemisphere. Separate the hemispheres as widely as possible along the great longitudinal fissure. At the upper edge of this fissure there will now be seen the inner aspect of the antero-marginal and postero-marginal convolutions already described, and between the lower edge of these and the corpus callosum there lies a thick convolution—the gyrus fornicatus.

The *Gyrus Fornicatus* (Fig. 31).—This is comparable to a lobe,* rather than to a convolution. It is disposed in a great curve, or arch, from which it is named. It begins at the forepart of the under surface of the hemisphere, in front of the lamina cinerea, and here it is narrow and pointed. It bends round the anterior extremity (*genu*) of the corpus callosum, acquiring at its point of reflection a great increase in thickness. It passes backwards above the corpus callosum, and below the antero-marginal and postero-marginal convolutions. From the former body it is separated by the *fissure of the corpus callosum*, while the *calloso-marginal* fissure (*great limbic fissure* of Broca) separates it from the marginal convolutions above. In this part of its course the gyrus is distinctly divided into two tiers by a fissure that traverses it in its length. Posteriorly this fissure becomes very shallow, and the gyrus, losing its double character, turns round the posterior end (*splenium*) of the corpus callosum and reaches the tentorial surface of the hemisphere. At this point it becomes slightly constricted; and after being connected with the convolu-

---

* Paul Broca (*Anatomic comparée des circonvolutions cerebrales*) considers that this part of the hemisphere represents not merely a lobe, but several lobes—that it is, in fact, the equivalent of all the rest of the cerebral cortex. He accordingly divides the surface of the hemisphere primarily into two great divisions—the *great limbic lobe* (gyrus fornicatus) and the *convolutionary mass*.

248   THE ANATOMY OF THE HORSE.

tions of the postero-superior and postero-inferior lobes, it is directed forwards at the base
of the brain (Plate 33, Hipp. con.), between the crus cerebri and the outer olfactory
tract. Finally, it terminates in the nipple-like eminence already noticed (Unc. con.).

The whole of this great convolution corresponds very closely to the gyrus fornicatus of
human anatomy. Thus, the part which turns round the genu and rests on the upper
surface of the corpus collosum is the *callosal convolution*; the part from the splenium
to the side of the cerebral crus is the *hippocampal convolution;* and the nipple-like
eminence is the *uncinate convolution.*

The hippocampal part of the gyrus fornicatus has a small process which projects
forwards under the splenium ; and as the convolution curves forwards to emerge at the
side of the crus, it rests on the optic thalamus. By carefully raising the convolution from
the thalamus, there will be brought into view a fissure on the under aspect of the former.
This is the *hippocampal fissure*, and it projects the convolution into the lateral ventricle
as the *hippocampus.* Beyond this fissure the edge of the hippocampus is seen, margined
by a thin-edged white band—the *taenia hippocampi.* The hippocampus and its taenia
here form from the upper boundary of the *great transverse fissure* of the brain, by which the
pia mater of the hemisphere projects towards the interior of the cerebrum as the *velum
interpositum.*

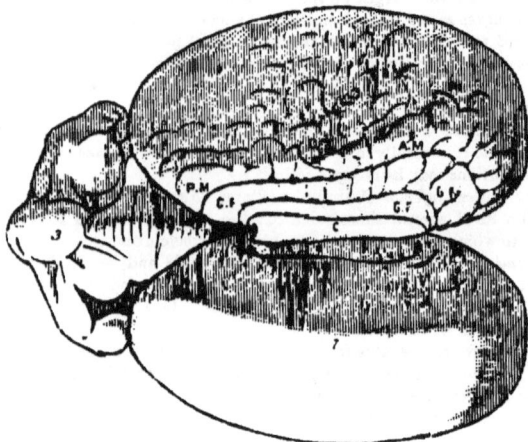

FIG. 31.

CORPUS CALLOSUM AND INNER FACE OF THE CEREBRAL HEMISPHERE.

1, 2. Right and left cerebral hemispheres; 3. Cerebellum; C. Corpus callosum; G. F. Gyrus
fornicatus (its callosal portion); C. M. Calloso-marginal fissure; A. M. Antero-marginal convolution; P. M. Postero-marginal convolution.

Directions.—With a large, thin-bladed, sharp knife a horizontal slice
should be removed from the top of one or both cerebral hemispheres,
down to the level of the corpus callosum.

The hemisphere will now be seen to contain both grey and white
matter. In the centre of the hemisphere the white matter forms a large
mass connected with that of the opposite side by the corpus callosum.
At the surface the mass sends a white core into each convolution. The

great sheet of grey matter on the surface of the hemisphere invests the white core in each convolution, and also extends across the bottom of each fissure.

The CORPUS CALLOSUM (Plate 35 and Fig. 31) is a great commissure of nerve fibres connecting the right and left hemispheres. It terminates behind in a thickened margin—the *splenium;* and in front it is abruptly bent downwards and backwards, the bend being named the *genu,* and the reflected portion the *rostrum.* The rostrum becomes narrower as it descends, and is connected to the optic commissure by the lamina cinerea. Along the middle line of its lower face the corpus callosum is connected posteriorly with the fornix, and anteriorly with the septum lucidum ; and on each side it forms the roof of a cavity in the hemisphere—the lateral ventricle. Nearly all the fibres of the corpus callosum have a transverse direction ; but on each side of the longitudinal middle line of its upper face there are a few longitudinal fibres termed the *striæ longitudinales,* or *nerves of Lancisi.*

Directions.—If the corpus callosum be now cut through in the longitudinal direction, a little to one side of the middle line, and dissected outwards, the lateral ventricle will be exposed. The corpus callosum, it will now be seen, is thickest at its posterior extremity, and thinnest at its middle.

The LATERAL VENTRICLES (Plate 35) are two in number, one in each hemisphere. They are separated from one another along the middle line by the fornix and septum lucidum, but beneath the former body they communicate through the foramen of Monro. The central portion of each cavity is termed the *body,* and its prolongations before and behind are termed respectively the *anterior* and the *descending cornu.*

On the floor of the *body* of the cavity the following objects will be noticed :—In front, a large pear-shaped grey eminence—the *corpus striatum ;* behind, another body of about the same size but white on its surface—the *hippocampus ;* between the corpus striatum and the hippocampus, a groove, in which there lies a red granular cord—the *choroid plexus.* Where the hippocampus bounds this groove, it is margined by a white band—the *tænia hippocampi ;* and if the choroid plexus be pulled gently backwards, another white band will be seen to margin the corpus striatum where it bounds the groove—this is the *tænia semicircularis.* [*]

The *anterior cornu* is occupied by the base of the corpus striatum. It curves downwards and forwards into the anterior part of the hemisphere, where it communicates with the cavity of the olfactory bulb.

The *descending cornu* contains the prolongations of the hippocampus and its tænia. It passes at first backwards and outwards, and then

[*] The optic thalamus and tænia semicircularis are generally enumerated among the objects visible in the body of the lateral ventricle. In the brain of the horse, however, the choroid plexus completely conceals from view the optic thalamus, and in most cases also the tænia semicircularis.

curves downwards, forwards, and inwards, terminating at the base of the hemisphere in the uncinate convolution. The ventricles are lined by a ciliated epithelium, which is continuous through the foramen of Monro with the lining of the 3rd ventricle.

The SEPTUM LUCIDUM is a thin, translucent partition between the two lateral ventricles. It is broadest in front, where it is attached to the rostrum, or reflected part of the corpus callosum. Its upper edge is attached to the corpus callosum, and its lower edge to the fornix ; and posteriorly these edges meet at an acute angle. The septum consists of white matter in its centre, with a layer of grey matter on each side. In man it contains a small isolated cavity—the 5th ventricle.

The FORNIX, or arch, is a mesially placed white band, consisting of a central part, or body, and two pairs of processes, or pillars. The *body* is flattened above and below, and broadest behind. Its upper face is adherent posteriorly to the corpus callosum, but in front it dips down and leaves beneath the forepart of the corpus callosum a space occupied by the septum lucidum. The under surface of the body rests on the velum interpositum, and at its anterior extremity arches over the foramen of Monro. The *anterior pillars* of the fornix are two white cords which descend in front of the foramen of Monro, being separated by a slight interval. Reaching the base of the brain, they turn on themselves, forming thus the corpus albicans, and they then enter the optic thalamus. The *posterior pillars* are broader and flatter, and not so well defined. Each in part bestows its substance on the surface of the hippocampus, and in part it descends along the anterior edge of that body as the *tænia hippocampi*, or *corpus fimbriatum*.

The CORPUS STRIATUM (Plates 35 and 36). This is the large grey body already noticed in the body and anterior cornu of the lateral ventricle. In shape it is pyriform, having its broad end directed forwards and inwards, and its tapering end backwards and outwards to the roof of the descending cornu. The body comprises two masses of grey matter, separated from each other by intermediate white fibres which curve upwards and outwards from the cerebral crus. The upper mass of grey matter—termed the *nucleus caudatus*—is that which projects into the lateral ventricle. The lower mass—the *nucleus lenticularis*—lies above the *quadrilateral space* already seen at the base of the hemisphere between the diverging olfactory tracts.

The TÆNIA SEMICIRCULARIS (Plate 36) is a narrow white band that extends between the corpus striatum and the optic thalamus. (Its relation to the thalamus will be better seen in the next stage of the dissection.)

The HIPPOCAMPUS is the curved eminence already noticed in the body and descending horn of the lateral ventricle. It rests on the optic thalamus, from which it is separated by the velum interpositum. The ventricular aspect of the body is white, but the surface that rests on the optic thalamus is grey. The hippocampus is to be viewed as a convolution of the cerebrum, being, in fact, an inward projection of the hippocampal convolution already noticed (page 248).

Directions.—The corpus callosum and septum lucidum should be cut away in order to see the upper surface of the fornix, which should then be divided transversely in its middle. The anterior part should be raised forwards and upwards to expose its anterior pillars and the foramen of Monro. The posterior part should be removed along with the hippocampus in order to bring into view the velum interpositum. The dissection will then assume the form of Plate 36; but the optic thalami and the pineal gland, there exposed, will be covered by the velum.

The VELUM INTERPOSITUM is a triangular fold of pia mater, continuous by its base with the pia mater on the hinder end of the cerebrum. Its apex lies at the foramen of Monro, and its lateral edges, fringed by the choroid plexuses, project towards the lateral ventricle through what is termed the *great transverse fissure of the cerebrum.* This is an arched cleft extending over the optic thalami, from the extremity of the descending horn on one side to the same point on the other. Above it is bounded centrally by the fornix, and on each side by the hippocampus and its tænia. The velum interpositum is, like the pia mater in general, a vascular membrane; and the choroid plexus of each side is a thickened and highly vascular portion of it. Along its centre the veins of Galen extend backwards, and unite to turn round the posterior extremity of the corpus callosum, and enter the inferior longitudinal sinus. The velum should now be raised from its apex backwards, when it will be seen to cover the optic thalami and the pineal gland, and care must be taken lest the latter be removed with it.

The OPTIC THALAMI. Each of these is a large grey-coloured body, superposed to the crus cerebri behind the corpus striatum, and in front of the corpora quadrigemina. Its upper surface is convex and covered by the velum interpositum. When followed outwards this surface changes its direction, looking backwards and downwards; and it there forms part of the boundary of the descending horn of the lateral ventricle. Inwardly the right and left thalami are opposed to one another along the middle line, and they include between them the 3rd ventricle. In front each thalamus is separated from the corpus striatum by a groove, in which will now be seen more distinctly the tænia semicircularis. Behind, another groove isolates the thalamus from the nates.

The PINEAL GLAND is a small, reddish, conical body, named from its resemblance to a pine cone. It stands by its base on the middle line between the optic thalami and the nates. From its base two white

bands—the *peduncles of the pineal gland*—extend forwards along the
groove between the two thalami; and at the foramen of Mouro each
peduncle unites with the anterior pillar of the fornix to descend to the base
of the brain, and concur in forming the corpus albicans. In structure
the body presents some resemblance to lymphoid tissue, but it also
contains some branched corpuscles which are possibly nerve cells.
Imbedded in it is a quantity of gritty calcareous matter termed the
acervulus cerebri, or brain-sand.

The THIRD VENTRICLE is a narrow space whose sides are formed by
the optic thalami. Its floor corresponds to the parts already examined
in the interpeduncular space, viz., the pons Tarini, corpus albicans, and
tuber cinereum. Its roof is formed by the velum interpositum covered
by the fornix. In front it is bounded by the lamina cinerea, and it here
communicates with the lateral ventricles by the foramen of Monro.
Posteriorly the aqueduct of Sylvius enters it from the 4th ventricle.
The cavity is crossed by three commissures: 1. The *Anterior Commissure* is a small white cord of nerve fibres stretching transversely between
the corpora striata at the anterior end of the cavity, and immediately in
front of the descending anterior pillars of the fornix. The fibres of the
commissure are traceable through the corpora striata into the white
matter of the hemispheres. 2. The *Middle (soft) Commissure* is composed of delicate grey matter cementing the inner surfaces of the thalami,
and apt to be more or less ruptured in handling the brain. 3. The
Posterior Commissure is white, like the anterior; and its fibres connect
the two thalami at the base of the pineal body, and immediately in front
of the nates.

The 3rd ventricle has a ciliated lining continuous with that of the
4th through the aqueduct of Sylvius, and with that of the lateral
ventricles through the foramen of Monro. In the fœtus the cavity communicates through the tuber cinereum and infundibulum with the
pituitary body.

The FORAMEN OF MONRO, or FORAMEN COMMUNE ANTERIUS, is the common
point of communication between the 3rd and lateral ventricles. It
might be described as a short vertical shaft ascending from the forepart of the 3rd ventricle, and opening under the fornix, which is thrown
over it like an arch. Beneath this arch the lateral ventricles communicate with one another and with the 3rd ventricle.

The CORPORA QUADRIGEMINA are two pairs of bodies superposed to the
crura cerebri behind the optic thalami. The anterior pair of bodies, or
nates, are larger than the posterior pair, or *testes*, from which they are
separated by a groove. Between the right and left nates there is a well-
defined groove, but the groove between the testes is faint or not observ-
able. The nates are grey on their surface, but the testes are white.
The bodies were named nates and testes from a fancied resemblance

to the hips and testicles of a man, but these terms are far from express-
ing the relative size of the two bodies.

The AQUEDUCT OF SYLVIUS, or ITER, is a tunnel which, commencing
posteriorly in the 4th ventricle, beneath the valve of Vieussens,
extends forwards beneath the corpora quadrigemina, and opens into the
hinder part of the 3rd ventricle.   It possesses a ciliated lining continuous with that of the ventricles which it connects.

OPTIC TRACTS and CORPORA GENICULATA.—The optic tracts have already
been seen at the base of the brain, where they form the anterior
boundaries of the interpeduncular space.   When followed backwards,
each tract will be found to turn round the crus cerebri, and join the
optic thalamus.   At the point of junction two eminences are placed, an
outer, or anterior, and an inner, or posterior.   These are named respectively
the *corpus geniculatum externum* and *internum*.   They are composed of
grey matter from which some fibres of the optic tract pass.   Other
fibres of the tract come directly from the optic thalamus, and others
from the corpora quadrigemina.

THE CRANIAL OR ENCEPHALIC NERVES (PLATE 33).

In the examination of the base of the brain, the roots of the cranial
nerves have already been noticed, but it will be advantageous to describe
them here as a series.   The cranial nerves are distinguished by special
names, and also by numerical designations.   It must be observed, however, at the outset that there are two different systems of enumeration
in use among anatomists, the first of which recognises twelve, and the
other nine, pairs of nerves.   This diversity of nomenclature is apt to lead
to confusion, but fortunately this confusion does not extend to veterinary
anatomy, in which, both at home and abroad, the first and more natural
of these methods is exclusively employed.   This system is also that
employed by human anatomists on the continent, but by British human
anatomists the number of cranial nerves is stated as nine pairs.   The
following table exhibits in the central column the special names of
the nerves, and in the side columns their numerical designations under
the two systems :—

1st pair	.	. Olfactory nerves .	. . .	. 1st pair.
2nd ,,	.	. Optic nerves .	. . .	. 2nd ,,
3rd ,,	.	. Oculo-motor nerves	. . .	. 3rd ,,
4th ,,	.	. Pathetic or Trochlear nerves	.	. 4th ,,
5th ,,	.	. Trifacial or Trigeminal nerves	.	. 5th ,,
6th ,,	.	. Abducent nerves .	. . .	. 6th ,,
7th ,,	.	. Facial nerves (Portio dura) .	.	·⎫ 7th ,,
8th ,,	.	. Auditory nerves (Portio mollis) .	.	·⎭
9th ,,	.	. Glosso-pharyngeal nerves .	.	·⎫
10th ,,	.	. Pneumogastric or Vagus nerves .	.	·⎬ 8th ,,
11th ,,	.	. Spinal Accessory nerves .	.	·⎭
12th ,,	.	. Hypoglossal nerves .	. .	. 9th ,,

The OLFACTORY or 1st nerve.   The fibres of this nerve leave the surface of the *olfactory bulb*, and pass through the foramina of the cribriform plate to reach the summit of the nasal chamber. They are there distributed in the olfactory division of the lining membrane of that chamber.

The OPTIC or 2nd nerve arises from the *optic chiasma* or *commissure*, and reaches the back of the orbit by passing through the optic foramen. Piercing the sclerotic and choroid tunics of the eyeball, its fibres radiate outwards and form one of the layers of the retina. As already seen, the optic chiasma is formed by the fusion of the optic tracts, each of which derives its fibres from the optic thalamus, corpora geniculata, corpora quadrigemina, and decussation of the pyramids. In the optic chiasma some of the fibres of each tract cross and are continued in the optic nerve of the opposite side. Some of the fibres of each optic tract, it is stated, cross in the chiasma and return to the brain by the opposite tract, while in the same way fibres pass from the one optic nerve to the other optic nerve.

The OCULO-MOTOR or 3rd nerve arises from the inner side of the crus cerebri by a number of bundles, the fibres of which are traceable to nerve cells in the corpora quadrigemina. The nerve leaves the cranium by the foramen lacerum orbitale, and reaches the orbit.

The PATHETIC, TROCHLEAR, or 4th nerve appears to arise in the valve of Vicussens (Plate 35). Some of its fibres are decussate with those of the opposite nerve, and the others are traceable to nerve cells of the locus cœruleus, or of the corpora quadrigemina. Emerging from the valve, the nerve winds round the crus cerebri, and appears in front of the pons. It leaves the cranium by the minute pathetic foramen, and reaches the back of the orbit. It is the smallest of the cranial nerves.

The TRIFACIAL, TRIGEMINAL, or 5th nerve springs out of the side of the pons by two roots. The outer and larger of these is termed the *sensory root;* and its fibres are traceable to cells of the grey matter of the medula, pons, and locus cœrulens, and possibly also to the cerebellum. This root near its origin expands into a large ganglion—the *Gasserian ganglion*, beyond which it divides into three branches, viz., the ophthalmic, superior maxillary, and inferior maxillary divisions. The inner or *motor root* of the 5th nerve is traceable to grey matter of the pons. It joins the inferior maxillary division of the sensory root. The *superior maxillary division* leaves the cranium by the foramen rotundum, the *ophthalmic division* by the foramen lacerum orbitale, and the *inferior maxillary division* by the forepart of the foramen lacerum basis cranii. The trifacial is the largest of the cranial nerves.

The ABDUCENT or 6th nerve. This nerve springs from the anterior part of the medulla, in line with the faint groove that limits outwardly the inferior pyramid. Some of its fibres issue from the groove between the pons and the medulla, while others penetrate the trapezium. The fibres of the nerve are traceable to a group of nerve cells in the medulla.

The nerve reaches the orbit by passing through the foramen lacerum orbitale.

The FACIAL or 7th nerve springs out of the medulla, close behind the pons, its fibres seeming to continue outwards the trapezium. Its rootlets are traceable to nuclei of grey matter in the medulla. The nerve is joined by a delicate filament—the *portio intermedia*—which appears between the roots of this and the next nerve. The 7th nerve enters the internal auditory meatus in company with the 8th nerve. Separating from that nerve, it passes along a canal in the petrous temporal bone—the aqueduct of Fallopius—from which it emerges by the stylo-mastoid foramen, under the parotid gland. Within the aqueduct of Fallopius the nerve forms a knee-shaped bend, and at that point it shows a minute ganglion—the *geniculate ganglion*—from which proceed the great and small superficial petrosal nerves (pages 189 and 214).

The AUDITORY or 8th nerve springs from the medulla, close behind the pons, and immediately external to the root of the 7th. It is here compounded of two roots—a superior and an inferior. The *superior root* (Plate 35) passes over the restiform body to the grey matter at the floor of the 4th ventricle. The *inferior root* springs out of the side of the restiform body, its fibres arising from nerve cells of that body or of the grey matter at the floor of the 4th ventricle, and possibly also from the cerebellum. The 8th nerve enters the internal auditory meatus, and penetrates to the internal ear.

The GLOSSO-PHARYNGEAL or 9th nerve springs out of the side of the medulla, a little behind the outer extremity of the trapezium. It is here compounded of two or three bundles, the outermost being in line with the roots of the next two nerves. The fibres emanate from nerve cells of the grey matter at the floor of the 4th ventricle. The nerve leaves the cranium by the posterior part of the foramen lacerum basis cranii, and at that point it shows a minute ganglion—the *petrous ganglion*, or the *ganglion of Andersch*, from which the nerve of Jacobson arises (page 269).

The PNEUMOGASTRIC, VAGUS, or 10th nerve is formed by a number of rootlets which spring from the side of the medulla, behind and in line with the outermost fibres of the 9th nerve. Its fibres arise from nerve cells of the medulla. The nerve passes out of the cranium by the posterior part of the foramen lacerum basis cranii, and is joined by the inner division of the 11th nerve. As the nerve passes through the foramen it presents an enlargement—the *upper ganglion*, or *ganglion of the root*. From this ganglion arises the *auricular branch* of the vagus, which penetrates to the aqueduct of Fallopius, where it anastomoses with the 7th nerve; afterwards emerging from the bone in company with that nerve, to be distributed to the mucous membrane of the external auditory process.

The SPINAL ACCESSORY or 11th nerve comprises two sets of roots—a *spinal* and a *medullary*. The *spinal* roots appear along the lateral column of the cervical part of the spinal cord, in which they arise from a group of nerve cells towards the middle of the grey crescent. By the union of these roots there is formed a cord which travels upwards between the superior and the inferior roots of the cervical spinal nerves, becoming thicker as it ascends. This cord enters the cranial cavity by the foramen magnum, and is then joined by the medullary roots. The *medullary roots* spring out of the side of the medulla oblongata, behind and in line with the roots of the 10th nerve, the fibres arising from nerve cells at the floor of the 4th ventricle. These roots join the spinal part of the nerve, which then leaves the cranium by the foramen lacerum basis cranii, along with the 10th nerve. In the foramen of exit the trunk of the nerve resolves itself into two portions—an internal and an external. The *internal portion* joins the 10th nerve ; the *external portion* is that which has already been seen in the dissection of the neck (page 151).

The HYPOGLOSSAL or 12th nerve is formed by the fusion of rootlets that spring from the lower face of the medulla, along the line that indicates the outer limit of the inferior pyramid. These roots are in series with the inferior roots of the spinal nerves ; and sometimes there is also present a superior root, in series with the superior roots of the same nerves, and provided with a minute ganglion. These roots arise from nerve cells of the medulla. The roots of the nerve perforate the dura mater, and unite in emerging from the cranium by the condyloid foramen.

CHAPTER VII.

DISSECTION OF THE EYEBALL.

Directions.—Let the student procure three or four eyes of the horse, or, failing these, of the ox. They should be excised from the orbit immediately after death, and as much as possible of the optic nerve should be preserved in connection with the eye. While an assistant holds the eye without squeezing it, the dissector should clean the optic nerve and the outer surface of the sclerotic with forceps and a sharp scalpel. One of the eyes so prepared should be completely frozen in a mixture of ice and salt, and it should then be bisected vertically with a large knife or fine saw. While still frozen, the section to which the optic nerve is attached should be fastened by a strong pin to a layer of solid paraffin at the bottom of a wide and shallow basin. It should be fastened with the cut surface upwards, the pin being passed vertically from the centre of that surface; and the vessel should then be filled with water. The remaining segment should be laid on the freezing mixture, with its cut surface upwards. By an examination of both segments, the student should make out the following points :—

The *Globe* or *Ball* of the eye approaches the spherical in form, as is expressed by these designations. On closer inspection, however, it will appear to be made up of two combined portions from spheres of different sizes. The posterior portion, forming about five-sixths of the ball, is a sphere of comparatively large size with a small segment cut off it in front ; and at this point there is applied to it the anterior portion, which, being a segment of a smaller sphere, projects at the front of the ball with a greater convexity than the posterior portion.

The eyeball consists of concentrically arranged coats, and of refracting media enclosed within these coats. The coats are three in number, viz., (1) an external protective tunic made up of the *sclerotic* and *cornea*, (2) a middle vascular and pigmentary tunic—the *choroid*, (3) an internal nervous layer—the *retina*. The sclerotic is the white opaque part of the outer tunic, of which it forms about the posterior five-sixths, being co-extensive with the larger sphere already mentioned. The cornea forms the remaining one-sixth of the outer tunic, being co-extensive with the segment of the smaller sphere. It is distinguished from the sclerotic by being colourless and transparent. The choroid coat will be

s

recognised as the black layer lying subjacent to the sclerotic. It does not line the cornea, but terminates behind the line of junction of that coat with the sclerotic, by a thickened edge—the *ciliary processes.* At

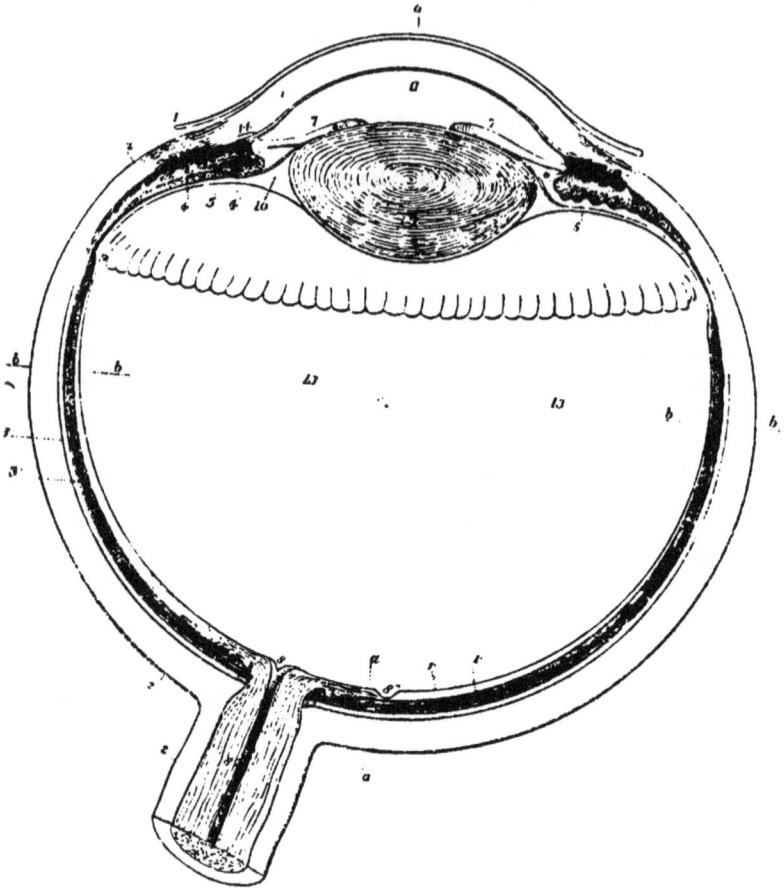

FIG. 32.

VIEW OF THE LOWER HALF OF THE RIGHT ADULT HUMAN EYE, DIVIDED HORIZONTALLY THROUGH THE MIDDLE. MAGNIFIED FOUR TIMES (*A. Thomson*).

1. The cornea ; 1'. Its conjunctival layer ; 2. The sclerotic ; 2'. Sheath of the optic nerve passing into the sclerotic ; 3. 3'. The choroid ; 4. Ciliary muscle, its radiating portion ; 4'. Cut fibres of the circular portion ; 5. Ciliary fold or process ; 6. Placed in the posterior division of the aqueous chamber, in front of the suspensory ligament of the lens ; 7. The iris (outer or temporal side) ; 7'. The smaller, inner, or nasal side ; 8. Placed on the divided optic nerve, points to the arteria centralis retinæ ; 8'. Papilla optica at the passage of the optic nerve into the retina ; 8''. Fovea centralis retinæ ; r. The nervous layer of the retina ; r'. The bacillary layer ; 9. Ora serrata, at the commencement of the ciliary part of the retina ; 10. Canal of Petit ; 11. Anterior division of the aqueous chamber, in front of the pupil ; 12. The crystalline lens, within its capsule ; 13. The vitreous humour ; a. a. a. Parts of a line in the axis of the eye ; b. b. b. b. A line in the transverse diameter.

the line of junction of the sclerotic and cornea, the *iris* passes across the interior of the eye. This, which may be viewed as a dependency of the choroid, is a muscular curtain perforated by an aperture termed the *pupil*. The retina will be recognised as a delicate glassy layer, lining the greater part of the choroid.

The refracting media of the eye are three in number, viz., (1) the *aqueous humour*—a watery fluid enclosed in a chamber behind the cornea ; (2) the *crystalline lens* (and its capsule)—a transparent soft solid of a biconvex form, and placed behind the iris ; (3) *the vitreous humour*—a transparent material with a consistence like thin jelly, and occupying as much of the interior of the eye as is subjacent to the choroid.

Directions.—Another eye should be cleaned like the first, and used for the more particular examination of the sclerotic and cornea.

The SCLEROTIC is a strong, opaque fibrous membrane which in great measure maintains the form of the eyeball, and protects the more delicate structures within it. Its anterior portion, which is covered by the ocular conjunctiva, is visible in the undissected eye, and is commonly known as the "white of the eye." In form it is bell-shaped, and the optic nerve pierces it behind like a handle. The point of perforation, however, is not exactly at the centre of the summit of the bell, but a little to its inner side. When the nerve is cut off close to the sclerotic, the nerve-bundles appear as if passing through the apertures of a sieve, and to this appearance the term *lamina cribrosa* is applied. The sheath of the nerve passes on to the sclerotic around the point of perforation. In front the rim of the bell becomes continuous with the cornea. The outer surface of the membrane receives the insertion of the muscles of the eyeball. The inner surface (which will afterwards be exposed) is of a light brown colour, and is connected to the choroid by fine processes of connective-tissue—the *lamina fusca*. The coat is thickest over the posterior part of the eyeball, and is thinnest a little behind its junction with the cornea.

Structure.—The sclerotic is composed of connective-tissue, there being a great preponderance of white fibres, but intermixed with these are some fine elastic fibres. The bundles of fibres, which are disposed both meridionally and equatorially, have a felted arrangement, but the surface fibres are mostly longitudinal. The texture of the sclerotic is only slightly vascular, the capillaries forming a wide-meshed network. It is most vascular just behind the cornea.

The CORNEA is the anterior transparent portion of the outer coat of the eyeball. It may be viewed as a part of the sclerotic specially modified to permit the passage of light into the interior of the eye. Its outline is elliptical approaching the circular, and its greatest diameter is transverse. At its periphery it joins the sclerotic by continuity of tissue ; and as the edge of the cornea is slightly bevelled, and has the fibrous sclerotic carried for a little distance forward on its outer surface, the cornea is generally said to be fitted into the sclerotic like a watch-glass into its rim. The *venous canal of Schlemm* runs circularly around the eyeball at the line of junction of the sclerotic and cornea. The anterior surface of the cornea is exquisitely smooth, and is kept moist by the lachrymal secretion. Its posterior surface forms the anterior boundary of the chamber in which the aqueous humour is contained. The cornea is of uniform thickness; and, as will afterwards be proved in removing it, it is very difficult to cut, being of a dense, almost horny consistence. When its normal convexity is disturbed, the cornea becomes opaque.

Structure.—Save a few capillary loops at its margin, the cornea is without vessels. Its structure comprises the following layers, which are enumerated in order from the anterior to the posterior surface :—

1. The *Anterior Epithelium* is a stratified, pavement epithelium, continuous at the margin of the cornea with the conjunctival epithelium.

2. The *Anterior Elastic Lamina* (Bowman's membrane). This is a structureless, elastic layer. It is extremely thin in the eye of the lower animals, but is better developed in the human eye.

3. The *Substantia Propria.* This, which forms the main thickness of the cornea, is composed of fibrous connective-tissue arranged in lamellæ parallel to the surfaces of the cornea. Between adjacent lamellæ there is left a network of spaces and branching canals, in which are found the branched *corneal corpuscles.*

4. The *Posterior Elastic Lamina* (Descemet's membrane) is a thick, structureless, elastic layer.

5. The *Posterior Epithelium* is a single layer of polygonal cells.

Directions.—A strong pin should now be passed through the optic nerve, and used to fasten the eye beneath the surface of water in a wide and shallow vessel, as already directed in the case of the frozen section. While one hand steadies the eye beneath the water, an incision is to be made with the other through the cornea, using for the purpose a very sharp scalpel. As soon as the incision is made, some of the aqueous humour will escape into the water, and may possibly be recognised by a slight inky discoloration, which is due to a *post-mortem* disintegration of the pigmented epithelium lining the cavity in which the humour is contained. Still keeping the eye under water, one blade of a pair of small scissors should be introduced within the incision, and the cornea should be excised immediately in front of its junction with the sclerotic. The iris will by this means be exposed, and the next step must be to remove a portion of the sclerotic so as to expose the sub-jacent choroid. Beginning at its anterior edge, it may be incised back-wards towards the optic nerve, snipping it bit by bit with the point of the scissors. Another incision may then be made parallel to the first, and about half an inch from it. The piece of sclerotic between the incisions may then be raised and turned backwards by destroying the slender processes, nerves, and vessels that connect it to the choroid. At the anterior edge of the piece of choroid thus exposed, and immediately behind the rim of the iris, there will be seen a whitish zone—the *ciliary body*, or *annulus albidus*.

The AQUEOUS HUMOUR occupies a chamber which is bounded in front by the posterior surface of the cornea; and behind by the capsule and suspensory ligament of the lens, and by the ends of the ciliary processes. It is across this chamber that the iris extends, and the chamber is some-times described as being divided by the iris into two compartments, viz., an anterior, in front of the iris; and a posterior, behind it. In the living eye, however, the posterior surface of the iris contacts with the lens-capsule, so as to leave only a narrow chink behind the attachment of the curtain to which the term posterior chamber may be applied. The aqueous humour is composed of water with a small proportion of common salt in solution.

The IRIS is a muscular pigmented curtain extended across the interior of the eye, and having about its centre an aperture termed the *pupil*. By variations in the size of this aperture, the amount of light trans-mitted to the retina is regulated. It varies somewhat in colour, but is most frequently of a yellowish-brown tint. Its anterior surface, which shows some lines con-verging to the pupil, is bathed by the aqueous humour, as is also its posterior surface immedi-ately internal to its attachment. The greater part of the posterior sur-face, however, is in con-tact with the capsule of the lens, and glides on it

FIG. 33.

CHOROID MEMBRANE AND IRIS EXPOSED BY THE REMOVAL OF THE SCLEROTIC AND CORNEA (*Quain* after Zinn).

a. One of the segments of the sclerotic thrown back; *b.* Ciliary muscle; *c.* Iris; *e.* One of the ciliary nerves; *f.* One of the vasa vorticosa or choroidal veins.

during the movements of the curtain. The circumferential border is attached within the junction of the sclerotic and cornea. The inner border circumscribes the pupil, which varies in outline according to its size. When much contracted, the pupil is a very elongated ellipse, the long axis of which is in the line joining the nasal and temporal canthi of the eyelids; but when it is extremely dilated, the ellipse approaches the circular in form. Appearing at the upper margin of the pupil, there are generally two or three little sooty masses termed the *corpora nigra*. These are little dependent balls of the *uvea*, or pigmentary layer covering the back of the iris.

Structure.—This comprises a connective-tissue stroma, muscular tissue, and an anterior and a posterior epithelium.

The *Stroma* is a framework of connective-tissue, the fibres having a radial arrangement, and the corpuscles being branched and pigmented. The pigment varies in shade from yellow to dark brown or almost black.

The *Muscular Tissue* is of the non-striated variety, and its fibres are arranged in two sets, viz., (1) the *sphincter of the pupil*, a narrow band around the pupil, and close to the posterior surface of the curtain ; (2) the *dilator of the pupil*, whose fibres begin at the attached edge of the curtain, and extend radially inwards to end in the sphincter. The size of the pupil is regulated by the state of contraction of these two muscles. When the action of the sphincter preponderates, the aperture is contracted ; when that of the dilator preponderates, the pupil is dilated.

The *Anterior Epithelium* is continuous at the attached edge of the iris with the posterior epithelium of the cornea. It is a single layer of pigmented cells.

The *Posterior Epithelium*, or *Uvea*, comprises several layers of cells similarly pigmented ; and, as before stated, the *corpora nigra* are small dependent portions of it. In the eyes of albinos the iris is devoid of pigment ; and occasionally in the horse and dog the pigment is only present in spots, and the animal is then said to be " wall-eyed."

Vessels.—The arteries of the iris are derived from the ciliary branches of the ophthalmic. They form at the circumference of the iris a larger circle, from which radial vessels pass inwards and form around the pupil a smaller circle. The veins have a similar disposition, and terminate in those of the choroid.

In the fœtus the pupil is closed by a vascular transparent membrane —the *membrana pupillaris*, which disappears before birth.

The CILIARY MUSCLE. This is a zone of non-striated muscular tissue which forms the outer layer of the ciliary body, and lies behind the circumferent edge of the iris. It consists (1) of an outer *radiating* set of fibres, which arise from the inner surface of the sclerotic close behind its line of junction with the cornea, and pass backwards to be inserted into the choroid and ciliary processes ; and (2) of an inner *circular* set, which surround the rim of the iris. When the radiating fibres contract, they pull forward the choroid coat and ciliary processes, and allow the lens to bulge forwards by slackening its tense suspensory ligament. This is the mechanism by which the eye is accommodated for near objects.

The CHOROID COAT. This is a bell-shaped, dark membrane which lines the sclerotic. Its outer surface, when exposed by the removal of the sclerotic, has a shaggy appearance due to the *tunica fusca* which unites the two coats. Between the two the ciliary vessels and nerves pass forwards. The inner surface of the choroid is lined by the layer of pigmented hexagonal cells belonging to the retina. Behind it is pierced by the optic nerve; and in front it is continued as the ciliary processes, which form, as it were, the rim of the bell.

Directions.—In the eye prepared to expose the iris and choroid, a segment of the former and of the ciliary muscle should be carefully and delicately removed with scissors, so as to lay bare a number of the ciliary processes. This is to be done while the eye remains immersed in water.

The CILIARY PROCESSES. These form a fringe around the slightly inverted rim of the choroid. They number upwards of a hundred, and each projects on the inner side of the rim, as a small swelling separated by depressions from the adjacent processes. The outer surface of each is covered by the ciliary muscle; the inner surface rests in a depression on the suspensory ligament of the lens; behind each is continuous with the texture of the choroid; and in front it terminates in a rounded end which bounds in part the so-called posterior chamber of the aqueous humour, behind the peripheral part of the iris.

Structure.—The choroid possesses a *stroma* of connective-tissue with ramifying corpuscles containing brown or black pigment—*melanin*. This stroma is lined internally by a structureless layer—the *lamina vitrea*, and it supports the vessels of the choroid. The arteries—which are derived from the ciliary branches of the ophthalmic—and the veins lie together in the outer part of the stroma, while the capillaries lie in its deeper part and form there the *tunica Ruyschiana*. The smaller veins converge in whorls—the *vasa vorticosa*—to join four or five principal trunks. The ciliary processes have the same structure as the choroid. Each contains a rich plexus of tortuous vessels. The branched cells at the anterior end of each process are without pigment. Over a considerable area on the inner surface of the choroid the pigment is absent; and there the choroid shines with a peculiar iridescent, metallic appearance termed the *tapetum lucidum*. In the eyes of albinos the choroid is entirely free from pigment.

The *Ciliary Nerves* are efferent branches of the lenticular ganglion. They perforate the sclerotic in company with the ciliary arteries, and run forwards between the sclerotic and cornea. They give branches to the cornea and ciliary muscle, and terminate in the iris. They contain sensory fibres, which are derived from the ophthalmic division of the 5th nerve; motor branches to the ciliary muscle and sphincter muscle of the pupil, which come from the third nerve; and motor fibres to the dilator muscle of the pupil, which are derived from the sympathetic system.

Directions.—In the immersed eye from which the cornea and part of the sclerotic have been removed, the portion of choroid exposed is to be torn away with two pairs of forceps from the subjacent retina. The inner surface of the membrane will be seen, through the transparent vitreous humour, in the submerged half of the eye that was frozen.

The RETINA is the most delicate of the coats of the eyeball. It is formed by the radiation of the optic nerve on the inner surface of the choroid, and like that coat it is bell-shaped. Its external or choroidal surface is covered by a layer of hexagonal pigment cells, which were at one time referred to the texture of the choroid. Its inner surface is moulded on the vitreous humour. This surface shows a little to the inner side of the summit of the bell, or of the antero-posterior axis of the eyeball, a disc-like elevation—the *papilla optica*, which is the point at which the optic nerve begins to expand. In the centre of this spot the *arteria centralis retinæ* appears, and divides into branches which radiate on the inner surface of the retina. The nervous structures of the retina terminate at a wavy line—the *ora serrata*—behind the ciliary processes; but the retina is continued beneath these processes in the form of an epithelial layer—the *pars ciliaris retinæ*, which forms the edge of the bell.

In the human eye a yellow spot—the *macula lutea*—is placed a little external to the papilla optica, and almost exactly in the antero-posterior axis of the eyeball. This is not present in the eye of the horse or in any mammal lower than the quadrumana.

The perfectly fresh retina is translucent and of a pale pink colour, but it speedily becomes opaque. In consistence it is delicate and jelly-like.

Structure.—Ten distinct layers are described as composing the thickness of the retina. These enumerated from within to without are as follows :—

The *Membrana Limitans Interna.*—This, although appearing as a distinct line in a transverse section, is not a distinct stratum, but merely the inner limiting line of a sustentacular framework—the radial fibres of Müller—which pervades and supports the nervous elements in the other layers of the retina.

2. The *Layer of Nerve fibres.*—This layer results from the radiation of the optic nerve, whose fibres at their point of entrance into the eyeball lay aside their medullary sheath.

3. The *Layer of Nerve Cells.*—This is a single layer of multipolar nerve cells.

4. The *Inner Molecular Layer* is a thick stratum of fibres and intermediate granular matter.

5. The *Inner Nuclear Layer* contains spindle-shaped or bipolar nerve cells with distinct oval nuclei and only a small amount of protoplasm. The inner and outer poles of the cells are continued through the 4th and 6th layers respectively.

6. The *Outer Molecular Layer* repeats the structure of the inner molecular layer.

7. The *Outer Nuclear Layer* contains spindle-shaped cells with conspicuous nuclei and a small amount of protoplasm, the poles of the cells being prolonged as in the case of the similar elements in the inner nuclear layer.

8. The *Membrana Limitans Externa.*—This is the outer boundary of the sustentacular framework of fibres already mentioned.

9. The *Layer of Rods and Cones*, or the *bacillary layer*, is composed of two different kinds of elements. The longer elements, the rods, extend vertically between the 8th and 10th layers; the cones are much shorter than the rods, and do not reach so far as the next layer.

10. The *Pigmented Epithelium.*—This is a layer of polygonal pigmented cells, generally six-sided.

Directions.—The third eye should be transversely divided with a sharp scalpel, about half an inch behind the junction of the sclerotic and cornea. This should be done with the eye immersed in water. The posterior half, after removal of the vitreous humour, should be used for the better examination of the inner surface of the retina. The lens should be removed for examination from the anterior half. In the eye already used for the display of the retina, that coat should be in part removed, so as to display the vitreous humour with the lens imbedded in its anterior part. By a combined examination of all the preparations, the following points regarding the lens and vitreous body may be made out.

The LENS is situated behind the pupil, and is contained within a capsule of its own.

The *Capsule* is a close-fitting, firm, transparent membrane, which is four or five times thicker on the front than on the back of the lens. The anterior surface of the capsule forms the posterior boundary of the cavity in which the aqueous humour is contained, and the iris in its movements glides on it. At its periphery the suspensory ligament of the lens blends with it. The posterior surface is in contact with the vitreous humour.

The lens is a transparent solid body of a biconvex shape, the convexity of its posterior surface being considerably greater than that of the anterior. It is maintained in a depression on the front of the vitreous humour by a *suspensory ligament*. This ligament, which is also known as the *zonula of Zinn*, arises behind and beneath the ciliary processes, where it is connected with the hyaloid membrane of the vitreous humour. It passes over the rim of the lens, and blends with the anterior part of the lens-capsule. Behind the rim of the lens the ciliary processes rest on the outer surface of the ligament; and when these are removed, the ligament is there seen to have a fluted or plaited appearance, each plait fitting into the depression between two processes. At this same point the inner surface of the zonula forms the outer boundary of a triangular chink which runs round the lens behind its rim. This is the *canal of Petit*, which is bounded in front by the lens-capsule, behind by the hyaloid membrane of the vitreous humour, and outwardly by the zonula.

Structure.—When removed from its capsule, the lens is found to be soft and pulpy in its outer portion, but its density increases in passing from the surface to the centre. Both its surfaces show some faint white lines radiating from the central point of the surface. The number of these lines varies in the adult, but in the fœtus they are three in number, and each line on the posterior surface is in position midway between two of the anterior lines.

A lens that has been hardened in spirit or by boiling may be broken down into concentric laminæ like the coats of an onion. Each of these laminæ is composed of long riband-shaped fibres. These *lens-fibres* when examined microscopically are seen to have finely serrated edges by which adjacent fibres are interlocked.

The fœtal lens is nearly spherical, it is of a reddish colour, and not quite transparent. In the young adult it is distinctly biconvex, firm, colourless, and transparent. With advancing age it tends to become flatter, denser, less transparent, and of a yellowish colour.

The VITREOUS HUMOUR occupies four-fifths of the interior of the eyeball. It is globular in form, with a depression in front for the lodgment of the lens. It is colourless, transparent, and of a consistency like thin jelly. It is enveloped by a delicate capsule—the *hyaloid membrane*, which is connected in front with the suspensory ligament of the lens, and ends by joining the capsule behind the lens. .

Structure.—The vitreous humour is composed of branched connective-tissue corpuscles in a jelly-like matrix.

CHAPTER VIII.

THE EAR.

THE organ of hearing consists of three divisions—the external, the middle, and the internal ear. The first of these comprises the osseous external auditory process, and the trumpet-like organ which collects the waves of sound and transmits them along that process to the middle ear. It is described at page 159.

The middle and the internal ear are cavities excavated in the substance of the petrous temporal bone. From their situation and the minuteness and intricacy of their parts, their dissection is extremely difficult. The student is therefore recommended to study the anatomy of these parts on the models and special dissections to which he is likely to have access, and by the aid of the fuller description given in systematic text-books. At the same time, an outline description will be here given, which the student may illustrate to himself by procuring two or three petrous temporal bones and dissecting them after they have been decalcified in a hydrochloric or chromic acid solution.

THE MIDDLE EAR.

The MIDDLE EAR—called also the TYMPANUM, or drum of the ear—is a cavity of the petrous temporal bone. It contains air, and across it there stretches a chain of minute bones, which transmit the sound waves from the outer to the inner ear. The inner wall of the chamber is formed by that portion of bone in which the divisions of the internal ear are excavated, and it shows the following objects :—The *promontory*—a projection, or bulging, which corresponds to the first turn of the cochlea. Above the promontory, the *fenestra ovalis*—an opening which is closed by the base of the stapes (the innermost of the auditory ossicles). Below the promontory, another opening—the *fenestra rotunda*, which is closed by a thin membrane. A pin passed through the fenestra ovalis, would enter the vestibular division of the internal ear ; if passed through the fenestra rotunda, it would penetrate the scala tympani of the cochlea. The outer wall of the chamber is formed mainly by the *membrana tympani*. This is a thin, translucent membrane which forms the septum between the tympanum and the outer ear.

The rim of the membrane is fixed in a groove of the bone. The membrane is slightly cupped towards the outer ear; while its inner surface is convex, and has the handle of the malleus (the outermost ossicle) attached to it. The surfaces of the membrane are inclined so that the outer surface looks somewhat downwards, and the inner upwards. In structure the membrane comprises (1) a middle fibrous stratum, the fibres being arranged both radially and circularly, with (2) an outer and (3) an inner epithelial covering. The roof and the floor of the tympanum present nothing of interest. The former is the more extensive. The anterior extremity of the chamber shows a fissure by which air is admitted from the Eustachian tube. Through this opening also the mucous lining of the cavity is continuous with that of the Eustachian tube. The posterior extremity, and part of the floor and outer wall communicate with the cellular spaces of the mastoid protuberance.

The *Auditory Ossicles.*—There are three of these, viz., the malleus, the incus, and the stapes.

The MALLEUS, named from its resemblance to a hammer, is the largest bone. It possesses a *head*, a *handle*, and two *processes*. The *head* is articulated by a synovial joint to the stapes. The *handle* is fixed on the inner surface of the membrana tympani. The *long process* is slender, and projects forwards to be fixed in a slit of the petrous temporal. The *short process* is a mere projection of the root of the handle, and is fixed to the membrana tympani.

The INCUS is named from its supposed resemblance to an anvil, but it has more likeness to a human bicuspid tooth. It presents a *body* and two *processes*, or *crura*. The *body* has a saddle-shaped articular facet for the malleus. The *short process* is directed backwards to be fixed to the wall of the tympanum. The *long process* curves downwards and inwards to terminate in a rounded point—the *orbicular process*, which articulates with the head of the stapes.

The STAPES is stirrup-shaped. It is the smallest bone, and possesses a *head*, a *neck*, a *base*, and two *crura*. The *head* is depressed for articulation with the orbicular process, and is succeeded by the slightly constricted *neck*. The *base* is a thin plate which closes the fenestra ovalis. The *crura* are slender rods of bone connecting the base and the neck.

Muscles of the Ossicles.—These are two—the tensor tympani and the stapedius. (The so-called *laxator* tympani is now believed to be a ligament.)

The TENSOR TYMPANI *arises* from the petrous temporal bone near the Eustachian orifice, and it is *inserted* by a slender tendon into the handle of the malleus near its root.

Action.—To tense the membrana tympani.

The STAPEDIUS *arises* within the *pyramid*—a small process of bone at the back of the tympanum. Issuing from the pyramid, it is *inserted* into the neck of the stapes. Its tendon of insertion contains a small nucleus of bone.

Action.—To regulate (diminish the excursions of) the movements of the stapes.

BLOODVESSELS. The arteries of the tympanum are derived from the *tympanic artery*, a branch of the internal maxillary artery.

NERVES. The *chorda tympani* branch of the 7th nerve enters the cavity of the tympanum from the aqueduct of Fallopius; and passing across the membrana tympani it leaves the cavity by the styloid foramen. The sensory nerves of the tympanum are derived from the *tympanic branch* (Jacobson's nerve) of the glosso-pharyngeal.

The *Nerve to the Stapedius* is a branch of the 7th.

The *Nerve to the Tensor Tympani* comes from the 5th, through the otic ganglion.

THE INTERNAL EAR.

The INTERNAL EAR, called also, from its complexity, the LABYRINTH, consists of a series of chambers, or passages, in the petrous temporal bone, and of certain fluids and soft textures contained within these passages. The chambers, with the wall of condensed bone tissue which immediately surrounds them, constitute the *osseous labyrinth*; the contained soft structures form the *membranous labyrinth.* The osseous labyrinth consists of three divisions: —the *vestibule*, the *cochlea*, and the *semicircular canals*, and each of these contains a division of the membranous labyrinth.

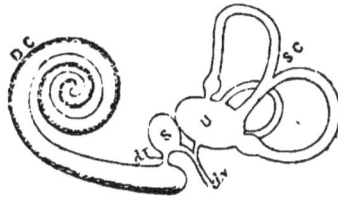

FIG. 34.

DIAGRAM OF THE MEMBRANOUS LABYRINTH.
DC. Ductus cochlearis; *dr.* Ductus reuniens; S. Sacculus; U. Utriculus; *dv.* Ductus vestibuli; SC. Semicircular canals. (*Turner,* after *Waldeyer*).

The VESTIBULE. This is the central division of the labyrinth. It lies between the inner wall of the tympanum and the internal auditory meatus. In front it communicates with the scala vestibuli of the cochlea, and the semicircular canals open into it behind by five openings. On its outer wall, which separates it from the tympanum, is the fenestra ovalis, closed by the base of the stapes. On its inner wall in front there is a depression—the *fovea hemispherica*— placed over the meatus auditorius internus, and pierced by minute foramina for the passage of the filaments of the auditory nerve. Behind the fovea hemispherica is a small slit which leads into the *aqueductus vestibuli.* The roof of the vestibule shows another depression—the *fovea hemi-elliptica.*

Contained immediately within the osseous vestibule there is a quantity of limpid, serous fluid—the *perilymph*, which surrounds the parts of the *membranous labyrinth* here found. These are two delicate sacs—the *saccule* and the *utricle.*

The *Saccule* is the anterior and smaller of the two sacs, and is lodged in the fovea hemispherica. It contains a fluid termed the *endolymph.* It communicates with the membranous canal of the cochlea by a minute tube—the *canalis reuniens,* and with the utricle by a Y shaped tube—the *ductus vestibuli,* the stem of which ends blindly in the aqueduct of the vestibule.

The *Utricle,* placed above and behind the saccule, is lodged in the fovea hemi-elliptica. Like the saccule, it contains endolymph. It communicates, as aforesaid, with the saccule; and the five openings of the membranous semicircular canals open directly into it.

The interior of both saccule and utricle is elevated into a ridge—the *crista acoustica,* in which are distributed the terminal filaments of the vestibular division of the auditory nerve. On this crest are certain peculiar cells, each having a peripheral hair-like process which projects into the endolymph, and a central process which is probably continuous with a filament of the auditory nerve. Here are also found the *otoliths,* which are minute calcareous particles imbedded in a jelly-like material.

The SEMICIRCULAR CANALS are placed behind the vestibule. They are three in number and are distinguished as *superior, posterior,* and *external.* The two first have a vertical direction, while the latter is nearly horizontal. Each canal opens into the vestibule by a dilated extremity, termed the *ampulla.* The non-ampullated end of the external canal opens by an independent orifice into the vestibule, while the non-ampullated ends of the other two canals have a common opening into the same cavity. The three canals have thus five openings into the vestibule, and three of these openings are ampullated.

The *Membranous Semicircular Canals.*—Contained immediately within the osseous canals is a quantity of *perilymph,* which surrounds the membranous canals. Each of these repeats the form of the osseous canal in which it is lodged; and they communicate with the utricle by five openings, three of which are ampullated. The membranous canals contain endolymph, and the ampullated end of each is raised inwardly into a ridge, or *acoustic crest,* having hair cells, otoliths, and nerve terminations similar to those of the saccule and utricle.

The COCHLEA is named from its resemblance to a snail's shell. It has the form of a slightly tapering tube wound spirally two and a half times around a central axis—the *modiolus.* It is thus somewhat conical in form, the base lying inwards near the internal auditory meatus, from which point the axis of the cone is directed outwards, forwards, and downwards to the apex. Projecting half way into the tube of the cochlea is a lamina, or shelf, of bone—termed the *osseous spiral lamina.* The tube is thus imperfectly divided into two passages, termed respectively the *scala tympani* and the *scala vestibuli.* The separation between these two passages is rendered more complete, and a third passage is marked off, by certain membranous structures. These are the *basilar membrane* and *Reissner's membrane.* The *basilar membrane* stretches from the free edge of the osseous spiral lamina to the outer wall of the tube, where it joins a thickening of the lining of the tube, termed the *spiral ligament.* *Reissner's Membrane* is much more delicate, and stretches from the *crista spiralis* at the free edge of the osseous spiral lamina, obliquely upwards and outwards to the wall of the tube.

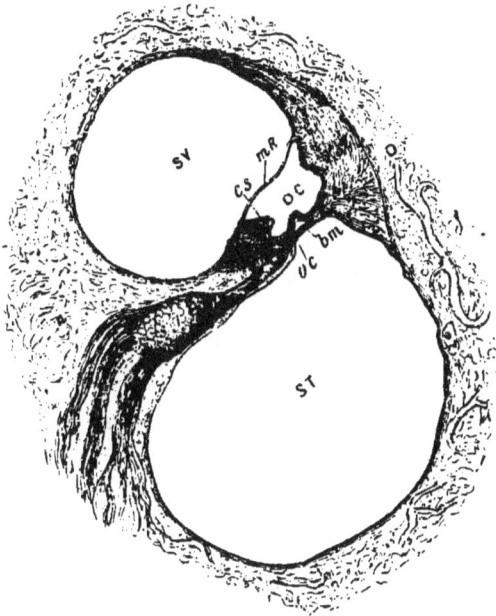

FIG. 35.

TRANSVERSE SECTION THROUGH THE TUBE OF THE COCHLEA.

m. Modiolus; O. Outer wall of cochlea; SV. Scala vestibuli; ST. Scala tympani; DC. Ductus cochlearis; *m*R. Membrane of Reissner; *bm.* Basilar membrane; *sc.* Crista spiralis; *sl.* Spiral ligament; *sg.* Spiral ganglion of auditory nerve; *oc.* Organ of Corti (*Turner*).

The tube is thus divided into three passages, viz., the *scala tympani,* the *scala vestibuli,* and the *scala intermedia.*

The *Scala Tympani* is the largest of the three passages, and is separated from the other two by the osseous spiral lamina and the basilar membrane. At the base of the cochlea it begins at the fenestra rotunda, by which, in the dried bone, it communicates with the tympanum.   At the apex of the cochlea it communicates with the scala vestibuli by a small opening—the *helicotrema*.

The *Scala Vestibuli* is separated from the preceding by the osseous spiral lamina, and from the scala intermedia by Reissner's membrane. At the apex of the cochlea it communicates with the scala tympani by the helicotrema, and at the base it opens freely into the osseous vestibule. Like the vestibule, it therefore contains perilymph, and this passes also by the helicotrema into the scala tympani.

The *Scala Intermedia* is the smallest but the most important of the three passages.   It is the true *membranous cochlea*, and is called also the *ductus cochlearis*.   It is separated from the scala vestibuli by the membrane of Reissner; and from the scala tympani mainly by the basilar membrane, but partly by the osseous spiral lamina near its free edge. At the base of the cochlea it communicates by the slender *canalis reuniens* with the sacculus, and it thus contains endolymph.

The terminal filaments of the cochlear division of the auditory nerve are distributed in the substance of the basilar membrane; and on that surface of the membrane which is directed towards the scala intermedia, there occurs a peculiar arrangement of cells, termed the *organ of Corti*.

The *Organ of Corti*.—When the basilar membrane is examined in transverse section, it is seen to support about the centre of the surface directed towards the scala intermedia a double row of elongated rod-like cells, termed *Corti's rods*.   The rods of the two rows, where they rest on the basilar membrane, are separated by a slight interval; but they incline towards each other and meet at the opposite extremity, so as to enclose a minute canal—the *canal of Corti*.   On the outer side of the external row of rods, the basilar membrane supports four or five rows of shorter cells, the free extremity of each of which bears a tuft of stiff, hair-like processes.   In the same way the membrane supports a single row of hair-bearing cells on the inner side of the inner rods of Corti.   On either side these *hair-bearing cells* are succeeded by cells which become progressively shorter and pass into the general columnar cell lining of the scala intermedia.   A delicate cellular membrane—the *membrana reticularis*—is spread over the outer hair-bearing cells. Through apertures in this membrane, the tufts of hair-like processes project, in a manner comparable to tufts of grass springing through the interstices of a wire net.   Still another membrane—the *membrana tectoria* —springs from the edge of the osseous spiral lamina between the lines of origin of the basilar and Reissner's membranes, and passes outwards over the organ of Corti.

The AUDITORY (8th) CRANIAL NERVE.   This nerve enters the internal auditory meatus in company with the 7th, which passes into the aqueduct of Fallopius.   The 8th divides into two branches, one for the cochlea, the other for the vestibule and semicircular canals.   The filaments of the latter branch penetrate the minute foramina seen at the bottom of the internal auditory meatus, and are finally distributed in the saccule, utricle, and ampullated ends of the membranous semicircular canals. The cochlear branch penetrates the modiolus, and in its passage detaches twigs which pass outwards in the osseous spiral lamina to reach the basilar membrane. Within the spiral lamina there are numerous ganglion cells placed on the course of the nerve fibres.

CHAPTER IX.

DISSECTION OF THE PERINÆUM IN THE MALE.*

UNDER this section there will be described not only the perinæum proper, but also the scrotum, testicle, prepuce, and penis. The dissection of all these must precede that of the hind limb and abdomen, and it should therefore be begun without delay.

THE PERINÆUM.

Position.—Place the animal on the middle line of its back, and draw its hind legs upwards and outwards by ropes running over pulleys fixed to the ceiling. The posterior extremity of the trunk should be level with, or project slightly over, the end of the table on which the subject rests. Empty the posterior part of the rectum, and stuff it with tow saturated in some preservative solution. A stitch should then be put through the edges of the anus.

Surface-marking.—The deep boundaries of the perinæum are those of the outlet of the pelvis (page 341), but its superficial boundaries are as follows :—Above it is limited by the root of the tail, on each side it is bounded by the semimembranosus muscle, and inferiorly it is continued without any limit into the cleft between the thighs.

On the middle line below the root of the tail is the anus. This forms an eminence more pronounced in the young, than in the old, animal. The integumental covering of the eminence is thin, puckered, and hairless; and it is generally dark-pigmented. Passing between the rectum and the root of the tail on each side, and most distinct when the latter is forcibly elevated, there is a projection caused by the so-called suspensory ligament of the rectum. Beneath the anus there can be seen or felt a longitudinal prominence formed by the urethra; and on the middle line of this, there is a median raphe which is prolonged between the thighs.

Directions.—Make a mesial incision through the skin for a length of six inches below the anus. Carry this incision round the sides of the anus, and up to the root of the tail. Make another incision transversely from one tuber ischii to the other. These incisions will enable

* The description of the perinæum in the female is incorporated with that of the pelvis.

sufficient skin to be raised as four triangular flaps. Around the anus there is a quantity of fat, whose amount varies with the condition of the subject, but is greater in the young, than in the old, animal. In this fat the perineal nerves are to be followed.

PERINEAL CUTANEOUS NERVES.—1. *Hæmorrhoidal Branch of 5th Sacral Nerve.* This nerve will be found emerging at the hinder edge of the coccygeal origin of the semimembranosus, and curving downwards and inwards at the root of the tail. It supplies the skin there, and gives some twigs downwards to the skin of the anus.

2. *Hæmorrhoidal Nerve.* The trunk of the hæmorrhoidal nerve, which cannot be reached at present (page 343), divides between the sacro-sciatic ligament and the retractor ani. Its branches are as follows :—1. A branch appears at the inner side of the coccygeal origin of the semimembranosus, and is distributed at the side of the anus. 2. External to the preceding a branch perforates the semimembranosus; and descending over the tuber ischii, it is distributed at the side of the penis. 3. About an inch or two below the anus a branch appears near the middle line, and descends over the urethra.

3. *Pudic Nerve.* Ascending on the side of the anus, beneath the branches of the hæmorrhoidal nerve, are some twigs from the pudic nerve. They terminate in the skin and the sphincter ani.

PERINEAL FASCIA. The lower part of the perinæum is covered by two layers of fascia, viz., a superficial and a deep. The *superficial layer* is attached laterally to the fascia covering the muscles on the inside of the thigh, towards the anus it loses its aponeurotic character and becomes cellular, and inferiorly it blends with the dartos. The *deep layer* is reflected upwards at each side of the penis, while above and below it loses its distinctness and becomes cellular.

Directions.—These layers of fascia should be removed, and the parts should be cleaned after the manner of Plate 37. Beneath the deep layer a branch of the pudic nerve will be found descending on the accelerator urinæ muscle. The transversus perinæi, if present (it was absent in the subject from which the Plate was taken), will be found concealing the internal pudic artery, and may be removed on one side.

The INTERNAL PUDIC ARTERY. This vessel is a branch of the internal iliac artery (Plates 46 and 47). It descends obliquely along the side of the pelvis, on the inner side of the sacro-sciatic ligament or within its texture. At the small sacro-sciatic foramen it passes backwards and inwards to turn round the ischial arch. It penetrates the urethral bulb, immediately resolving itself into a number of branches that supply the erectile tissue of that body. Its position should be particularly noted with reference to the operation of lithotomy, in which, by making a mesial incision, the urethra may be opened without danger of wounding the artery.

In this part of its course the vessel gives off small branches to the anus and to the erector penis muscle.

The INTERNAL PUDIC VEIN accompanies the artery.

The SPHINCTER ANI EXTERNUS. The fibres of this muscle are of the striped variety, and they are circularly disposed around the anus. Above the anus the fibres are fixed at the root of the tail, and below it they unite to form a pointed slip inserted into the perineal fascia. The muscle should be removed in order to expose the next.

The SPHINCTER ANI INTERNUS. This is comprised between the outer muscle and the mucous membrane. Its fibres are circularly disposed like those of the external sphincter, from which they differ in being of the non-striped variety. They are, in fact, nothing more than the last of the circular muscular fibres of the rectum; and in the horse they are not aggregated in the form of a ring, as they are in man.

Action of the sphincters.—To maintain the anus closed except during the passage of excreta.

The RETRACTOR ANI (*Levator ani* of human anatomy). This muscle is red like the external sphincter. It arises (but this cannot be seen at present) from the superior ischiatic spine, and from the inner surface of the sacro-sciatic ligament over the small sacro-sciatic foramen. Its fibres pass upwards and backwards, and terminate in tendinous slips that are insinuated beneath the anterior edge of the external sphincter.

Action.—During the passage of fœces the anus is carried backwards and everted, and the action of this muscle is to carry the anus forwards and invert it after the act of defœcation.

The RETRACTOR PENIS. This muscle descends at the side of the rectum, immediately in front of the external sphincter, and under cover of the termination of the retractor ani, which must therefore be raised and turned forwards. The fibres of the muscle are non-striped, and they form a narrow riband which arises from the 1st and 2nd or 2nd and 3rd coccygeal bones. The right and left bands meet below the rectum, for which they thus form a kind of sling. They are then prolonged downwards on the middle line of the corpus spongiosum, on which they are lost near the extremity of the penis.

Action.—To retract the penis within the prepuce when erection passes off.

The SUSPENSORY LIGAMENT of the RECTUM (Plate 46). This, although denominated a ligament, is composed of non-striped muscular tissue. It is derived from the longitudinal muscular fibres of the rectum, which it leaves in front of the external sphincter; and passing upwards, it becomes inserted into the 4th and 5th coccygeal vertebræ. It forms at the root of the tail a prominence which has already been referred to.

The TRANSVERSUS PERINÆI. This muscle is not constantly present. It arises from the tuber ischii, behind the origin of the erector penis;

and it passes transversely inwards to terminate on the middle line over the urethra, being confounded with its fellow of the opposite side, and with the first fibres of the accelerator urinæ.

Action.—To dilate the bulbous part of the urethra.

The ACCELERATOR URINÆ. These muscles (right and left) cover the sides and lower face of the urethra from the ischial arch to the free extremity of the penis. Along the inferior median line of that tube the right and left muscles are joined by an intermediate fibrous raphe. From this raphe the fibres pass round the urethra on each side, with a slightly forward inclination, and are lost on the upper aspect of the tube, but without reaching the middle line. Beneath the anus the first fibres of the muscle seem to arise from the retractor penis muscle, but elsewhere the retractor is superficial to the intermediate raphe of the right and left muscles.

Action.—The muscles of opposite sides always act together; and when they do so, they diminish the calibre of the urethra and expel its contents. In this way they are instrumental in the ejaculation of semen. In micturition the muscles ordinarily do not come into play until the close of the act, when they empty the urethra from behind to before. The necessity for this action exists because the expelling power of the bladder is lost as soon as its own cavity is emptied.

The ERECTOR PENIS. This is a thick, dark-red muscle covering the crus penis. Its fibres arise from the inferior ischiatic spine (of the tuber), and they terminate on the crus.

Action.—To aid in erecting the penis by compressing the crus and thus retarding the return of blood from the cavernous body of the penis.

THE SCROTUM (PLATE 37).

Position.—Let the subject remain in the dorsal position, but unfasten the rope from one of the hind limbs, and allow the trunk to incline to the same side. The loose limb should be fastened backwards out of the way.

Directions.—Grasp the neck of the scrotum close to the wall of the abdomen, so as to tighten the skin over the testicle, and then tie a piece of soft cord round the constricted neck of the scrotum. This will facilitate the dissection of the different layers of the pouch.

The scrotum is the bag, or pouch, in which the testicles are suspended. It is laminated, and comprises the following layers:—

1. The SCROTAL INTEGUMENT. This is continuous with the surrounding skin, of which it is a modified portion. It is thin, with short fine hairs, and numerous sebaceous glands, whose secretion renders it moist. It is traversed mesially by a raphe, continuous posteriorly with the median raphe of the perinæum. The scrotal integument forms a single bag for the two testicles.

2. The DARTOS.   If a portion of skin be removed from the scrotum, it will expose a reddish-yellow layer, composed of connective-tissue with many elastic fibres and a considerable quantity of involuntary muscular tissue.   This is the dartos, and, like the remaining tunics of the testicle, it forms two distinct pouches, one for each testicle.   In the mesial plane, over the median raphe, the right and left pouches are applied together and form the *septum scroti;* but superiorly they separate to allow the penis to pass between them.   Traced upwards, the dartos is continuous around the external abdominal ring with the subcutaneous fascia.   Under the contraction of the muscular tissue of the dartos, the scrotum becomes firm and wrinkled; during relaxation the scrotum is smooth and pendulous.

3. The SPERMATIC FASCIA, continuous with the tendon of the external oblique tendon.

4. The CREMASTERIC FASCIA, continuous with the internal oblique muscle.

5. The INFUNDIBULIFORM FASCIA, continuous with the transversalis fascia.

6. The TUNICA VAGINALIS REFLEXA, a layer of serous membrane continuous with the peritoneum of the abdominal cavity.

In Plate 37 these layers are semi-diagrammatically represented as succeeding each other like the coats of an onion.   The dissector will probably be unable to discriminate layers 3, 4, and 5.   The tunica vaginalis reflexa, he will recognise as a semitransparent layer which, when cut through, takes him into a smoothly lined pouch in which the testicle lies free. This is the *sac of the tunica vaginalis*, a diverticulum of the peritoneal cavity.

In the fœtus the testicles make their first appearance in the sublumbar region, close behind the kidneys.   As development proceeds, they descend through the abdominal wall into the scrotum ; and hence the correspondence between the coverings of the testicle and the layers that compose the wall of the abdomen.

If the dissector will now lay hold of the testicle, and endeavour to drag it out of the opening which he has made in its coverings, he will bring into view the spermatic cord.   The testicle, he will observe, is covered by a glistening serous membrane, the *tunica vaginalis propria*, which he can trace upwards on the cord.   This spermatic cord contains the vessels, nerves, and excretory duct (vas deferens) of the testicle, which structures descend through the abdominal wall by an oblique passage termed the *inguinal canal*.   In the upper part of this canal, which is not to be exposed at present, the tunica vaginalis propria of the cord is continuous with the tunica vaginalis reflexa.

The CREMASTER MUSCLE is continuous with the cremasteric fascia already described.   It is a bright red muscle, placed in the cord beneath its serous covering.   It passes upwards through the inguinal canal, in which its connections will be observed at a later stage.

The SPERMATIC VESSELS.   The *spermatic artery* is an important vessel from the hæmorrhage to which it may give rise in castration.   It is placed in the anterior part of the cord, and in a well-injected subject its remarkably convoluted disposition will be evident without dissection. The *spermatic veins* accompany the artery.   They are large and tortuous.

The VAS DEFERENS is the excretory duct of the testicle, and is placed at the posterior part of the spermatic cord, where it may be seen and felt as a thick, firm tube.

Directions.—The student, having identified these different elements of the cord, may practise the operation of castration by any one of the common methods, taking care to sever the spermatic cord just above the epididymis, at the upper border of the testicle.   The cord is to be left in the inguinal canal.

THE TESTICLE AND EPIDIDYMIS (PLATES 46 AND 47).

The TESTICLE is the gland that secretes the *semen*—the male fertilizing fluid.   In form it is ovoid.   Its faces, right and left, are smooth and rounded; its inferior border is slightly convex and free; its upper edge is nearly straight, and is related to the epididymis.   Its anterior extremity shows below the globus major of the epididymis a small cystlike body—the *pedunculated hydatid of Morgagni.*

The EPIDIDYMIS is made up of the convolutions of the excretory tube of the testicle.   It presents anteriorly an enlargement termed the *globus major*, and posteriorly a lesser enlargement termed the *globus minor*, the intermediate part being called the *body.*   At the globus minor the tube loses its convoluted disposition, and is continued as the vas deferens, which, as already seen, becomes one of the constituents of the spermatic cord.

STRUCTURE.   The testicle has for its most external investment the *tunica vaginalis propria.*   This, as already explained, is a serous membrane which passes on to the testicle from the cord, and is continuous with the peritoneum at the upper opening of the inguinal canal.   It is, as it were, the visceral part of a serous membrane, the *tunica vaginalis reflexa*—the inner lining of the bag in which the testicle lies free— being the parietal portion of the same membrane.   This covering is thin and transparent, and closely adherent to the next covering—the *tunica albuginea.*   The *tunica albuginea* is a complete envelope of dense, lamellated connective-tissue, containing some fibres of non-striped muscular tissue.   Towards the upper and anterior part of the testicle, a strong process from the tunica albuginea passes into the interior of the gland.   This is termed the *corpus Highmori*, or *mediastinum testis;* and

between it and the inner surface of the tunic, numerous trabeculæ pass,
forming a framework for the gland, and dividing it into a number of
conical compartments, or *lobules*, which lodge the *seminal tubules*. On
the inner surface of the tunica albuginea, and on its trabeculæ, the
bloodvessels are distributed, forming the *tunica vasculosa*.

Each seminal tubule begins either with a blind extremity, or by anastomosing with an adjacent tubule. The tubes are highly convoluted
until they approach the mediastinum, where they unite to form a series
of straight tubes—the *tubuli recti*, which enter the mediastinum and
form in it a network—the *rete testis*. From this network arise a number
of tubes termed the *vasa efferentia*, which perforate the tunica albuginea
above the anterior end of the testicle. On leaving the gland, these
become convoluted, forming little masses known as the *coni vasculosi*;
and they then unite with one another until there results a single excretory tube, whose convolutions make up the globus major, body, and
globus minor of the epididymis. The seminal tubules are composed of
a membrana propria and an epithelial lining. The epithelium is
arranged in several layers, and through the agency of the innermost
cells—*spermatoblast cells*—the spermatozoa of the semen are produced.
The tubuli recti and rete testis are lined by a single layer of columnar
epithelium. The tubes of the vasa efferentia and epididymis have a
wall that contains non-striped muscular fibres, and they possess a columnar ciliated lining.

THE PREPUCE.

The prepuce, vulgarly called the "sheath," is the involution of skin
which lodges the free portion of the penis when that organ is non-
erect. In this condition it consists of two layers—an external, similar
to the surrounding integument, with which it is continuous; and an
internal, which is intermediate in texture between skin and mucous
membrane. The latter layer is smooth, destitute of hair, and provided
with numerous preputial glands, which secrete a strong-smelling sebaceous
material. This material facilitates the protrusion of the penis during
erection; and, ordinarily, it accumulates in considerable amount within
the prepuce. These two layers are continuous with one another at the
orifice of the preputial cavity, and at the posterior end of the cavity the
inner layer is continuous with the investment of the penis. Towards
the orifice of the preputial cavity, two rudimentary tubercle-like teats
are sometimes found. Lay hold of the extremity of the penis, and pull
it forcibly forwards, at the same time pulling the prepuce backwards.
This will obliterate the prepuce, as in erection, in which condition the
inner layer of the prepuce becomes a part of the covering of the penis.

Directions.—While the penis is pulled forwards out of the prepuce,
carry a mesial incision through the skin from the perinæum to the

entrance of the prepuce, and reflect the skin on each side for three or four inches.

SUSPENSORY LIGAMENTS of the prepuce. When the outer cutaneous layer of the prepuce is removed, there is exposed an elastic fibrous layer which descends into it on each side from the abdominal tunic. These are the suspensory ligaments of the prepuce.

VESSELS and NERVES. The *cutaneous nerves* of the prepuce and scrotum are branches of the *inguinal* nerve or nerves. One or more of these, derived from the 2nd and 3rd lumbar nerves, descend through the inguinal canal. The arteries are branches of the *subcutaneous abdominal artery*. This vessel, which is a branch of the external pudic artery, passes forwards a few inches from the middle line. The trunk of the artery is to be left undisturbed at present.

A rich plexus of veins exists in and around the scrotum. This plexus is drained by a comparatively small vein that accompanies the external pudic artery, and by a larger vessel which penetrates the gracilis to empty itself into the femoral vein.

THE PENIS.

Directions.—While the penis is pulled forwards, reflect the integumental covering from the upper face of its free portion, and follow backwards its dorsal vessels and nerves.

DORSAL ARTERIES of the penis (Plates 39 and 46). On each side there are two of these, distinguished as anterior and posterior. 1. The *anterior dorsal artery* of the penis is one of the terminal branches of the external pudic artery. It results from the bifurcation of that vessel immediately after its emergence from the inguinal canal, and after a course of a few inches it divides into an anterior branch which passes forwards on the free portion of the penis, and a posterior which passes backwards on the fixed portion, meeting and anastomising with the posterior dorsal artery. When the penis is non-erect, the anterior of these branches has a flexuous disposition, which permits it to be elongated without stretching when the organ becomes erect. 2. The *posterior dorsal artery* of the penis is a branch of the cavernous artery (from the obturator). It runs forwards on the dorsal aspect of the fixed portion of the penis, and anastomoses with the posterior division of the anterior dorsal artery. These arteries are mainly expended in branches to the cavernous and spongy portions of the penis, and they also give off some twigs to the prepuce.

DORSAL NERVES of the penis. These nerves, right and left, accompany the dorsal vessels on the dorsum, or upper surface, of the penis. Each is the continuation of the pudic nerve, which reaches the penis by turning round the ischial arch. In proceeding forwards along the penis, the nerves are disposed in a flexuous manner to allow them to be adapted without stretching to the varying length of the organ. They emit numerous branches to the cavernous and spongy portions of the penis, and terminate in the glans.

SUSPENSORY LIGAMENTS of the penis (Plate 46). These are two fibrous bands, right and left, which are attached superiorly to the tendon of origin of the gracilis, and below to the cavernous body of the penis.

Directions.—The penis may now be freed as far as its posterior extremity, and its surface cleaned of vessels, nerves, and connective-tissue. On one side the erector penis muscle should be removed, to lay bare the crus and expose the artery of the corpus cavernosum.

The ARTERY of the CORPUS CAVERNOSUM (Plate 46). This is a branch of the obturator artery, detached after the emergence of that artery from the obturator foramen. It passes backwards on the lower face of the ischium, and perforates the crus penis. It gives off as a collateral branch the posterior dorsal artery of the penis.

The PENIS (Plates 46 and 47) is the male organ of copulation. It begins at the ischial arch, where it is attached by its *crura* to the ischial tuberosities; and it terminates anteriorly in a free enlargement—the *glans*. It may be said to consist of a posterior fixed portion, and an anterior portion which is free and protrusible. The former portion extends from the ischial arch to the scrotum; the latter, when the organ is non-erect, is lodged in the prepuce, but during erection the prepuce becomes obliterated, and this part of the penis then projects freely in front of the scrotum.

The penis is compounded of three longitudinal and parallel columns, viz., two *corpora cavernosa* and a single *corpus spongiosum*. From the relationship of these to one another, the penis has been happily compared to a double-barrelled gun, the barrels being represented by the corpora cavernosa, and the ramrod by the corpus spongiosum.

The CORPORA CAVERNOSA. Each corpus cavernosum begins at the tuber ischii, to whose inferior ridge (inferior ischiatic spine) it is firmly attached under cover of the erector penis muscle. These constitute the roots, or *crura*, of the penis, and they converge towards each other and form a single mass which makes up the main thickness of the penis as far as the glans. The united corpora cavernosa have an upper flattened surface, or dorsum, along which the dorsal vessels and nerves pass. Their sides are smooth and slightly rounded, and inferiorly they form a shallow median groove for the corpus spongiosum (Fig. 44). Anteriorly they terminate bluntly in the glans.

The CORPUS SPONGIOSUM forms a much more slender column than the corpora cavernosa. It is traversed in the whole of its length by the extra-pelvic part of the urethra. This urethra, as will subsequently be seen, begins at the neck of the bladder, and its first few inches are intra-pelvic, being placed over the ischiatic symphysis. Turning round the ischial arch, the intra-pelvic urethra becomes directly continuous with the extra-pelvic portion, and from the point of continuity onwards the urethra is enveloped in a sheath of erectile tissue, which is the corpus spongiosum.   The corpus spongiosum forms at either of its extremities an enlargement.   The posterior enlargement, which is situated at the ischial arch, is termed the *bulb;* the anterior enlargement is the *glans penis.*   The glans forms the expanded free extremity of the penis, and it surrounds the blunt anterior end of the united corpora cavernosa.   During erection the enlargement assumes a shape resembling, somewhat, the rose of a watering-can, having a prominent ridge—the *corona glandis,* behind which there is a slight constriction—the *cervix.*   The front of this rose-like swelling presents a fossa from which the urethra projects for about half an inch as a free tube—the *urethral tube.*   Above the base of the urethral tube there is the opening of a double cavity—the *urethral sinus,* which generally contains some of the partially inspissated secretion of sebaceous glands that open into the cavity.   Inferiorly the corona glandis is interrupted on the middle line by the *suburethral notch.*

The corpus spongiosum as far as the glans is surrounded by the accelerator urinæ muscle.   Superiorly it fits into the groove on the lower aspect of the corpora cavernosa, and along its under aspect pass the retractor muscles of the penis.

Directions.—Immediately in front of the junction of its crura, the penis should now be amputated, that the structure of its component parts may be examined.

STRUCTURE of the corpora cavernosa.   The corpora cavernosa possess a strong envelope of white fibrous tissue, termed the *tunica albuginea.* This tunica albuginea, besides forming a common envelope to the united bodies, sends inwards an incomplete mesial septum between the two—the *septum pectiniforme.*   This septum when viewed laterally is seen to be perforated by numerous vertical slits, which give its processes a resemblance to the teeth of a comb; hence the name.   Besides the septum pectiniforme, numerous small trabeculæ pass into the interior of the corpora cavernosa, and by their anastomosis form a framework for these bodies.   The trabeculæ are composed of fibrous tissue with some bundles of non-striped muscular tissue.   Between the trabeculæ are innumerable intercommunicating spaces, placed between the capillaries and the small veins.   During erection the blood is poured into these spaces, and thus is brought about the increase in the size of the organ.   At other times the blood passes in the ordinary manner from the capillaries to the venous radicles.   In the crura and peripheral part of the cavernous bodies some of the small arteries terminate directly in these venous spaces.   The small arteries are imbedded in the trabeculæ, and when these are contracted, in the non-erect state, the arteries assume a coiled disposition, from which they receive the name *arteriæ helicinæ*.

STRUCTURE of the corpus spongiosum. The structure of the spongy body resembles, somewhat, that just described. It possesses an envelope of fibrous tissue with trabeculæ and a plexus of large veins. In its peripheral part, and in the bulb, it also contains true cavernous spaces, like those of the cavernous bodies but smaller.

STRUCTURE of the spongy (or extra-pelvic) part of the urethra. This should be laid open on its under aspect with scissors. The lumen of the tube is not uniform. At the ischial arch (this will not be seen at present) it presents a dilatation; and its calibre is again increased as it enters the glans, forming what is termed in man the *fossa navicularis*. The interior of the tube is lined by mucous membrane having simple columnar epithelium, except at its orifice, where it is stratified and squamous. The ducts of numerous small racemose glands open on the surface of the membrane. External to the mucous membrane the wall of the urethra is made up of non-striped muscular tissue, arranged as an inner circular and an outer longitudinal layer.

CHAPTER X.

DISSECTION OF THE ABDOMEN.

BEFORE this part can be begun in the male subject, the dissection of the perinæum (Chapter IX.) must be completed.

THE ABDOMINAL WALL.

Position.—The subject should be placed on the middle line of its back, or slightly inclined to one side, its limbs being drawn upwards and outwards by ropes and pulleys.

The MAMMARY GLANDS, or the UDDER. It is convenient to describe here these glands, since their dissection must precede that of the abdominal wall. They are organs peculiar to the female, occupying the position of the scrotum in the male. As regards their function, they may be viewed as an accessory part of the reproductive system, secreting the milk upon which the young animal subsists for some time after birth. It is only during the period of lactation that they become fully developed, and therefore a subject suited for the satisfactory display of their structure seldom presents itself in the dissecting-room.

The glands are two in number, and are placed side by side on the middle line of the abdominal wall, in front of the pubes. They form here a single mass, with a wide and shallow mesial furrow between them. The term "udder" is used to include both glands. From the most prominent part of each, the *mamilla, teat,* or *nipple,* projects. This has the form of a short, flattened cone. Its free extremity is perforated by two or three orifices belonging to the large milk ducts by which the milk is extracted from the gland. The integumentary covering of both glands and teats is thinner than the surrounding skin, and it is generally black-pigmented. Moreover, the ordinary body hairs are absent over it, their place being taken by a fine down, except over the summit of the teat, where there are no hairs. It is richly provided with sebaceous and sudoriparous glands, whose secretion renders it moist.

When the cutaneous covering of the gland is reflected, there is exposed a second envelope, composed of yellow elastic tissue. This covering detaches a number of processes into the interior of the gland

between its main lobes, and on the mesial plane the elastic envelopes
of the two glands are applied together, and form a kind of intermediate
septum.   A few strong slips of the same texture descend into the gland
from the abdominal tunic, and play the part of suspensory ligaments.

The secretory structure of the gland is arranged on the racemose
type.   If a bristle be passed into one of the orifices seen at the extremity
of the teat, it will pass upwards by the large milk duct, and enter a
dilatation at the base of the teat, termed the *galactopherous* or *lactiferous
sinus.*   The secretion of milk during the period of lactation is constant,
and the liquid accumulates in these reservoirs, to be drawn off by the
young animal.   The milk ducts and the sinuses are lined by a mucous
membrane; and in the substance of the teat, between this mucous lining
and the external skin, there are some fibres of non-striped muscular
tissue, arranged both longitudinally and circularly.   The circular fibres
prevent the escape of the milk from the sinus.

The milk enters each sinus from a number of tubes which, when
traced into the substance of the gland, divide and subdivide ; and the
smallest ducts resulting from this subdivision lead up to the ultimate
acini of the gland structure.   These acini are lined by a secretory
epithelium by whose agency the milk is formed.

The arteries and veins of the glands are branches of the external
pudic vessels.   They undergo a great increase in size during lactation.
The nerves of the gland are branches of the inguinal nerves.

Directions.—Reflect the skin as shown in Plate 38.   If the dissector
of the fore limb be engaged with the pectoral region, the skin from the
posterior part of that region will be turned back in a piece with that
over the front of the abdomen.   If not, the dissector of the abdomen
must limit the skin which he is about to reflect, by an incision carried
outwards from the ensiform cartilage to the point of the elbow.   He
will be guided in the same way towards the hind limb.   Care must be
taken not to reflect the panniculus with the skin.

A slight degree of tympanitic distension of the intestines is favourable
for the dissection of the abdominal wall.   When excessive, however, as
it often becomes, it interferes with the dissection, and is almost certain
to rupture the diaphragm, or the abdominal wall before its dissection
can be completed.   This should be prevented by tapping the large
intestine with a canula and trochar, making the puncture at the most
prominent part.

CUTANEOUS NERVES.   In reflecting the skin, a multitude of small
nerves will be seen on its inner surface.   They are derived from the
intercostal nerves.

The SUBCUTANEOUS ABDOMINAL ARTERY (Plate 38).   Look for this
vessel near the middle line, in the region of the prepuce or mammary
gland.   It is one of the terminal divisions of the external pudic artery,

and is distributed to the scrotum and prepuce (skin of mammary gland in mare), superficial inguinal glands, and skin, terminating a little in front of the umbilicus.

The SUBCUTANEOUS ABDOMINAL VEIN runs in company with the artery.

The SUPERFICIAL INGUINAL LYMPHATIC GLANDS (Plate 38). These form a small group close to the subcutaneous abdominal artery, at the side of the prepuce.

The SUBCUTANEOUS THORACIC (SPUR) VEIN (Plate 38) will be found on the surface of the panniculus. The primary rootlets of the vein collect blood from the skin in front of the mamma or prepuce, and pass on to the surface of the panniculus, where they unite to form the trunk of the vein. This is at first lodged in a groove on the superficial aspect of the panniculus. It then perforates the muscle; and gaining its deep face, it passes forwards towards the axilla (Plate 1), where it joins the brachial vein. The course of this vein is usually distinctly visible in the living animal. From its position it is liable to be injured in deep spurring, and hence one of its names.

The PANNICULUS CARNOSUS (Plate 38). This is a thin extended sheet of muscular tissue, which is adherent to the deep surface of the skin over a large part of the abdomen and thorax, being continued also from the latter region over the outer aspect of the shoulder. The most posterior angle of the muscle is included in the fold of skin at the groin, but it does not reach the hind limb. From this angle the superior edge of the muscle (which will not at present be seen) slopes upwards with two or three wide sinuosities to near the spine in the dorsal region, while from the same point the posterior edge of the muscle slopes downwards and forwards to a second angle which is rounded and placed from three to six inches external to the umbilicus. The inferior edge extends from this latter angle forwards towards the elbow. Anteriorly the muscle is continued over the scapular region, and sends also an aponeurotic tendon between the fore limb and the chest-wall to be attached to the internal tuberosity of the humerus. The edges of the muscle are prolonged by a thin fascia which is attached superiorly to the vertebral spines, and below and behind is adherent to the abdominal tunic. The outer surface of the muscle is with difficulty separated from the skin, which indeed receives the insertion of its fibres. The muscular tissue of the panniculus is, as compared with striped muscles in general, of a pale colour.

Action.—It twitches the skin, and plays the part of a hand to the animal in removing offending insects.

Directions.—Begin at the lower edge of the panniculus and raise it upwards from the subjacent structures. This is easy over the abdominal tunic, but anteriorly it is closely adherent to the edge of the deep pectoral muscle. The panniculus is not to be removed, but raised as far as is necessary to bring the origin of the external oblique muscle of the abdomen into view. Notice on the inner surface of the muscle ramifying nerves, and anteriorly the spur vein accompanied by a small branch of the external thoracic artery.

PERFORATING NERVES. The nerves seen descending on the inner surface of the panniculus are perforating branches derived from the intercostal trunks, and from the last dorsal and first lumbar nerves. These perforating nerves appear along a curved line a few inches below the origin of the external oblique. They supply the panniculus, and give cutaneous twigs through it to the overlying skin.

A perforating branch from the 2nd lumbar nerve appears close to the bony prominence of the haunch, and descends to the skin on the front of the thigh. A perforating branch from the 3rd lumbar nerve appears below the same bony prominence, and two inches below the point of exit of the preceding nerve. It is accompanied by a branch of the circumflex iliac artery, with which it descends to the thigh, internal to the last described branch.

The SUBCUTANEOUS THORACIC NERVE (Plate 1). This will be found running horizontally backwards on the inner surface of the panniculus, behind the shoulder, and in company with the vessels of the same name. It comes from the brachial plexus.

PERFORATING VESSELS. Small un-named branches, mostly branches of the intercostal vessels, appear at the same points as the nerves.

The ABDOMINAL TUNIC (Plate 38). This is a great expansion of yellow elastic tissue which is spread over the inferior and lateral walls of the abdomen. It is nearly co-extensive with the external oblique muscle, to which it is adherent. It is thickest in its posterior part, near the linea alba; and becomes gradually thinner as it is traced outwards over the muscular part of the external oblique, and forwards beneath the posterior deep pectoral. Posteriorly it furnishes the suspensory ligaments of the prepuce, or analogous slips to the mammary gland. The tunic acts as an admirable elastic abdominal bandage, assisting the muscles to support the heavy abdominal viscera, and adapting the wall of the abdomen to the varying volume of its contents.

Directions.—The abdominal tunic must be entirely removed. This is an operation requiring time and care, for the tunic is intimately adherent to the tendon of the external oblique muscle, especially in its anterior half. Transverse incisions should be made through it, taking care not to cut the fibres of the subjacent tendon, which will be recognised by its different colour and texture. Then seize the cut edges of the tunic with the forceps, and tear it off in strips forwards and backwards. Proceed in this way until the whole of it has been torn away.

Muscles of the Abdominal Wall. On each side there are four of these, viz., the obliquus abdominis externus, the obliquus abdominis internus, the rectus abdominis, and the transversalis abdominis. They are stated in the order of their occurrence, the first being the most external. These muscles have not only to discharge the ordinary function of a muscle, but they have also to close in the abdominal cavity; and for this latter purpose, they are, with the exception of the rectus abdominis, peculiarly modified in form. Thus, the two oblique muscles and the transverse muscle have their tendons of insertion extended in the form of great fibrous or aponeurotic sheets, and the fibres in each of these tendons have a direction different from that of the others.

The LINEA ALBA is the white mesial raphe, or band, which extends from the ensiform cartilage to the pubes. It is fibrous in structure, and is formed by the meeting of the aponeurotic tendons of the right and left muscles. A little behind its mid point is a puckered cicatrix—the *umbilicus.*

The EXTERNAL ABDOMINAL RING (Plate 39). This is the lower orifice of the inguinal canal. It has the form of a slit in the tendon of the external oblique. The direction of the slit is oblique forwards and out-wards. The lips, or *pillars*, of the slit are simply fibres of the external oblique tendon. The inner angle or *commissure* is placed at the edge of the prepubic tendon. This prepubic tendon is a strong fibrous band by which the abdominal muscles get a common insertion into the anterior edge of the pubic bones, and from whose surface the pubio-femoral liga-ment of the hip-joint arises. The external abdominal ring gives passage in the male to the spermatic cord, the external pudic vessels, and the inguinal nerves. In the female it transmits merely the corresponding vessels and nerves.

The OBLIQUUS ABDOMINIS EXTERNUS (Plate 39). This consists of a muscular band at its antero-superior edge, and an aponeurotic tendon over the inferior and lateral parts of the abdomen. It *arises* by its muscular portion from the outer surface of the last fourteen ribs, and behind the last rib from the tendon of the latissimus dorsi. Its anterior slips of origin interdigitate with the serratus magnus. The muscular fibres are directed obliquely downwards and backwards, and are suc-ceeded by the aponeurotic tendon. The fibres of the tendon continue in the same direction, and become *inserted* into the linea alba, the prepubic tendon, and the external angle of the ilium ; while between the two last-mentioned points they are continued to form *Poupart's ligament.* Along the line between these two points the fascia of the inside of the thigh is inserted to the surface of the tendon, and it must be cut in order to expose the ligament. It will then be observed that from the prepubic tendon to the bony prominence of the haunch, the fibres of the external oblique tendon, instead of becoming inserted into bone, curve upwards and forwards and are lost to view. It is these reflected fibres that constitute the *ligament of Poupart* (Plate 40), which may be described

U

as having two extremities, two surfaces, and two edges. Its extremities are attached to the pubis and angle of the haunch respectively. Its anterior surface is concave, and directed towards the abdomen. This surface gives origin outwardly to fibres of the internal oblique muscle, and inwardly it forms the posterior wall of the inguinal canal. The posterior surface is convex, and forms an arch over the femoral vessels, the crural nerve, and the sartorius, iliacus, and psoas magnus muscles (Plate 13). Neither of the edges of the ligament has a distinct existence. The posterior or inferior edge is the line of continuity between the ligament and the tendon of the external oblique. At its anterior or superior edge the ligament becomes thin in texture, and disappears on the fascia covering the sublumbar muscles. All of these points cannot be made out at present, but they will become evident as the dissection proceeds.

Action of the external oblique muscle.—When the right and left muscles act in concert, they bend the trunk, and arch the back. If the spine is fixed, they pull the ribs backwards and assist in expiration. If both the spine and ribs are fixed, they compress the abdominal viscera, and assist in urination, defæcation, and parturition. If only one muscle acts, it bends the trunk or pelvis to the same side.

The INGUINAL CANAL is the oblique passage in the abdominal wall through which the testicle descends in the young animal, and in which the spermatic cord is lodged in the adult. The *external abdominal ring*, which has already been examined, is the lower opening of the canal. Its upper orifice, which will be seen at a later stage, is termed the *internal abdominal ring.* The direction of the canal is oblique downwards and inwards, and it is slightly curved with the concavity forwards. Introduce the finger into the canal and press on the posterior wall. This, it will be seen, is formed by the reflected portion of the external oblique tendon—in other words, by Poupart's ligament. Rotate the hand, and press the finger on the anterior wall, at the same time separating the edges of the external abdominal ring. The anterior wall will be seen and felt to be formed by muscular substance, viz., by the muscular part of the internal oblique.

The canal gives passage in the male to the spermatic cord, the external pudic vessels, and the inguinal nerves. In the female it is much smaller, and transmits the corresponding vessels and nerves.

The SPERMATIC CORD. See page 278.

The EXTERNAL PUDIC ARTERY (Plate 39) is one of the terminal divisions of the prepubic. In the inguinal canal it descends posterior and internal to the spermatic cord. After its emergence it divides into the subcutaneous abdominal artery, and the anterior dorsal artery of the penis. In the mare the latter branch is represented by the mammary artery.

The EXTERNAL PUDIC VEIN is proportionally smaller than the artery, which it accompanies.

The INGUINAL NERVES are derived from the 2nd and 3rd lumbar nerves, and are distributed to the prepuce, the scrotum, and the adjacent skin.

Directions.—Incise the external oblique tendon, from the external angle of the ilium to the edge of the prepubic tendon. Reflect Poupart's ligament towards the thigh, and hook it up after the manner of Plate 40. Then strip away the tendon of the external oblique from the subjacent internal oblique. This will be found easy in the region of the flank, where the tendon is related to the muscular part of the internal oblique ; but over the inferior part of the abdomen, and especially in front, where the tendons of the two muscles are applied to each óther, the operation is difficult, and in some parts impossible. In this proceeding the dissector has to guard against removing the thin tendon of the inner muscle along with the outer, and this he will best do by observing that the fibres of the inner tendon cross these of the outer at right angles, being directed downwards and forwards. Observe that anteriorly the two tendons are not simply in apposition, but actually interwoven—a disposition of tendons which is unique, and one which greatly increases the strength of the abdominal floor. The muscular portion of the external oblique should be raised as far as the lower extremities of the ribs. A better view of the inguinal canal and its contents will now be obtained.

The OBLIQUUS ABDOMINIS INTERNUS (Plate 40) consists of a fan-shaped fleshy portion situated in the flank, and an aponeurotic tendon spread over the abdominal floor. It *arises* from the external angle of the ilium, and from the adjacent part of Poupart's ligament. It is *inserted* into the prepubic tendon and the linea alba by the inferior edge of its tendon, and by tendinous slips into the four or five last costal cartilages. In front of the lower end of the fourth last intercostal space, the aponeurotic tendon has a free edge which ordinarily lies under concealment of the line of overlapping costal cartilages. When the abdomen is tympanitic, however, this edge is thrust outwards, and the transversalis muscle is exposed as in Plate 40. The posterior edge of the fan-like muscular portion lies in contact with Poupart's ligament ; and the inguinal canal, as already seen, passes between the two structures. The highest fibres of its muscular part are parallel to the edge of a small muscle—the *retractor costæ*—inserted into the last rib, under cover of the most posterior slip of the serratus posticus. This is described with the muscles of the back (page 96).

Action.—Similar to that of the external oblique.

Directions.—The internal oblique covers the transversalis and rectus abdominis muscles. The outer edge of the last may be seen through the thin tendon of the internal oblique, and through the same tendon the posterior abdominal artery may be seen if well injected

(Plate 40). The circumflex iliac artery is on the deep surface of its muscular portion. In order to see these connections of the muscle to the most advantage, incise the muscle along the line of junction of the muscular fan and the tendon. Raise the muscular portion carefully, and hook it back. Strip away entirely the aponeurotic tendon, using the scalpel where the tendon is firmly adherent to the rectus abdominis.

The CIRCUMFLEX ILIAC ARTERY. This is a branch of the external iliac artery, and will be better seen in the dissection of the sublumbar region (Plate 44). It has an *anterior division* whose branches are distributed to the internal oblique and transverse muscles in the flank, and a *posterior division* which, after giving some twigs to the oblique muscles, perforates them below the angle of the haunch, and descends to the thigh.

The POSTERIOR ABDOMINAL ARTERY (Plate 40). This is a branch of the prepubic artery, beginning at the inner side of the internal abdominal ring. It places itself on the abdominal aspect of the internal oblique muscle, crosses behind and internal to the ring, and runs forwards to enter the rectus abdominis, in which, about midway between the sternum and the pubis, it anastomoses with the anterior abdominal artery.

These arteries are accompanied by veins of the same names.

The RECTUS ABDOMINIS (Plate 40). This muscle extends in the form of a broad band from the sternum to the pubis, at the side of the linea alba. To a large extent it separates the internal oblique and transverse muscles, but beyond its outer border these muscles are in contact in the flank and below the extremities of the ribs. The muscle is widest about its centre, and it is crossed from side to side by a number (about a dozen) of white lines—*lineæ transversæ*, which are caused by as many tendinous intersections of its muscular substance. It *arises* from the lower face of the sternum, and from the five costal cartilages behind the 4th. It is *inserted* into the anterior border of the pubis by the prepubic tendon.

Action.—Similar to that of the oblique muscles.

NERVES (Plate 40). At the lower ends of the last ten intercostal spaces, the *intercostal nerves* are prolonged beyond the rim of overlapping cartilages to pass between the straight and transverse muscles, giving fibres to both and also some perforating twigs to reach the skin. The *last dorsal nerve* (behind the last rib) has a similar distribution. The inferior primary branches of the 1st and 2nd *lumbar nerves* are similarly prolonged after furnishing twigs to the oblique muscles in the flank.

Directions.—Cut the rectus abdominis transversely about the umbilicus, and reflect it forwards and backwards from the subjacent transversalis. Look for the anterior abdominal artery on its deep face.

The ANTERIOR ABDOMINAL ARTERY is one of the terminal branches of the internal thoracic artery. It appears at the side of the ensiform cartilage, where it turns round the 9th costal cartilage behind its tip. It runs backwards along the middle of the superior face of the rectus, giving off lateral branches, and terminating about midway between the sternum and pubis in branches which anastomose with those of the posterior abdominal artery. It is accompanied by a satellite vein.

The TRANSVERSALIS ABDOMINIS (Plate 40). This muscle consists of a fleshy band at its origin, and of an aponeurotic tendon over the abdominal floor. In both of these the direction of the fibres is transversely downwards and inwards towards the linea alba. It *arises* by its fleshy portion from the lower extremities or cartilages of the asternal ribs (last ten), meeting here the origin of the diaphragm; and from the transverse processes of the lumbar vertebræ. It is *inserted* by the inner edge of the aponeurotic tendon into the ensiform cartilage and the linea alba. The posterior edge of the tendon is thin and ill-defined. The inner surface of the entire muscle is related to the parietal peritoneum, there being interposed, however, a very thin layer of connective-tissue representing the *fascia transversalis* of man. Slender branches from the intercostal or asternal vessels run on the peritoneal surface of the muscle.

Action.—Similar to that of the oblique muscles.

Directions.—The abdominal cavity will be exposed by the removal of the transverse muscle and its peritoneal lining. If only one side of the abdominal wall has been dissected, the other side may now be used for the better display of things not satisfactorily made out in the first; and particularly, a portion of the abdominal wall in front of Poupart's ligament should be turned back in its entire thickness, so as to expose its peritoneal aspect and the internal abdominal ring.

The INTERNAL ABDOMINAL RING (Plate 44) is the abdominal opening of the inguinal canal. As seen from the abdominal side, its posterior or outer edge is prominent, and corresponds to the edge of the muscular part of the internal oblique; while the opposite boundary of the ring is flattened over the sublumbar muscles covered by the continuation of Poupart's ligament.

The student can now see the direct continuity between the peritoneum and the tunica vaginalis, the latter membrane passing directly into the inguinal canal, and forming a well-defined edge on the posterior and outer side of the entrance. It is by this opening that a portion of intestine or mesentery sometimes passes into the inguinal canal, or onwards into the scrotum, constituting an inguinal or scrotal hernia.

The PREPUBIC ARTERY (Plate 44). This vessel arises from the femoral artery at the brim of the pubis, forming a short common trunk with the deep femoral branch. It crosses to the edge of the internal oblique, and divides into the external pudic and posterior abdominal arteries. The former enters the inguinal canal at a point internal to the internal abdominal ring. The latter passes behind the ring, and crosses it on the inner side. Both branches have already been followed, but the relation of the posterior abdominal artery to the ring should now be specially noted, as, in consequence of its position, an incision for the relief of a strangulated hernia must be made outwards to avoid wounding the vessel.

The SPERMATIC CORD. The various structures which compose the spermatic cord meet at the internal abdominal ring. The *vas deferens* is seen turning inwards to enter the pelvis, and projecting the peritoneum to form a small band, or *frænum*, for itself. The *vessels* and *nerves* of the cord are to be left undisturbed, so that they may be followed to their source at a later stage.

The CREMASTER MUSCLE (Plate 44). The fibres of this muscle are now seen at their origin from the iliac fascia, where they are close to the muscular fibres of the internal oblique. They pass into the inguinal canal, where, separating but remaining connected by intermediate areolar tissue, they constitute the cremasteric covering of the cord and testicle. When the muscle contracts, it twitches the testicle upwards by shortening the spermatic cord.

THE CAVITY OF THE ABDOMEN.

Boundaries of the Cavity.—The abdomen is the largest of the visceral cavities of the body. It is placed behind the thorax, from which it is separated by the diaphragm; posteriorly it is directly continuous with the cavity of the pelvis; laterally and inferiorly it is enclosed by muscular, tendinous, and elastic textures making up what is generally termed the abdominal wall; and superiorly it is bounded by the lumbar portion of the spine clothed by the sublumbar muscles.

Contents of the Cavity.—The cavity is occupied mainly by the gastro-intestinal part of the alimentary tube, and its associated glands—the liver and the pancreas. Besides these, it lodges the spleen and the kidneys. In the female it contains the ovaries and the uterus (in part), and in the male the vas deferens passes through it.

Divisions of the Cavity.—As a matter of convenience in describing the position of its contained organs, the cavity is arbitrarily divided into the following nine areas:—

left hypochondriac	epigastric	right hypochondriac
left lumbar	umbilical	right lumbar
left iliac	hypogastric	right iliac

This subdivision is quite arbitrary, the boundaries between these areas being certain imaginary planes. Thus, the three anterior regions are separated from the three middle regions by a transverse vertical plane passing through the lower end of the 15th rib, and the three middle regions are separated from the three posterior regions by another transverse vertical plane passing through the external angle of the ilium (angle of the haunch). Again, each of these three regions—anterior, middle, and posterior—is further subdivided into a central and two lateral regions, this subdivision being effected by two vertical and parallel longitudinal planes, each passing through the centre of Poupart's ligament.

Directions.—The intestines of the horse, owing to their unwieldy size, and generally also to the weight of their contents, are extremely inconvenient to dissect. From the following description and the accompanying plates, the student should first learn how the intestinal tube is divided. He should then, with as little disturbance of the different intestines as possible, observe how they are disposed within the abdominal cavity.

The INTESTINES (Plates 41 and 42). The intestinal tube begins at the pyloric orifice of the stomach, and it terminates on the surface of the body, at the anus. It is primarily divided into *small* and *large intestines*, and each of these is naturally or arbitrarily divided into segments.

The *Small Intestine* comprises the first portion of the tube, and in a horse of medium size it measures about seventy-two feet in length. As is expressed by its name, it is of smaller calibre than the large intestine. Moreover, it is distinguished from nearly every part of the large intestine by having a smooth and regular contour when distended. The first two feet of the tube occupies a fixed position, and is termed the *duodenum*. It received this name because in man its length is about equal to the breadth of twelve fingers. The remainder of the small intestine has a comparatively loose mode of suspension; and it is arbitrarily divided into *jejunum* and *ileum*, the former succeeding the duodenum, and measuring about thirty feet, the latter comprising the remainder of the tube—about forty feet. These terms are borrowed from human anatomy, where the term *jejunum* was applied in consequence of that portion of the intestine being generally found empty in the dead body, while the *ileum* was so designated on account of its convoluted disposition.

The *Large Intestine* is, for the most part, of vastly greater calibre than the small; and, unlike the latter, it has when distended, not a smooth, but a bosselated, surface. In a medium-sized animal it is about twenty-five feet in length. It is subdivided—and in a much more natural fashion than the small intestine—into *cæcum*, *colon*, and *rectum*, the colon being further subdivided into *double* and *single* colon.

When the muscles which enclose the abdomen below and on each side have been removed, it most commonly happens that only the large intestines are exposed, and consequently their examination must precede that of the small intestines.

The CÆCUM is the first of the large intestines. In an animal of medium size it measures about three feet in length, and when moderately distended it has a capacity of about four gallons. At one of its extremities it is curved, forming what is termed the *crook of the cæcum*, while the opposite extremity tapers to a blind point, from which the bowel is named. The bowel has a puckered appearance, which is most evident when it is distended. This is owing to the longitudinal muscular fibres of its wall being not uniformly distributed as they are in the small intestine, but collected into bands, which shorten the bowel by throwing it into folds. The terminal portion of the ileum (small intestine) joins the cæcum on the concave side of the crook, and a few inches above the point of communication is the orifice by which alimentary matters are passed on to the colon. The crook of the cæcum is fixed in the right sublumbar region by means of loose cellular tissue, and it is in contact with the right kidney and the pancreas. On its inner side it adheres by cellular tissue to the termination of the double colon, and the duodenum passes round it on the outer side. The remaining portion of the bowel extends downwards and forwards through the right hypochondriac region, terminating by its blind point in the epigastrium. The first portion of the large colon, which lies to its inner side, extends in the same direction, and the peritoneum in passing from the one bowel to the other forms a fold which has been termed the *meso-cæcum*. As the cæcum is not adherent to the abdominal parietes except in the neighbourhood of its crook, it admits of some displacement; and the student must therefore be prepared to find it deviating somewhat from the course just described.

The DOUBLE or LARGE COLON. This bowel is termed double because when taken out of the abdomen it is arranged in the form of two parallel portions; but in order that it may be accommodated within the cavity, it has again to be doubled, so that in its natural disposition it presents four portions, which receive numerical designations. In an animal of medium size its length is about ten feet, and its capacity about sixteen gallons. It is puckered like the cæcum, and from the same cause.

The 1st division of the bowel begins at the crook of the cæcum, by an orifice of communication which is comparatively small. It extends downwards and forwards through the right hypochondriac region, bulging laterally into the umbilical region; and on reaching the epigastrium, the bowel becomes bent on itself, forming what is termed, from its relation to the ensiform cartilage of the sternum, the *suprasternal flexure*. The angle of this flexure forms the point of separation between the 1st and 2nd portions of the double colon.

The 2nd division, beginning at the suprasternal flexure, runs backwards on the left side of the abdomen, occupying the hypochondriac, umbilical, and lumbar regions; and on approaching the entrance of the pelvic cavity, the bowel forms in the iliac or hypogastric region a second flexure—the *pelvic flexure*, the angle of which marks the point of separation between the 2nd and 3rd portions. The 1st and 2nd portions of the double colon have extensive contact with the abdominal wall, and they conceal the other two divisions of the bowel, which lie above them (in the natural standing posture).

The 3rd division, beginning at the pelvic flexure, extends forwards along the left side of the abdomen, through the same areas as the 2nd portion, being closely bound to it, and lying immediately above it. On reaching the epigastric region, a third flexure is formed, in contact with the diaphragm, liver, and stomach, and from these relations named the *diaphragmatic* or *gastro-hepatic flexure*. This will be brought into view by grasping and pulling backwards the suprasternal flexure, above which it lies.

The 4th portion begins at the angle of the diaphragmatic flexure, and passes backwards on the right side of the cavity, lying above the 1st division, and closely bound to it. On reaching the inner side of the crook of the cæcum, to which it is adherent, it suddenly becomes much reduced in calibre, and is continued as the small or floating colon.

The pelvic flexure of the colon should now be seized and carried forwards, so as to place the bowel in the position shown in Plate 41. It will now be observed that the bowel is quite unattached except at its beginning and termination, where it adheres to the pancreas and the crook of the cæcum. In this disposition the suprasternal and diaphragmatic flexures are obliterated, and the 1st and 4th portions are seen to be closely adherent to one another, and, in like manner, the 2nd and 3rd portions, except just at the pelvic flexure, where, in the angle of the flexure, a small space is bridged over by a racket-shaped piece of peritoneum. It will be noticed also that the intestine varies greatly in calibre at different points. Its greatest diameter is in its 4th portion, and its smallest about the centre of the 3rd. This narrow portion of the intestine is further distinguished from the rest by being not puckered, but plain, when distended.

The SMALL or FLOATING COLON succeeds the double colon. It is much narrower than that bowel, indeed it does not greatly exceed in calibre the small intestine, from which, however, its coils are readily distinguished by their puckered appearance. In a medium-sized animal it is about ten feet in length. It is disposed within the abdomen after the manner of the small intestine, being suspended at the free edge of a dependency of the peritoneum, termed the *meso-colon* or *colic mesentery*. It has a convoluted disposition, and occupies the left lumbar and iliac regions. Its last coil passes into the pelvic cavity, and is continued as the rectum.

The RECTUM is the terminal portion of the intestine, and is about two

feet in length.   It derives its name from its approximately straight course
through the pelvic cavity, in connection with which it will be more fully
described.

Directions.—The coils of the jejunum and ileum should be arranged
in the left flank after the manner of Plate 41.   To get a view of the
duodenum, the cæcum should be thrown across the abdomen, with its
point towards the left side.   The duodenum will then be seen encircling
the crook of the cæcum on its outer side.   Should the large intestine
contain much ingesta, that should be evacuated through an incision across
the pelvic flexure of the double colon and another at the point of the
cæcum.   When the ingesta has been expelled, the bowels should be
moderately inflated, and the cut ends ligatured.

The DUODENUM (Plate 44) is the first segment of the small intestine.
Its length is about two feet, but it cannot be very well seen in its
entirety at this stage of the dissection.   It begins at the pyloric aperture of the stomach, where it is related to the posterior surface of the
liver.   It curves upwards and backwards across the lower face of the
right kidney, and then sweeping round the crook of the cæcum to its
outer side, it crosses the spine behind the anterior mesenteric artery, and
is continued as the jejunum.   It is maintained in position by a narrow
band of peritoneum, and in this fixity of position it is distinguished from
the rest of the small intestine.

The JEJUNUM and ILEUM.   These comprise the remaining portion of
the small intestine, of which about thirty feet is arbitrarily apportioned to the former, and the remainder (about forty feet) to the latter.
They are arranged in the form of numerous coils, which occupy the
iliac, umbilical, and hypogastric regions.   The coils are attached to
the free edge of a fold of peritoneum called the *great mesentery;* and
inasmuch as this mesentery is of considerable breadth, they may move
from place to place within the above-mentioned areas.   When distended,
they have not a puckered, but a smooth, surface.   The terminal part of
the ileum joins the crook of the cæcum, into which it projects for a little
distance, after the manner of a tap into a barrel; and at the point of
entrance there is a valvular arrangement—the *ileo-cæcal valve,* to prevent regurgitation from the cæcum into the ileum.

The PERITONEUM is the lining membrane of the abdominal and pelvic
cavities.   It belongs to the class of serous membranes, and, like all such
membranes, it consists of a *parietal* and a *visceral* division, these being
portions of one great sac.   The parietal part is that which lines the
abdominal walls, or parietes; the visceral part invests the solid and
hollow organs, or viscera, of the abdominal cavity.   In virtue of this
membrane, all the free surfaces that present themselves when the
abdominal wall is removed, have a smooth and shining appearance.
The surface of the membrane is covered by a layer of endothelial cells,

and these rest upon a layer of vascular connective-tissue. The object of the membrane is to facilitate the movements of the different abdominal organs on each other and on the walls of the cavity, and especially to facilitate the vermicular or peristaltic movements of the intestines. For this purpose the surface of the membrane is kept moist by a sparing amount of serous fluid, which gives to the membrane its glistening aspect.

To trace the exact disposition of the peritoneum in the horse is very difficult, in consequence of the unwieldy character of the intestines. When the student has the opportunity he should examine the membrane in a foal, in which the different organs can be manipulated with ease.

The parietal and visceral peritoneum, as has already been stated, form portions of one great sac, and the various abdominal viscera are external to this sac. The sac of the peritoneum, it must be observed, encloses not an actual, but merely a potential, cavity; the inner surface of every portion of the sac being in contact with the same surface of another portion. To facilitate the understanding of this, let the student imagine the cavity of the abdomen (including the pelvis) as having its natural form, but deprived of all its contents, and completely lined by peritoneum, which, for simplicity's sake, he may suppose to be elastic. The continuity of the membrane, and the fact that it formed a close sac would then be apparent. Now let him imagine a simple tube of intestine extending between this membrane and the spinal column, that is, *outside* the serous sac. Conceive next this tube of intestine let gradually down, until it extends through the cavity about its centre. In this descent the intestine would first surround itself with peritoneum; and then, as it sank farther, it would stretch the membrane so as to form a kind of sling passing upwards to the point from which it started. The membrane would now have lost its simplicity, for it would have a parietal division continuing to line the abdominal walls, and a visceral portion surrounding the tube of intestine. Moreover, these two portions would be continuous with each other along the sling-like portion suspending the tube. Lastly, imagine the tube of intestine to grow and branch, so as to completely fill up the abdominal cavity, and obliterate the space between the parietal and visceral peritoneum. This, of course, would not destroy the continuity of the serous sac, although it would complicate it so that its continuity would be difficult to trace.

All the organs, then, that actually project into the abdominal cavity get a more or less complete investment of visceral peritoneum; and, in the case of each organ, this visceral covering is traceable on to a neighbouring organ, or on to the walls of the abdomen. Where organs are contiguous to each other or to the abdominal parietes, the peritoneum may pass directly from the one organ to another or to the abdominal

parietes ; but, at other times, the connection between the parietal and
visceral peritoneum is traceable along bands or folds analogous to the
sling-like membrane that was formed in the imaginary case.   These
folds constitute the various mesenteries, omenta, and peritoneal ligaments that will hereafter be described.

Although there is but a single peritoneal sac, this sac is so disposed
that it forms two compartments, termed respectively the greater and
lesser cavities of the peritoneum, the latter being also known as the
cavity of Winslow.   The greater cavity is that which is exposed when
the inferior wall of the abdomen is removed, the lesser cavity is situated
behind the stomach, and is separated from the greater cavity mainly by
the omentum.

The *Great* or *Gastro-colic Omentum.*—Passing backwards among the
intestines, on the left side of the abdomen, there will have been noticed
a large lace-like membrane, which is the *great omentum, epiploon,* or web.
In order to examine its connection, the cæcum and double colon should
be thrown backwards over the right flank, and the coils of the single
colon arranged over the left flank.   The coils of small intestine should
at the same time be gathered backwards and to the right.   The omentum is composed of two layers of peritoneum, which include between
them vessels, and a varying quantity of fat.   This fat is deposited
mainly along the course of the vessels, leaving, except in obese subjects,
intervening transparent areas that are free from fat ; and it is from this
arrangement that the membrane possesses a lace-like appearance.   The
two layers of the omentum may be distinguished as superficial and
deep.

When the superficial layer is traced backwards, it is seen to pass on
to the terminal part of the double colon (4th part) and initial part of
the single colon, covering the posterior aspect of these where they extend
across the roof of the abdominal cavity.   Behind these it passes backwards along the roof of the abdominal cavity, from which it descends to
envelop the small intestine, forming the great mesentery, and the floating colon, forming the colic mesentery.   To the right, again, it passes
directly on to the cæcum and the double colon ; and after enveloping
these intestines, it returns to the abdominal wall, to pursue its backward course to the pelvis.   When followed forwards, the superficial
layer reaches the convex curvature of the stomach, and the initial dilatation of the duodenum ; and it passes over the anterior surfaces of
these organs as visceral peritoneum.   Passing off the duodenum and
stomach, it next forms the anterior layer of the gastro-hepatic omentum, and thus reaches the posterior surface of the liver at the portal
fissure.   From that point it descends over the posterior surface of the
liver as visceral peritoneum, and turns round the inferior edge of the
gland to gain its diaphragmatic surface.   It ascends on this surface ;

and where the liver and diaphragm are united, it passes from the former to the latter, on which it descends to the inferior wall of the abdomen. Along this it passes until it enters the pelvis, where it becomes continuous with the same layer already followed backwards along the roof of the abdomen. In the male it is to be observed that the parietal peritoneum of the abdominal floor passes into the inguinal canal, and forms the tunica vaginalis of the testicle, the sac of which is a simple diverticulum of the great peritoneal sac. Returning again to the omentum, it will be noticed that its superficial layer, towards the left side, in passing forwards to gain the convex curvature of the stomach, encounters the spleen. Passing round that organ, it gives to it a visceral covering, and then continues its course to the stomach. The portion of omentum between the spleen and the left sac of the stomach is termed the *gastro-splenic omentum.*

Now make a transverse opening about the centre of the great omentum, and introduce the hand through the opening. The hand is now in what is termed the *cavity of Winslow,* and the deep layer of the omentum is exposed. When this layer is traced forwards, it is seen to reach the convex curvature of the stomach, where, separating from the superficial layer, it passes over the posterior surface of the stomach, and initial dilatation of the duodenum. From these, again, it passes as the posterior layer of the *gastro-hepatic omentum,* and reaches the liver at the portal fissure. There it separates from the other layer of the gastro-hepatic omentum, and ascends on the liver. It turns round the superior edge of the gland, and passes from its anterior face to the diaphragm, on which it ascends to the spine. The deep layer of the omentum is now to be followed in the backward direction. It is seen to reach the terminal part of the double colon, and the initial part of the single colon; and, separating there from the superficial layer, it passes over the anterior aspect of these portions of intestine, and is reflected forwards on the under surface of the pancreas. It turns round the anterior edge of that gland, covers for a little distance its upper face, and then passes on to the spine, where it meets the same layer advancing in the opposite direction. It is thus seen that the deep layer of the omentum, when traced in the antero-posterior direction, forms a continuous layer; and at first sight it does not appear to be continuous with the remainder of the peritoneum. As already stated, however, the peritoneum forms a single sac, and the before-mentioned layer is continuous with the remainder of the serous membrane at a narrow opening termed the *foramen of Winslow.* To find this opening, pass the dorsal aspect of the left forefinger along the posterior surface of the lobulus caudatus of the liver, close to the spine; and insinuate the point of the finger onwards towards the left (of the subject). At the same time pass the right hand up to the spine in the cavity of Winslow, and insinuate the forefinger towards the right, above and behind the pylorus. The tips of the forefingers of opposite hands can thus be made to meet, showing the continuity of the larger sac of the peritoneum, in which the left hand is, with the smaller sac, or cavity of Winslow, in which the right hand is. Perhaps the simplest way to get an understanding of the relationship of the two cavities, is to imagine the deep layer of the omentum to be suppressed. In that condition, the anterior aspect of the double and single colon at their point of junction, the pancreas, the posterior surface of the stomach and initial dilatation of the duodenum, the upper parts of the liver and diaphragm, and the roof of the abdomen for a short space behind the hiatus aorticus would be without a serous covering. It may be supposed that to supply this deficiency, a pouch of the great sac of peritoneum has to be made. This pouch is made at the foramen of Winslow, the peritoneum being there thrust outwards towards the right, and expanded until it forms what has already been traced as the deep layer of the omentum. The foramen will be observed to have the following boundaries :—the base of the lobulus caudatus in front, the 4th part of the double colon behind, the free edge of the gastro-hepatic omentum below, and the posterior vena cava and right pillar of the diaphragm above.

The *Great Mesentery* is the membrane that suspends the small intestine. Like the omentum, it is composed of two layers of peritoneum. These layers leave the spine at the root of the anterior mesenteric artery, being there continuous with the parietal peritoneum ; and they descend, one on each side of the branches of that artery, until they reach the intestine. At the concave edge of the bowel the two layers separate ; and after encircling the tube as visceral peritoneum, they meet and become continuous at its convex or free border. Where the mesentery suspends the first part of the jejunum, it is continuous with the peritoneal frænum of the duodenum ; and at its opposite extremity, where it envelops the termination of the ileum, it passes on to the cæcum. At the latter point it will be observed that the two layers of mesentery do not become continuous around the convex border of the ileum, but are prolonged beyond that, so that the terminal portion of the small intestine is included in the mesentery some distance from its free edge.

The *Colic Mesentery.*—This is the membrane that suspends the single or floating colon. It is composed of two layers of peritoneum, which leave the roof of the abdomen along a line extending from the root of the anterior mesenteric artery to the inlet of the pelvis. These two layers include between them the posterior mesenteric artery and its branches ; and after enveloping the single colon, they become continuous at its free edge. At its anterior extremity the colic mesentery is continuous with the great omentum and with the great mesentery, and at the pelvic inlet it is continuous with the meso-rectum.

The *Uterine Broad Ligaments.*—These are the double peritoneal folds that suspend the uterus, ovaries, and Fallopian tubes. Each ligament leaves the roof of the abdomen in the lumbar region, and descends to the concave edge of the cornu, and to the side of the upper face of the body, of the uterus. At these points the layers of the ligament separate, and pass on to the uterus as its visceral covering. The ligaments are widely apart in front; but as they are traced backwards, they become narrower and nearer to each other. The Fallopian tube is sustained between the two layers of each ligament at its anterior edge, and here the fimbriated extremity of the tube opens into the sac of the peritoneum. In the female, therefore, the peritoneum does not form a shut sac. Stretching between the ovary and the uterine cornu is a cord of non-striped muscular tissue—the ligament of the ovary—which forms the free edge of a small secondary fold of peritoneum. This forms with the adjacent part of the broad ligament a pocket-like cavity. On the outer side of the broad ligament another secondary fold extends as far as the internal abdominal ring, and contains a layer of non-striped muscular tissue corresponding to the round ligament of the human uterus. Besides some scattered fibres of non-striped muscle, the layers of the broad ligament include between them the uterine and ovarian vessels and nerves.

The other peritoneal ligaments will be described in connection with the organs to which they belong.

PERITONEAL POCKETS. The peritoneum, in passing from one organ to another, forms several remarkable pockets, one of which has been mentioned above in connection with the ligament of the ovary. The exact position of the others will now be indicated. So far as I am aware, these have not hitherto been described. Nevertheless, they possess considerable interest, since, in the human subject, a coil of intestine has been known to become incarcerated in a similar pocket of peritoneum.

1. The entrance to the first of these pockets will be found immediately in front of the base of the lobulus caudatus, which separates it from the foramen of Winslow. It is bounded by the anterior end of the right kidney, and by the lobulus caudatus and upper part of the right lobe of the liver. It extends inwards to near the spine between the diaphragm and the upper part of the right lobe of the liver.

2. Another pocket will be found a little to the left of the root of the anterior mesenteric artery, the entrance to it being on the anterior surface of the mesentery suspending the first few inches of the jejunum. The pocket is bounded in part by this piece of mesentery, and in part by a peritoneal fold passing between the jejunum and the first part of the single colon.

3. Other two pockets will be found at the termination of the small intestine. Turn the point of the cæcum backwards and to the right,

and pull upon the terminal part of the jejunum. On each side of the point at which the latter perforates the cæcal crook, there will be found a recess, the posterior (in this position) being the deeper.

4. Another considerable pocket will readily be found on the concave side of the cæcal crook, being formed by the peritoneum in passing between the cæcum and the beginning of the double colon.

5. Another but much smaller pocket will be found in the cavity of Winslow, above and in front of the first few inches of the single colon.

Directions.—For the display of the mesenteric vessels and the sympathetic nerve, the intestines should first be disposed after the manner of Plate 41. When well injected, the arteries require but little dissection, and they are closely accompanied by the veins and nerves. The arteries of the cæcum and colon should be taken where most conspicuous, and traced in both directions. Each of these vessels must be carefully dissected up to its point of origin, but only two or three of the arteries of the small intestine need be fully dissected. The whole intestinal tube with the exception of a short piece of the duodenum next the stomach, is supplied by the anterior and posterior mesenteric arteries, which are branches of the abdominal aorta. The first supplies the whole of the small intestine except the piece of duodenum specified; and it also supplies the cæcum, the large colon, and a few inches of the beginning of the small colon. The remainder of the small colon, and the rectum are supplied by the posterior mesenteric artery.

The ANTERIOR MESENTERIC ARTERY (Plate 41) comes off from the inferior aspect of the aorta at the 1st lumbar vertebra. It is only about an inch and a half in length, but it has a large calibre; and in old horses it often shews aneurismal dilatation. It divides into three terminal branches, which from their direction are distinguished as *left*, *right*, and *anterior*. The *left* distributes its branches to the whole of the small intestine except a few inches at the beginning of the duodenum and about two feet at the end of the ileum; the *right* supplies the terminal portion of the ileum, the entire cæcum, and the double colon as far as the pelvic flexure; and the *anterior* is distributed to the double colon beyond the pelvic flexure, and to the first few inches of the single colon. It is an assistance to the memory to study the different branches in the order of their distribution to the intestine, taking first those that supply the most anterior segment of the tube.

1. The *Left Branch* of the anterior mesenteric artery is no sooner detached than it splits up into about fifteen or twenty arteries, which pass between the layers of the mesentery to supply the small intestine. Indeed, the left branch can scarcely be said to exist, for these arteries of the small intestine seem to spring from a common point of the anterior mesenteric trunk. As each artery approaches the intestine, it bifurcates, each branch inosculating with the corresponding branch of an adjacent

artery to form an arch. From the convexity of these arches smaller vessels pass to each side of the intestine, and anastomose round it. At the anterior part of the tube two sets of superposed arches are formed before the ultimate vessels to the intestine are detached. The branch which is most anterior in point of distribution anastomoses with the duodenal branch of the cœliac axis, while the one which is most posterior anastomoses with the ileo-cæcal artery from the right branch of the anterior mesenteric.

2. The *Right Branch* of the anterior mesenteric artery divides into four vessels, viz., the ileo-cæcal, the superior cæcal, the inferior cæcal, and the direct colic arteries.

a. The *Ileo-cæcal Artery* (Plate 41, for *ilio-cæcal* read *ileo-cæcal*) supplies the terminal portion of the ileum (about two feet in length), and inosculates with the last of the arteries from the left branch.

b. The *Superior Cæcal Artery*, in the present inverted position of the intestines, passes beneath the termination of the ileum to run along one of the longitudinal muscular bands of the cæcum. It sometimes gives off the ileo-cæcal artery as a collateral branch, and at the point of the cæcum it anastomoses with the next vessel. It gives off branches right and left to the walls of the cæcum.

c. The *Inferior Cæcal Artery*, in the present position of parts, passes above the termination of the ileum to run along another of the muscular bands of the cæcum. Besides collateral branches to the main portion of the bowel, it gives off the *artery of the arch*, which follows the concavity of the cæcal crook and terminates on the beginning of the double colon.

d. The *Direct* or *Right Colic Artery.*—This is a large vessel, receiving the first of these designations because the course of its blood stream is the same as that of the alimentary matters in the bowel. It supplies, by right and left collateral branches, the 1st and 2nd portions of the double colon, and anastomoses at the pelvic flexure with the retrograde colic artery.

3. The *Anterior Branch* of the anterior mesenteric artery divides after a very short course into two vessels of unequal size, viz., the retrograde colic artery and the first artery of the small colon.

a. The *Retrograde* or *Left Colic Artery*, much the larger of the two, supplies successively the 4th and 3rd portions of the double colon, running parallel to the direct colic artery, but carrying its blood in a direction counter to the course of the alimentary matters in the intestine.

b. The *First Artery of the Small Colon* supplies a short piece at the beginning of that bowel. It is included between the layers of the colic mesentery, and anastomoses with the first branch of the posterior mesenteric artery.

Directions.—To display the posterior mesenteric artery, the small

x

colon must be spread out over the left flank after the manner of Plate 42.

The POSTERIOR MESENTERIC ARTERY is a much smaller vessel than the anterior. It is a branch of the abdominal aorta, from which it is given off at the 4th lumbar vertebra. It passes in a curved direction between the layers of the colic mesentery and meso-rectum, and terminates near the anus in vessels which supply the end of the intestinal tube. From the convexity of its curve, which is directed downwards, about twelve or fourteen branches pass to supply the small colon (except a few inches at its beginning) and the rectum. The branches which supply the first half of the small colon divide and form arches by anastomosis in the mesentery, close to the bowel; but the more posterior branches do not anastomose until they perforate the intestinal wall.

The *Intestinal Veins.*—The blood which is brought to the intestines by the arteries just considered is carried away by vessels belonging to the *portal system.* These veins for the most part run in close company with the arteries, and receive the same names.

The ANTERIOR MESENTERIC VEIN is a very large vessel having tributaries which correspond almost exactly to the divisions of the artery of the same name. It joins the splenic and posterior mesenteric veins to constitute the vena portæ.

The POSTERIOR MESENTERIC VEIN has its roots in the hæmorrhoidal veins around the termination of the rectum, which veins, on the other hand, communicate with the internal pudic vein. After receiving blood from the walls of the rectum and small colon, the posterior mesenteric vein forms by union with the splenic a very short trunk which joins the anterior mesenteric to form the vena portæ.

LYMPHATIC VESSELS of the INTESTINE.—In an ordinary dissecting-room subject the lymphatic vessels will not be visible unless the animal is emaciated and has been killed shortly after a meal, in which case the mesenteric vessels may be seen without dissection. They will be recognised as vessels with very thin walls and milky contents, coursing between the layers of the mesentery, from the intestine towards the anterior mesenteric artery. The lympathic vessels of the small intestine are called *lacteals.*

The LYMPHATIC GLANDS of the INTESTINE are very numerous. Those of the small intestine are chiefly aggregated in the form of a cluster of about thirty included between the layers of the mesentery, near the anterior mesenteric artery; but a number are placed lower down in the mesentery, along the course of the ileo-cæcal artery. The glands of the cæcum are distributed in the form of two chains along the track of the superior and inferior cæcal arteries, and numerous glands are similarly placed on the colon along the course of the direct and retrograde colic arteries. Those of the small colon and rectum are, for the most part, placed on the wall of the bowel, at the edge of the mesentery ; but a few are included between the layers of the colic mesentery. The lacteals from the small intestine and the lymphatic vessels from the large intestine traverse these various groups of glands on their course towards the receptaculum chyli.

The SYMPATHETIC NERVE. This nerve forms on the aorta, in front of the anterior mesenteric artery, a great network termed the *Solar plexus.* The solar plexus is at present concealed by the pancreas, but the student has to notice the anterior and posterior mesenteric plexuses, which are wholly or in part derived from it.

The *Anterior Mesenteric Plexus* comprises numerous nerves already met in dissecting the branches of the anterior mesenteric artery. The nerves interlace around the arteries, and pass with them to gain the bowel, where they further interlace before penetrating its wall.

The *Posterior Mesenteric Plexus.*—The branches of this plexus run in company with the divisions of the artery of the same name. Its nerves are derived in part from the aortic plexus, which is a backward continuation of the solar plexus, and in part from roots furnished by the lumbar cord of the sympathetic. The ultimate branches are distributed in the wall of the small colon and rectum.

Directions.—The intestinal mass is now to be removed in the following manner. The ropes must be unfastened from the left limbs of the animal, while those on the right limbs are to be lengthened until the subject inclines considerably over to the left side. Two ligatures a few inches apart are to be passed round the duodenum where it encircles the crook of the cæcum, and the bowel is then to be cut across between the ligatures, the object of which is to keep the contents from escaping. Where the small colon joins the rectum, at the entrance to the pelvis, the bowel is to be served in the same way, and the colic mesentery is to be cut along its point of origin at the spine. Both large and small intestines are then to be thrown as far as possible outwards over the left flank. The next step must be to take the scalpel and carefully sever the connective-tissue adhesions between the cæcal crook and colon on the one hand, and the sublumbar region and pancreas on the other. In doing this, the dissector must cut close to the wall of the bowel, and take especial care not to take away any portion of the pancreas, which will be recognised by its dark colour. The operation will be favoured by the weight of the intestines, which tends to tear these connections. When the cæcum and colon have been freed, it will be found that strong resistance to the removal of the intestines is still offered by the mesentery, or rather, by its included vessels. These must therefore be cut near the spine, and the entire mass will then slip over the left side, the omentum being cut or torn from its attachment to the colon. The intestines should now be spread out on a table ; and when the student has refreshed his memory regarding their form and connections with one another, he must proceed to examine their structure. This should be done by taking a short piece of the gut, slitting it up, and pinning it with its mucous surface downwards on a block of wood.

STRUCTURE OF THE SMALL INTESTINE. The wall of the bowel is made up of four layers, viz., serous, muscular, submucous, and mucous.

1. The *Serous Layer*, the most external, is a part of the visceral peritoneum. It reaches the bowel by the mesentery, whose two layers separate at the concave border of the intestine, and pass round each side to meet and become continuous on its convex or free border. It is closely adherent to the subjacent muscular layer, which it completely covers except at the line of separation of the two layers of the mesentery, where the vessels enter. It must be stripped off to expose the next coat.

2. The *Muscular Coat* is made up of two distinct sets of fibres: 1. Longitudinal fibres, which are most external, and form a thin layer uniformly spread along the wall. 2. Circular fibres, thicker than the preceding, and also spread over every part of the wall. These fibres are of the pale, non-striated variety.

3. The *Submucous Coat* is composed of loose areolar tissue uniting the muscular and mucous layers. In the duodenum it contains the glands of Brunner, which have the racemose type of structure, and are about the size of a hemp seed. Their ducts pass through the mucous membrane, and open on its free surface.

Directions.—A few feet of the jejunum and about the same length of the ileum should be taken and slit up along the line of attachment of the mesentery. After the pieces have been gently washed, they should be spread on a flat surface with the peritoneal coat downwards.

4. The *Mucous Membrane* forms an inner lining to the intestine. It is a soft, velvety-looking membrane which, when healthy and fresh, has a pinkish-yellow colour. When a piece of intestine is floated in water, the mucous membrane is seen to be studded with short, thread-like projections, to which the velvety appearance of the membrane is due. These are the *intestinal villi*. Each of them may be regarded as an upheaval of the mucous membrane, containing in its interior microscopic blood and lymph vessels, some non-striped muscular fibres, and a framework of lymphoid tissue. The villi are important agents in the absorption of nutrient particles from the contents of the bowel. They are found throughout the whole of the small intestine, but are more numerous in the jejunum than in the ileum. The free surface of the mucous membrane, including the villi, is formed by a single layer of columnar epithelium with goblet cells interspersed. Contained within the substance of the membrane are numerous microscopic tubular glands—the *glands fo Lieberkühn*—whose mouths open on the free surface. The *solitary glands* are small spherical bodies about the size of a mustard seed.

They are covered by the epithelium, and occur throughout the whole intestine, but are more numerous in the ileum than in the jejunum. They are composed of lymphoid tissue. The *glands of Peyer*, or, as they are commonly called, *Peyer's patches*, are circular or oval patches formed by the aggregation of solitary glands. They are more numerous in the ileum than in the jejunum, their total number being about one hundred. They are distributed along the convex or free border of the intestine, and hence it was directed that the bowel should be opened along the attachment of the mesentery, so as to leave the patches intact.

Directions.—The cæcum, with the first few inches of the double colon and a like length at the end of the small intestine, should be separated from the rest of the intestinal mass. After the serous and muscular coats have been observed on the inflated cæcum, the bowel should be slit open on the *convex* side of its crook, the incision being extended to its point. The mucous surface is to be gently washed; and in connection with its study, the student is to examine the two orifices found on the concave side of the crook.

STRUCTURE of the LARGE INTESTINE. Throughout nearly the whole of its length, the wall of the large bowel is made up of four coats, similar to those of the small intestine.

FIG. 36.

VERTICAL SECTION THROUGH THE WALL OF THE DUODENUM, SHOW-ING THE GLANDS OF BRUNNER (*Turner*).

V. Intestinal villi; L. Layer of glands of Lieberkühn; B. A Brunner's gland, *d.* its excretory duct; S.M. Submucous coat; M. Muscular coat.

1. The *Serous Coat* is derived from the peritoneum, but it forms here a less complete investment than in the case of the small intestine, considerable areas of the wall being without this covering. Thus, it is absent where the cæcum and double colon adhere to the pancreas and abdominal parietes in the sublumbar region; it is also absent where these two intestines adhere to each other, and where the parallel portions of the double colon come into contact; and lastly, as will be seen in the dissection of the pelvis, the terminal part of the rectum is without a peritoneal covering.

2. The *Muscular Coat* consists of two distinct layers—an external longitudinal and an internal circular. Throughout nearly the whole extent of the large intestine, the longitudinal fibres are not uniformly distri-

buted over the wall, but are collected into distinct bands, the areas between the bands being provided only with circular fibres. When these bands contract, they shorten the intestine, and throw the wall of the bowel between them into alternate ridges and furrows. The number of these bands is different at different points. The cæcum has four. The colon in its 1st part has also four. Three of these disappear on the 2nd part, so that at the pelvic flexure there is only a single band, on the concave side of the flexure. This single band is continued along the 3rd part, and near the diaphragmatic flexure other two bands originate. The 4th part has three bands. The single colon has two bands, one on each curvature, and these are continued on the first half of the rectum, but are lost on its terminal half, as will be seen in the dissection of the pelvis. The inner layer of circular fibres is uniformly distributed.

3. The *Submucous Coat* is a layer of loose areolar tissue uniting the muscular and mucous coats.

4. The *Mucous Coat* lines the cavity of the bowel. Its surface is covered by a single layer of columnar epithelium, and in its deeper part it contains *solitary glands* and *glands of Lieberkühn*, similar to those of

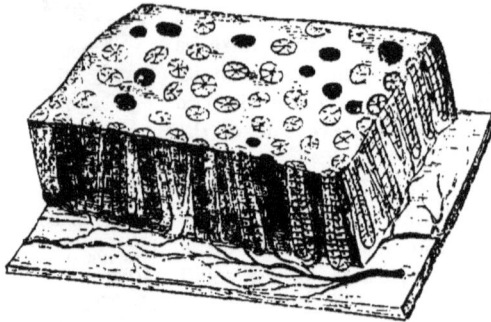

FIG. 37.

DIAGRAMMATIC VIEW (MAGNIFIED) OF A SMALL PORTION OF THE MUCOUS MEMBRANE OF THE COLON (*Allen Thomson*).

A small portion of the mucous membrane cut perpendicularly at the edges is shown in perspective; on the surface are seen the orifices of the crypts of Lieberkühn or tubular glands, the most of them lined by their columnar epithelium, a few divested of it and thus appearing larger; along the sides the tubular glands are seen more or less equally divided by the section; these are resting on a wider portion of the submucous tissue, from which the blood-vessels are represented as passing into the spaces between the glands.

the small intestine. No Peyer's patches are found in it; and it is without villi. The foldings of the wall of the bowel produced by the longitudinal muscular bands involve all the coats, and the interior therefore shows the alternately ridged and furrowed appearance already seen on the exterior.

Orifices of the Crook.—These are the apertures of communication with the ileum, and with the large colon. At its termination the ileum projects slightly into the interior of the cæcum, and beneath the mucous membrane surrounding the orifice, there is developed a ring of muscular fibres. This fold of mucous membrane with its included muscular fibres constitutes the *ileo-cæcal valve.* The opening of communication with the colon is considerably larger than the preceding, above which it is placed.

Directions.—The student must now return to the parts left within the abdominal cavity, where, without further dissection, he will be able to examine the stomach, spleen, pancreas, and liver. Should the stomach be nearly empty, the ligature should be untied from the cut end of the duodenum, and by means of bellows that intestine and the stomach should be moderately inflated. To permit this, it will not be necessary to ligature the œsophagus. At the present stage the above-mentioned organs may be studied as regards their form, situation, and relations, their structure being postponed for future consideration.

The STOMACH (Plates 43 and 44) is the most dilated segment of the alimentary tube. When moderately distended, it will be seen to have the following configuration. It possesses an anterior and a posterior surface, both being smoothly rounded. It has a concave or lesser curvature, which is turned upwards and to the right; and a convex or greater curvature, which is directed downwards and to the left. The left extremity of the organ is much the larger, and is termed the *cardiac extremity,* or the *fundus.* The smaller right end is termed the *pylorus.* The stomach occupies the epigastric and left hypochondriac regions, and it will be observed to have the following connections. The anterior surface is related to both the liver and the diaphragm, and in the natural position looks upwards as well as forwards. The posterior face looks downwards as well as backwards, and before the removal of the intestines was related to these, and chiefly to the gastro-hepatic flexure of the double colon. The smaller curvature is fixed to the liver by means of the gastro-hepatic omentum. If, in the present inverted position of the animal, the greater curvature be pulled backwards, so as to separate the anterior surface from the liver and diaphragm, the œsophagus will be found entering the stomach at its lesser curvature, about midway between the central point of that curvature and the extremity of the fundus. The greater curvature is related in its left half to the spleen, and throughout the rest of its extent to the intestines, particularly to the suprasternal flexure of the colon, now removed. The right extremity, or pylorus, is directly continued into the duodenum, a slight constriction being the only outward mark of their separation. The left or cardiac extremity extends to the left beyond the insertion of the œsophagus, and is related to the pancreas and base of the spleen. The stomach is retained in position by continuity with the œsophagus and duodenum, and by certain folds of peritoneum, viz., the gastro-phrenic ligament, and the gastro-hepatic, gastro-splenic, and gastro-colic omenta. The *gastro-phrenic ligament* extends from the diaphragm to the stomach, around the œsophageal insertion. The *gastro-hepatic omentum* passes between the lesser curvature and the posterior fissure of the liver. The *gastro-splenic omentum* passes from the cardiac extremity to the spleen. The *gastro-colic* or *great omentum* is continuous with the preceding, and passes in the form of a loose fold from the greater curvature. It extends backwards and downwards, and then curves upwards to the roof of the abdominal cavity; and, as has already been explained (page 300), it separates the greater and lesser cavities of the peritoneum; and, inasmuch as in man it hangs downwards to float upon the intestines, it has been termed the *Epiploon*. Even in emaciated subjects, it contains between its layers a considerable quantity of fat.

The DUODENUM (Plates 43 and 44). A better opportunity to examine this part of the intestine is now afforded. Commencing in the epigastrium, at the pyloric orifice of the stomach, it ascends across the posterior face of the right lobe of the liver, in passing into the right hypochondrium. It then curves backwards in the right lumbar region, beneath the right kidney; and sweeping round the crook of the cæcum, it crosses the spine and is continued as the jejunum. Its calibre is greatest just beyond the pylorus, and at this point it presents, when inflated, a small dilatation like a miniature stomach with its greater curvature superior. Throughout the whole of its course it is retained in position by a narrow band of peritoneum formed by the serous membrane as it passes to envelop the bowel. The right extremity of the pancreas rests against the duodenum, a few inches from the pylorus, and at that point the wall of the bowel is perforated by the bile and pancreatic ducts.

The SPLEEN (Plates 43 and 44) is a bluish-purple solid organ placed in close proximity to the left sac of the stomach. In the horse it has a scythe-shaped outline. It presents an external face, which is slightly convex; an internal face, which is slightly concave and narrower than the outer; an anterior thick border; and a posterior border, which is sharp. Its surfaces are widest above, where they terminate in the base of the organ, and below they taper to the apex. The spleen is situated in the left hypochondriac region, and has the following relations :—Its outer surface is related to the diaphragm; its inner surface contacts with the double colon ; its anterior border is penetrated by the vessels and nerves of the organ, and is related to the greater curvature of the stomach ; its posterior border is free, and is included between the intestines and the diaphragm ; its base is related to the pancreas and left kidney. The spleen is retained in the left hypochondrium by the *gastro-splenic omentum*, and by a special *splenic ligament*. The *gastro-splenic omentum* forms a loose connection between the left half of the greater curvature of the stomach and the anterior border of the spleen. The *splenic ligament* is a fold of peritoneum developed at the base of the organ, and formed by the serous membrane in passing from around the anterior end of the left kidney to envelop the spleen.

The PANCREAS (Plate 44) is a body having a lobulated structure and a very irregular shape. It is placed across the roof of the abdominal cavity, its central portion underlying the last dorsal vertebræ. Its upper face is applied to the aorta, the cœliac axis, the vena cava, the pillars of the diaphragm, and the right kidney, and is partly covered by peritoneum. Its lower face towards the right is adherent to the crook of the cæcum and the termination of the double colon, while to the left it is covered by peritoneum. Its anterior border is related to the stomach, the duodenum, and the liver. Its posterior border is related about its centre to the anterior mesenteric artery. Its right extremity, or *head*, is in contact with the duodenum; while the left extremity, or *tail*, is related to the base of the spleen. The entire thickness of the gland is perforated by the portal vein, which passes from its lower to its upper surface through what is named the *pancreatic ring*. The gland possesses two excretory ducts, both of which leave it at its right extremity. The main duct is named the *duct of Wirsung*, and it perforates the wall of the duodenum about six inches from the pylorus, and close by the point of entrance of the bile duct. The *accessory duct* is much smaller, and penetrates the bowel at a point opposite the entrance of the duct of Wirsung. The healthy fresh pancreas has a greyish-yellow colour; but when decomposition sets in after death, this speedily changes to an almost black hue.

The LIVER (Plates 43 and 44) is the largest gland in the body. It forms the bile and discharges it into the duodenum. In health it has a reddish-brown colour and a moderately firm consistence. In form it is not comparable to any common object, and its irregularity of shape makes its description somewhat difficult. It should be observed, in the first place, that inferiorly the rim of the organ is deeply indented, or notched, and two of the largest of these notches serve to partially divide the gland into its three main lobes, viz., a right, a left, and a middle, or *lobulus quadratus*. Of these the middle lobe is always the smallest, and its inferior border shows two or three minor indentations. The left lobe is generally the largest, but sometimes it is less than the right. The liver possesses a fourth lobe, in the form of a small projection of liver substance about the size of two or three of the human fingers, and situated at the upper part of the right lobe. This is the homologue of the *lobulus caudatus* of the human subject.

The exact form of the liver will be more distinctly seen when it has been removed from the body; but while it remains *in situ*, the student may endeavour to make out the following points:—Viewing the organ as a whole, it may be described as having an anterior and a posterior surface, and a circumference divisible into an upper and a lower border. The anterior surface is closely applied to the diaphragm, and is convex. The posterior vena cava, in descending from the spine to the foramen dextrum, passes between this surface and the diaphragm; and its course is marked on the liver by a vertical groove, which may be termed the *anterior fissure*. The posterior surface, when the organ is *in situ*, is concave; but when the liver is removed from the body, this surface, like the anterior, is slightly convex. It presents the *portal fissure* (L. *porta*, a gate), by which the portal vein, hepatic artery, bile duct, and hepatic nerves and lymphatics enter the liver. The upper border shows about its centre a rounded notch for the reception of the short abdominal portion of the œsophagus. The lower border shows the sharper and deeper indentations dividing the liver into its three principal lobes, and the lesser indentations that partially subdivide the lobulus quadratus.

The liver is situated in the epigastric and right and left hypochondriac regions. Its most important relations, besides those already mentioned, are as follows:—The anterior surface is applied to the diaphragm, the right lobe, which has the highest point of contact, being related to the most superior part of the muscular rim on the right side, the lobulus quadratus corresponding to the phrenic centre, while the left lobe touches the lowest point reached by the liver, and lies against the lower part of the muscular rim on the left side and the adjacent part of the tendinous centre. The posterior surface is related to the stomach, the duodenum, the gastro-hepatic flexure of the double colon, the pancreas, and the right kidney, the latter slightly indenting the upper part of the right lobe.

If an attempt be made to pull the liver from its position, it will be found that this is opposed by certain folds of peritoneum which pass between it and the abdominal parietes. These are the ligaments of the liver, and they are named as follows:—

The *Right Lateral Ligament* passes between the right lobe and the adjacent part of the phrenic rim.

The *Left Lateral Ligament* attaches the left lobe to the phrenic centre.

The *Falciform* or *Suspensory Ligament* attaches the lobulus quadratus to the diaphragm and to the abdominal floor a little to the right of the linea alba. Its posterior edge is concave and free, and contains the shrivelled remains of the umbilical vein—the so-called *round ligament*.

The *Ligament of the Caudate Lobe* is a small peritoneal fold passing between the anterior end of the right kidney and the lobulus caudatus. The *Coronary Ligament.*—If all the preceding ligaments be cut, and an attempt made to pull the liver out of position, it will be found that the gland is still firmly attached to the diaphragm by its anterior face. This adhesion takes place over an area that is traversed by the anterior fissure lodging the vena cava, and the peritoneum in passing between the gland and the phrenic centre on each side of this area constitutes the coronary ligament.

Directions.—The cœliac trunk and its branches must now be prepared; and, coincidently with this, the bile duct, portal vein, and solar plexus must be dissected. The portal vein will be found passing through the pancreas to the transverse fissure; and emerging from the fissure, below the vein, is the bile duct, which passes to open into the duodenum close by the principal pancreatic duct. The cœliac axis is concealed by the pancreas, which must be carefully raised by dissection at its anterior border, and pulled backwards. The same dissection will expose the semilunar ganglia and the solar plexus, whose branches are to be traced in company with the arteries. In dissecting the vessels, the student will meet the lymphatic glands of the stomach, spleen, and liver.

LYMPHATICS. The glands of the stomach form two groups, viz., (1) a few large glands situated at the lesser curvature, and (2) a number of smaller glands placed at the greater curvature. The glands of the liver also form two groups, viz., (1) a number situated in the posterior fissure, and (2) a group, between the portal vein and the pancreas. The glands of the spleen are placed on the course of the splenic vessels. The lymphatic vessels emanating from the stomach, liver, and spleen traverse these groups of glands; and after anastomosing with each other, they pass to the thoracic duct.

The BILE DUCT. This is the main duct for the conveyance of the bile from the liver to the intestine. It is formed at the portal fissure of the liver, by the union of secondary branches from the three principal lobes, and it passes between the layers of the gastro-hepatic omentum to penetrate the wall of the duodenum, about six inches from the pylorus. The excretory apparatus of the horse's liver has the peculiarity—shared by a few other animals—of being without a gall-bladder.

The CŒLIAC AXIS (Plates 43 and 44) is a collateral branch of the abdominal aorta, arising from the inferior face of that vessel between the pillars of the diaphragm. It is less than an inch in length, and it divides into three branches: the gastric trunk, the hepatic artery, and the splenic artery.

The HEPATIC ARTERY is directed obliquely forwards, downwards, and to the right, to gain the posterior fissure of the liver, which it penetrates in company with the portal vein and the bile duct. At first imbedded in the pancreas, it then passes over the duodenum, and reaches its destination by passing between the layers of the gastro-hepatic omentum. It crosses the posterior vena cava, from which it is separated by the foramen of Winslow. It gives off the following collateral branches :—

1. *Pancreatic Branches.*

2. The *Right Gastro-omental Artery,* which is, at its origin, of larger volume than the continuation of the parent trunk, crosses behind the duodenum ; and placing itself in the texture of the great omentum, it is carried round the greater curvature of the stomach to inosculate with the left gastro-omental artery. It gives off the pyloric and duodenal arteries, besides numerous omental and gastric branches. The *pyloric artery* is detached from the right gastro-omental artery near its origin, and sometimes it is a branch of the hepatic artery. It supplies the pylorus and the initial dilatation of the duodenum. The *duodenal artery* is detached from the right gastro-omental artery before that vessel crosses the duodenum ; and following the lesser curvature of the duodenum, in the narrow serous band that fixes the bowel, it meets, and inosculates with, the first artery from the left branch of the anterior mesenteric artery. The *omental branches* of the right gastro-omental are small and unimportant. The *gastric branches* pass from the concave side of the parent artery ; and bifurcating at the greater curvature, they are distributed to the right sac of the stomach on both its surfaces, where they anastomose with branches of the pyloric and gastric arteries.

The GASTRIC TRUNK is the central of the three terminal branches of the cœliac axis. After a course of a few inches downwards and forwards, it bifurcates to form the anterior and posterior gastric arteries.

The *Anterior Gastric Artery* reaches the anterior surface of the stomach by crossing the lesser curvature immediately to the right of the œsophagus.

The *Posterior Gastric Artery* descends to the lesser curvature of the stomach, where it divides into branches distributed on the posterior aspect of the organ.

The *Pleuro-œsophageal Artery* is a vessel constantly present, but variable as regards its origin. It may arise from the gastric trunk or one of its branches, or from the splenic artery. Passing through the foramen sinistrum along with the œsophagus, it enters the thoracic cavity, and there anastomoses with the œsophageal arteries, supplying the pulmonary pleura at the base of the lung.

The SPLENIC ARTERY is considerably larger than the gastric trunk or the hepatic artery. Under cover of the pancreas, it passes outwards between the left kidney and the cardiac extremity of the stomach. Reaching the spleen, it descends along the anterior border of that organ, beyond which it is continued as the left gastro-omental artery, From its convex side it gives off many large *splenic branches;* and from its concave side it emits *gastric branches,* which pass in the gastro-splenic omentum to reach the great curvature of the stomach, where they bifurcate to be distributed to both surfaces of the left sac.

The *Left Gastro-omental Artery* is the continuation of the splenic artery beyond the tip of the spleen. It passes in the texture of the great omentum to meet, and inosculate with, the right gastro-omental artery, advancing in the opposite direction. Besides *omental branches,* it emits *gastric branches,* which bifurcate at the great curvature of the stomach to be distributed to both its surfaces.

The PORTAL VEIN (Plates 43 and 44) is the trunk which collects the blood from the stomach, intestines, spleen, and pancreas, and conveys it to the liver, where, as will afterwards be described in connection with the liver structure, the vessel comports itself after the manner of an artery. The vessel is formed behind the pancreas, by the junction of the *anterior mesenteric vein* with a short trunk resulting from the union of the *posterior mesenteric* and *splenic veins.* It gains the upper face of the pancreas by passing through its substance, the perforation being termed the *pancreatic ring;* and descending in the gastro-hepatic omentum to the posterior fissure of the liver, it penetrates the substance of the gland in company with the bile duct and hepatic artery.

Anterior and *Posterior Mesenteric Veins,* satellites of the arteries of the same names, have already been described; but there is no venous trunk corresponding to the cœliac axis, the companion veins of the divisions of that artery behaving as follows :—

The *Splenic Vein* is the upward continuation of the *left gastro-omental vein.* After receiving the *posterior gastric vein,* it becomes one of the roots of the vena portæ, previously forming a short trunk by union with the *posterior mesenteric vein.*

The *Anterior Gastric Vein* joins the vena portæ in the posterior fissure of the liver.

The *Right Gastro-omental Vein* is continuous with the left vein of the same name, in the texture of the great omentum, opposite the middle of great curvature of the stomach. It receives gastric, omental, duodenal, pyloric, and pancreatic branches, all of which run in company with the arteries of the same names; and then, above the pancreas, it joins the portal vein.

The ŒSOPHAGEAL NERVES. These nerves are the backward continuations of the vagus, pneumogastric, or 10th cranial nerves. They reach the abdominal cavity by passing through the foramen sinistrum of the diaphragm, in company with the œsophagus and the pleuro-œsophageal branch of the gastric artery. The inferior nerve forms at the lesser curvature of the stomach a plexus whose filaments pass mainly to the right sac; while the superior, after giving branches to the left sac, joins the solar plexus.

The SPLANCHNIC NERVES. On each side there are two splanchnic nerves—a great and a small. Both are formed by efferent branches of the dorsal portion of the sympathetic gangliated cord, and they reach the abdomen by passing between the diaphragm and the psoas parvus muscle. The great splanchnic nerves terminate in the semilunar ganglia; the small nerves pass directly to the solar plexus, or they may be continued to the renal or the suprarenal plexus.

The SEMILUNAR GANGLIA are the largest in the body. They are placed one at each side of the lower face of the aorta, between the coeliac and anterior mesenteric arteries. Each receives the great splanchnic nerve of its own side, and the two ganglia communicate by transverse branches across the lower face of the aorta. The efferent branches which proceed from them form the solar plexus.

The SOLAR PLEXUS is an intricate network of nerves and ganglia. It is joined on each side by the lesser splanchnic nerve, and by the terminal filaments of the superior œsophageal nerve. From the plexus nerves pass to the abominal viscera, and in doing so they run in company with arteries. There is thus : a *cœliac plexus*, whose branches reach the liver, pancreas, spleen, and stomach, by accompanying the divisions of the hepatic, splenic, and gastric arteries ; a *renal* and a *suprarenal plexus*, which pass to the kidneys and suprarenal bodies ; an *aortic plexus*, continued backwards on the aorta to join the posterior mesenteric plexus; and an *anterior mesenteric plexus*, already described.

Directions.—The form, situation, and relations of the kidneys, and the course of the ureters should now be examined. Without displacing the kidneys, the fat and peritoneum is to be stripped from their lower face, their vessels being carefully cleaned at the same time. The ureter will be found passing backwards from the notched inner border of each kidney, and it is to be followed backwards to the entrance to the pelvis. In close relation to each kidney is its suprarenal body, which, to prevent displacement, may be transfixed in position with a long pin.

The KIDNEYS (Plates 44 and 47) are the two glandular bodies that secrete the urine. Each kidney occupies a position at the side of the vertebral column, on the inferior aspect of the loins, and at the roof of the abdominal cavity. In all except the most emaciated subjects, the kidneys are surrounded by a quantity of adipose tissue, which is so abundant in fat animals as to completely isolate them from surrounding objects. The most common shape of the mammalian kidney is so well known that it is popularly used as a descriptive term, objects having a similar form being described as "kidney-shaped." Each kidney possesses two surfaces, two borders, and two extremities. The inferior surface is convex ; the superior, which is concealed at present, is almost flat. The outer border is convex ; while the inner is concave, presenting a well-marked notch termed the *hilus*. From this hilus the ureter issues, and in its neighbourhood the renal vessels and nerves pass into or out of the kidney. The extremities are anterior and posterior, and both are rounded. It will at once be noticed, however (Plate 47), that although this description applies to both kidneys, they are far from being identical in shape. The right kidney has an outline somewhat like the " heart " of playing-cards, while the left has a decided resemblance to a haricot bean. The right has the longest transverse, but the shortest antero-posterior, diameter. The right is nearly symmetrical on each side of a line drawn from the hilus to the middle of the outer border; but if such a line be drawn on the left, the part in front of the line will be considerably smaller than the part behind it.

Furthermore, it will be noticed that the two kidneys differ in situation, and in relations. The right kidney is the more anterior; and taking their relation to the skeleton, the difference may be expressed thus: the right extends from the middle of the third last intercostal space at its upper end to a point beneath the 2nd lumbar transverse process ; the left extends from the second last intercostal space to the 3rd lumbar transverse process.

The right kidney is related by its upper face to the psoas muscles and to the rim of the diaphragm ; by its lower face it contacts with the pancreas and the crook of the cæcum, and is partly covered by peritoneum ; its inner border is margined by the posterior vena cava, and is in contact with the right suprarenal capsule in front of the hilus (sometimes behind); its outer border is in contact with the duodenum; its anterior extremity is in contact with the right lobe of the liver (which is slightly depressed for it) and with the lobulus caudatus, and to the latter a small fold of peritoneum passes from the lower face of the kidney.

The left kidney has the same relations on its upper face as the right ; its lower face is covered by peritoneum, and is related to the small intestines ; its inner border is margined by the aorta, and is related to the left suprarenal body in front of the hilus; its anterior border is related to the left extremity of the pancreas ; and the anterior half of its outer or convex border is related to the base of the spleen.

RENAL ARTERIES. Each kidney receives blood from a large vessel— the renal artery—which is a branch of the aorta. Springing at a right angle from the parent trunk, the artery passes towards the hilus, where it divides into a number of branches that penetrate the kidney substance. The left artery is short, and passes directly to the hilus ; the right is longer, and passes between the psoas parvus muscle and the vena cava to reach its destination. It is also generally a little anterior to the left in its point of detachment, both being a little behind the trunk of the anterior mesenteric artery.

The Renal Veins are as large in proportion as the arteries. They join the posterior vena cava, the right vein being the shorter and passing directly from the hilus, while the left crosses the lower face of the aorta behind the anterior mesenteric artery.

The Renal Plexus of nerves interlace around the artery, and enter the kidney with its branches.

The Suprarenal Capsules (Plates 44 and 47). These are two small solid bodies found in close relation to the kidneys, the right being between the vena cava and the inner border of the right kidney, and the left between the aorta and left kidney. They have an irregular elongated shape, and a slatey-brown colour. They are highly vascular, receiving branches from the mesenteric or renal arteries, and having veins that enter the posterior vena cava or the renal veins. They have also a rich nervous supply, receiving the suprarenal plexus—an offset from the solar plexus. Their substance consists of a *cortical* and a *medullary* portion. Each possesses a fibrous capsule continuous with an internal trabecular framework. The interspaces of these trabeculæ contain nucleated polyhedral or branched cells, which in the medulla and innermost layer of the cortex frequently enclose yellowish-brown pigment. They are without ducts, and their function is not well known.

The Ureters. The ureter is the tube which conveys the urine from the pelvis of the kidney to the urinary bladder. On the left side it has the following course. Beginning at the hilus, it is directed backwards and inwards across the lower face of the kidney to place itself at the side of the aorta, over (under, in the natural position) the psoas parvus muscle. Here it is crossed obliquely by the spermatic artery. It next curves a little outwards, crossing over the circumflex iliac artery and the artery of the cord, runs at the outer side of the external iliac artery, and then crosses it very obliquely to enter the pelvis. The right ureter has similar relations, except that it passes at the side of the vena cava instead of the aorta.

Directions.—Pin each ureter in position immediately behind the kidney, and then cut it across. Remove carefully the liver, stomach, duodenum, spleen, pancreas, and kidneys, by cutting the œsophagus and the various ligaments, vessels, and cellular adhesions which retain these organs in position. Put them in carbolic solution, or procure fresh organs, to serve for the examination of their structure. In the meantime proceed to dissect the sublumbar region.

THE SUBLUMBAR REGION.

Directions.—Under this heading there will be described the abdominal aorta and the vena cava, with their branches ; the inferior primary branch of the last dorsal nerve, and the corresponding branches of the first four lumbar nerves; the aortic plexus and gangliated lumbar cord of the sympathetic nerve ; and, lastly, a group of muscles, comprising the iliacus, psoas magnus, psoas parvus, quadratus lumborum, and lumbar intertransverse muscles. The great arterial and venous trunks are mesially placed, and the aortic plexus is on the great artery. The other structures enumerated are the same on both sides of the body. One side may be used for the nerves and the arterial and venous branches, the other being reserved for the muscles. In the mare, after the ovarian and uterine vessels have been examined, the broad ligaments must be cut to allow of the ovaries and uterus being pushed into the pelvic cavity, where they are to remain until they can be dissected along with the other reproductive organs of the female.

LYMPHATIC GLANDS. In cleaning the nerves and vessels, the following groups of lymphatic glands will be found :—1. *Sacral* glands, between the right and left internal iliac arteries at the entrance to the pelvis. 2. *Internal Iliac* glands, between and around the roots of the external and internal iliac arteries on each side. 3. *External Iliac* glands towards the point of bifurcation of the circumflex iliac artery. 4. *Lumbar* glands, on the lower face of the aorta around the roots of the posterior mesenteric and spermatic arteries. These various groups of glands are placed on the course of the lymphatic vessels of the hind limb, pelvis, and spermatic cord ; and the efferent vessels from the most anterior group (lumbar) pass to enter the receptaculum chyli.

The POSTERIOR AORTA (Plates 44 and 45). The abdominal portion of this great artery appears close to the spine, between the two pillars of the diaphragm, the opening being termed the *hiatus aorticus*. It passes backwards across the lumbar vertebral bodies, resting on the left pillar of the diaphragm and the inferior common ligament. At the 5th lumbar vertebra it terminates in four branches, two diverging to each side. These are the external and internal iliac arteries. On its right side the aorta is related to the vena cava. On the left it is related to the psoas parvus, the left lumbar sympathetic cord, and the left kidney and suprarenal capsule ; and the left ureter is beside or in actual contact with it. Besides the external and iliac arteries, which are described as its terminal branches, it gives off the following :—

1. *Phrenic Branches* (two or three) to the pillars of the diaphragm. They are given off at the hiatus aorticus.

2. *Lumbar Arteries.* There are six or seven of these on each side. The last comes from the lateral sacral artery, the second last from the internal iliac, and the others from the aorta. These last arise from the upper aspect of the vessel, and divide into two branches—a superior for the skin and muscles over the lumbar vertebræ, giving also a spinal twig through the intervertebral foramen; and an inferior which passes outwards in the intertransverse spaces to the flank, where it anastomoses with the circumflex iliac artery in supplying the abdominal muscles.

Y

3. The *Middle Sacral Artery* is an extremely slender vessel, and not always present. Search for it in the angle between the internal iliacs. Arising from the summit of that angle, it passes mesially backwards on the sacrum. It is of interest as representing the large vessel which in some animals continues the aorta to the coccygeal region.

4. The *Cœliac Axis* is detached as soon as the aorta passes through the hiatus aorticus.

5. The *Anterior Mesenteric*—the largest of the branches—is detached at the 1st lumbar vertebra.

6. The *Renal Arteries*, right and left, arise from the sides of the aorta at the articulation between the 1st and 2nd lumbar vertebræ.

7. The *Spermatic Arteries*, right and left, come off a few inches behind the renals, viz., between the 3rd and 4th lumbar vertebræ, and one generally a little in advance of the other. As seen in the dissection, each passes obliquely backwards and outwards over the ureter and circumflex iliac artery to gain the internal abdominal ring, where it joins the other constituents of the spermatic cord. In the cord it has a remarkably tortuous disposition; and, although a long vesssel, it detaches no branches of any size until it reaches the testicle. In the mare it is represented by the *ovarian artery*, which passes in a tortuous manner between the layers of the uterine broad ligament to reach the ovary. It gives off a uterine branch to the uterine horn.

8. The *Posterior Mesenteric Artery.* This vessel is usually detached at the 4th lumbar vertebra, a little behind the origin of the spermatics, but this relationship may be reversed.

The EXTERNAL ILIAC ARTERY is, speaking generally, the vessel of supply to the hind limb. It is regarded as a terminal branch of the aorta, and it has its root at the body of the 5th lumbar vertebra. It descends with a curved course at the pelvic inlet, and at the anterior border of the pubis it is directly continued as the femoral artery. It is placed immediately beneath the peritoneum, and each is related on its outer side to the psoas parvus, sartorius, and iliacus muscles, the tendon of the first of these separating it from the great crural nerve. On the inner side it is related successively to the common iliac and external iliac veins, the former separating it from the internal iliac artery. Its branches are:—

1. The CIRCUMFLEX ILIAC ARTERY. This is a large artery detached from the outer side of the external iliac close to its origin. It passes outwards across the psoas muscles; and at the outer edge of the psoas magnus it divides into an anterior and a posterior branch. The former is distributed in the flank, beneath the internal oblique muscle; and the latter perforates the oblique muscles near the bony angle of the haunch, and descends to the thigh.

2. The ARTERY of the CORD (Plate 44). This is a slender vessel arising close to the preceding, or it may come from the aorta itself.

Parallel and internal to the spermatic artery, it passes to the internal abdominal ring, and is distributed to the spermatic cord.

In the mare it is represented by the *uterine artery.* This, which is a much larger vessel, passes between the layers of the broad ligament to reach the uterus, being distributed to the body of that organ, and anastomosing anteriorly with the uterine branch of the ovarian artery, and posteriorly with the vaginal artery.

3. The PREPUBIC ARTERY (Plates 45 and 46). This vessel arises at the anterior border of the pubis, and marks the limit of the iliac and femoral arteries. It forms, at its origin, a short common trunk with the deep femoral artery. It is about two or three inches in length, and it passes on the anterior face of Poupart's ligament to the posterior edge of the internal oblique, where, at the inner side of the internal abdominal ring, it divides into the posterior abdominal and external pudic arteries.

The INTERNAL ILIAC ARTERY. This vessel may be described as the vessel for the supply of the pelvic walls and contents. It will be described with the pelvis.

The POSTERIOR VENA CAVA (Plates 43, 44, and 45). This great venous trunk is formed to the right of the termination of the aorta, by the union of the two common iliac veins. It passes forwards along the right side of the lumbar vertebral bodies, until it reaches the upper border of the liver. Here it descends in the anterior fissure of the liver, being included between that organ and the diaphragm. Passing through the foramen dextrum, it enters the thorax. It is related on its left side to the aorta; and on its right to the psoas parvus, ureter, kidney, suprarenal capsule, and lumbar sympathetic cord of the same side. It receives the following branches:—

1. *Lumbar Veins*, exactly corresponding to the arteries.

2. *Phrenic Veins*, or sinuses (2). These begin in the muscular rim of the diaphragm, and converge to the foramen dextrum, where they join the vena cava. They are distinctly visible without dissection in the tendinous centre of the diaphragm (Plate 45).

3. *Spermatic Veins* (*Ovarian* in the mare). The right and left veins often unite before joining the vena cava.

4. *Renal Veins*, the left longer than the right.

5. *Hepatic Veins*. These join the vena cava while it lies in the anterior fissure of the liver. They discharge the blood of the portal system of veins, after it has circulated in the liver.

ILIAC VEINS. There are *external* and *internal iliac* veins, with branches corresponding in all respects to the divisions of the homonymous arteries. The external and internal iliac veins of each side, however, unite and form a short trunk termed the *common iliac* vein, which is placed in the angle of separation between the external and internal iliac arteries. The right and left common iliac veins unite to form the posterior vena cava.

The RECEPTACULUM CHYLI. Separate the aorta and vena cava at the origin of the anterior mesenteric artery, and look above them for this. It is the dilated commencement of the thoracic duct. It is formed by the union of a variable number of large lymphatic vessels, and it is continued forwards by the duct. This is a thin-walled vessel of small calibre which passes into the thorax between the pillars of the diaphragm, being generally to the right of the aorta.

LAST DORSAL and FIRST TWO LUMBAR NERVES (Plates 44 and 45). The inferior primary branches of these nerves appear at the outer edge of the psoas magnus, the last dorsal being close behind the last rib, and the other two issuing in series behind it. These nerves have already been followed in the dissection of the abdominal wall, where they are distributed in the region of the flank to the abdominal muscles, panniculus, and skin. These lumbar nerves also furnish cutaneous branches to the inside and front of the thigh.

3RD LUMBAR NERVE (Plates 44 and 45). The inferior primary branch of this nerve will be found in front of the circumflex iliac artery, emerging from between the psoas magnus and parvus muscles, after having penetrated the substance of the latter. It accompanies the posterior division of the circumflex iliac artery to the front of the thigh, where it is expended in cutaneous branches. Before it emerges, it gives branches to the psoas and quadratus muscles.

INGUINAL NERVES. There is considerable variation in the mode of formation of these, but that figured in Plates 44 and 45 is probably as common as any other. A nerve is there seen passing obliquely backwards over the circumflex iliac artery. It is formed by the union of two branches which emerge at the inner side of the psoas parvus, these being from the 2nd and 3rd lumbar nerves respectively. It divides into three sets of branches, viz. :—

1. *Muscular*, to the internal oblique.

2. *Cremasteric*, to the cremaster muscle.

3. *Inguinal*, which descend in the inguinal canal to supply the scrotum, prepuce (mammary gland in the female), and surrounding skin.

The LUMBO-SACRAL PLEXUS (Plate 48). This is the plexus of nerves for the supply of the hind limb. Like the corresponding brachial plexus, the inferior primary branches of five nerves compose it, viz., the 4th, 5th, and 6th lumbar, and the 1st and 2nd sacral nerves. There is a loop of communication between the first of these and the 3rd lumbar, which to that extent also enters into the formation of the plexus. The majority of its branches fall to be dissected with the pelvis, and a complete account will then be given of it (page 349). In the meantime only the most anterior of its branches will be dissected.

1. *Branches* to the *psoas magnus* and *iliacus*. These are derived from the 4th lumbar root of the plexus, or from the loop between that and the 3rd.

2. The *Anterior* or *great crural nerve* (Plate 45). This is a large nerve which derives its fibres from the 4th and 5th lumbar roots, and from the loop between the 3rd and 4th. Emerging between the psoas magnus and parvus, it descends at the outer side of the external iliac artery, but separated from it by the tendon of the last-named muscle. It rests on the iliacus and psoas muscles, and crosses their common termination to end in a fasciculus of branches for the extensors of the leg. In this course it is covered by the sartorius muscle. It gives off as a branch the *internal saphenous nerve*, whose origin is about opposite the ilio-pectineal eminence.

The AORTIC PLEXUS of the SYMPATHETIC NERVE (Plate 45). This is the backward continuation of the solar plexus. Its branches interlace around the aorta behind the kidneys, and unite with the posterior mesenteric plexus. It receives some of the efferent filaments of the lumbar ganglia.

THE POSTERIOR MESENTERIC PLEXUS is formed around the root of the artery of the same name. It is united in front with the aortic plexus, and receives efferent branches from the lumbar ganglia. Three sets of branches pass from it:—

1. Branches following the divisions of the posterior mesenteric artery.

2. Branches accompanying the spermatic artery, and forming the *spermatic plexus.*

3. Pelvic branches to join the *pelvic plexus.*

The SYMPATHETIC GANGLIATED CORD in the loins. This is the backward continuation of the dorsal cord. Beginning between the psoas parvus and the diaphragmatic crus, it extends backwards to the lumbo-sacral articulation, where it is directly continued by the sacral division of the cord. Each nerve will be found on the inner aspect of the psoas parvus muscle of the same side, the left nerve being related inwardly to the aorta; and the right for the greater part of its course to the vena cava, but for a short distance in front to the aorta. Six fusiform, greyish ganglia stud the cord, and from these proceed the various branches of the cord. These branches are :—

1. Communicating branches with the inferior primary divisions of the lumbar spinal nerves.

2. Branches to the aortic and posterior mesenteric plexuses.

ILIAC FASCIA. This is the name given to the aponeurotic layer which covers the inferior face of the psoas magnus and iliacus muscles. It is densest and most adherent at the side of the pelvic inlet, and becomes more cellular as it is traced forwards and backwards. It is adherent inwardly to the psoas parvus tendon, and outwardly to the bony prominence of the haunch. Poupart's ligament is adherent to its inferior face, and it gives origin to the sartorius and cremaster muscles.

The PSOAS MAGNUS (Plates 44 and 45). This muscle is broad and flattened anteriorly, and thick and pointed behind, where it rests in a depression of the iliacus. It *arises* from the last two dorsal vertebræ and the under surfaces of the last two ribs at their upper part; also from the lumbar vertebræ except the last, covering their transverse processes. It is *inserted*, in common with the iliacus, into the internal trochanter of the femur.

Action.—It is a flexor and an outward-rotator at the hip-joint. When the hind limbs are fixed, the two muscles will arch the loins, or the single muscle will incline the trunk to the same side.

The Psoas Parvus (Plate 45). This is a smaller and more tendinous muscle than the preceding, to whose inner side it is placed. It *arises* from the bodies of the last three or four dorsal and all the lumbar vertebræ. It is *inserted* into the ilio-pectineal eminence (of the ilium) at the side of the pelvic inlet.

Action.—To flex the pelvis on the loins when both musles act; or to incline it laterally when a single muscle acts. If the pelvis be fixed, it will execute the same movements on the loins.

Directions.—Raise the outer edge of the psoas magnus, and remove it except its fibres of origin beneath the rim of the diaphragm and its conical tendon of insertion, as in Plate 45.

The Iliacus (Plates 45 and 16). This is a powerful fleshy muscle which, when the psoas magnus muscle is in position, appears to consist of an outer and an inner portion. In reality, however, it is a single mass, with a deep groove in it for the terminal tendon of the psoas magnus. It *arises* from the entire iliac surface of the ilium, from its external angle, and from the sacro-iliac ligament. It is *inserted* into the inner trochanter of the femur, in common with the psoas magnus.

Action.—The same as the psoas magnus.

The Quadratus Lumborum (Plate 45). This muscle lies under cover of the great psoas muscle. Its most external and strongest fasciculus *arises* from the sacro-iliac ligament. It is *inserted* by this same fasciculus into the tips of the lumbar transverse processes, and into the hinder edge of the last rib. From the main fasciculus others pass inwards to the lumbar transverse processes, and to the under surfaces of the three last ribs, close to the spine.

Action.—To assist in bending the loins to the side of the acting muscle. Both muscles, by fixing the last ribs, will enable the diaphragm to act to more advantage.

Lumbar Intertransverse Muscles (Plate 45). These are thin muscular and tendinous strata connecting the edges of adjacent lumbar transverse processes.

Action.—To assist in bending the loins to the side on which the muscles act.

Directions.—An examination of the diaphragm will complete the dissection of the abdomen. Define its pillars attaching it to the lumbar vertebræ, and clean the edges of its foramina. At its periphery, under the costal cartilages, follow the asternal vessels (Plate 45).

The ASTERNAL ARTERY. This is one of the divisions of the internal thoracic artery (Fig. 7, page 120). It passes from the thorax to the abdomen by perforating the rim of the diaphragm about the 9th chondro-costal joint. As here seen, it passes backwards at the rim of the diaphragm, in the interval between it and the origin of the transversalis abdominis, and terminates at the lower extremity of the 13th intercostal space. It has three sets of branches, viz., (1) ascending branches, which anastomose with the intercostal arteries of the spaces crossed; (2) internal branches to the diaphragm; (3) descending branches, which run on the peritoneal surface of the transversalis abdominis muscle.

The ASTERNAL VEIN accompanies the artery.

The DIAPHRAGM (Plate 45) is the muscle which serves as a partition between the thoracic and abdominal cavities. In outline it has some resemblance to the heart of playing-cards, the point being at the ensiform cartilage, and the base at the spine. Its general direction is oblique downwards and forwards. Its anterior or thoracic surface (Plate 22) is convex, covered by pleura, and related to the bases of the lungs. Its posterior surface is concave, covered for the greater part by peri-toneum, and related to the liver, stomach, spleen, and intestines. It consists of a muscular rim, two muscular pillars, or crura, and an aponeurotic centre.

The *Fleshy Rim* is composed of soft muscular fibres, the lowest of which are attached outwardly to the upper face of the ensiform cartilage about one inch behind its junction with the sternum (Fig. 7, page 120). From this mid point the line of attachment of the rim rises on each side, the fibres taking origin from the cartilages of the last ten ribs, or from the ribs themselves above the chondro-costal joints. On each side these fibres meet, or are separated by only a narrow line from, the fibres of the transversalis abdominis at their origin ; and along the line of separa-tion the asternal artery runs. The muscular fibres are all directed from these points of origin inwards, where, along a denticulated line, they terminate in the tendinous centre.

The *Pillars*, or *Crura*.—These are right and left. The right is the largest, and arises by a strong tendon from the lumbar vertebræ, through the medium of the inferior common ligament. Its muscular fibres terminate in the tendinous centre, some of them diverging to the right, but without joining the muscular rim, while others descend to near the mid point. The left pillar has a similar origin from the left side of the lumbar vertebræ, and its fibres terminate in the tendinous centre, being sometimes continuous outwardly with the muscular rim.

The *Tendinous Centre* is pearly white, and composed of glistening fibres interlacing in various directions. By the descent of the pillars into it, it is partially divided into right and left halves, or leaflets.

Ligamenta Arcuata.—On each side of the pillars the rim of the diaphragm arches with a free edge over the apices of the psoas muscles, forming the so-called arcuate ligament.

Foramina of the diaphragm.

The *Foramen Sinistrum.*—This is a slit between the fibres of the right crus, formed slightly to the left of the mesial plane, and a little below the spine. It transmits the œsophagus, the œsophageal continuations of the vagus nerves, and the pleuro-œsophageal branch of the gastric artery.

The *Foramen Dextrum.*—This is the aperture by which the posterior vena cava passes from the abdomen to the thorax. It is formed near the middle of the tendinous centre, but a little to the right of the mesial plane. The margins of the opening are closely adherent to the wall of the vein, and here the phrenic sinuses empty.

The *Hiatus Aorticus.*—This is the opening between the right and left pillars, close to the spine. It gives passage to the posterior aorta, and to the initial portions of the thoracic duct and great azygos veins.

Between the crus and the psoas parvus on each side the gangliated cord of the sympathetic passes, and a little outward the great splanchnic nerve passes between the same muscle and the edge of the diaphragm. The asternal vessels penetrate the edge of the diaphragm at the 9th chondro-costal joint.

Action of the Diaphragm.—The diaphragm is the principal muscle of inspiration. When it contracts, it moves backwards, and thus increases the antero-posterior diameter of the thorax. In this action it pushes back the abdominal viscera, and causes the abdominal wall to descend. The movements of the diaphragm affect principally its periphery, any great backward movement of the tendinous centre being prevented by the posterior vena cava, which passes like a ligament between the centre and the heart.

STRUCTURE OF THE STOMACH.

The wall of the stomach comprises four layers, viz., serous, muscular, submucous, and mucous.

1. The *Serous Coat* is a smooth, glistening covering derived from the critoneum. It is united to the subjacent muscular coat by areolar tissue sometimes termed the *subserous* coat.

2. The *Muscular Coat.*—This can be best displayed on a stomach which has been boiled for a few minutes. If two such stomachs can be procured, one of them should be everted and moderately inflated, and then its mucous coat stripped off with fingers and forceps. From the other, similarly inflated, the peritoneum should be stripped off. The muscular fibres are disposed in three planes, viz., an outer longitudinal, a middle circular, and an inner oblique layer. Of these the

circular layer is found all over the organ, but the other layers are mainly confined to the left half. At the right extremity of the stomach the circular fibres are aggregated to form the sphincter-like *pyloric ring.* The fibres are of the non-striped variety.

3. The *Submucous Coat* is composed of areolar connective-tissue, in which the blood-vessels ramify before they pass into the next coat.

4. The *Mucous Coat.*—It is desirable to study this on the stomach of an animal recently killed. If possible, take such a stomach with about a foot of the duodenum and a few inches of the œsophagus attached, and fasten the duodenum to a tap. Let water flow into the organ, and it will be noticed that, even when the stomach is much distended, none of the water escapes by the orifice of the gullet, although that is unligatured. This is an instructive experiment, as showing the difficulty or impossibility of vomition in the horse. Now allow the contents of the stomach to escape by the duodenum ;

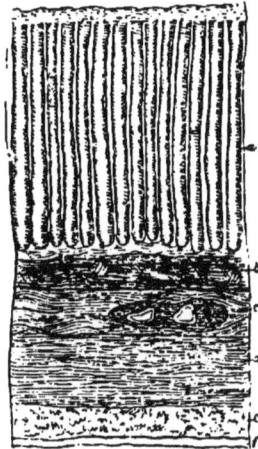

FIG. 38.

VERTICAL TRANSVERSE SECTION OF THE COATS OF A PIG'S STOMACH. 30 DIAMETERS (from *Kölliker*).

a. Gastric glands ; b. Muscular layer of the mucous membrane ; c, Submucous or areolar coat ; d. Circular muscular layer ; e. Longitudinal muscular layer ; f. Serous coat.

and either evert the organ and inflate it, or incise it along its convex curvature. It will at once be noticed that the mucous lining is not the same throughout. The left or *cardiac* half of the cavity is lined by a mucous membrane termed *cuticular ;* the right or pyloric half has a totally different lining, termed *villous.* The cuticular portion is pale, harsh, without true gastric glands, but possessed of a few mucous follicles, and covered on its free surface by a thick stratified squamous epithelium. It is, in fact, an extension of the œsophageal mucous membrane, which it resembles in all respects. Towards the middle of the stomach it is separated from the villous half by an abrupt, raised, and slightly sinuous line of demarcation—the *cuticular ridge.* The villous half is rosy, soft, and velvety (but without villi), thickly beset with gastric glands, and possessed of a single layer of columnar epithelium. The gastric glands are of the tubular variety, and by the aid of a lens numbers of them may be seen opening together into pits, or *alveoli*, of the mucous membrane. The cuticular portion is but slightly vascular, but the villous portion is richly supplied with blood-vessels. In the collapsed organ the mucous membrane is thrown into folds, or *rugæ.*

The ŒSOPHAGEAL ORIFICE, it will now be seen, is very narrow, and obstructed by the mucous membrane gathered into folds.

The PYLORIC ORIFICE is much larger, but capable of being completely closed by the pyloric ring of muscular fibres.

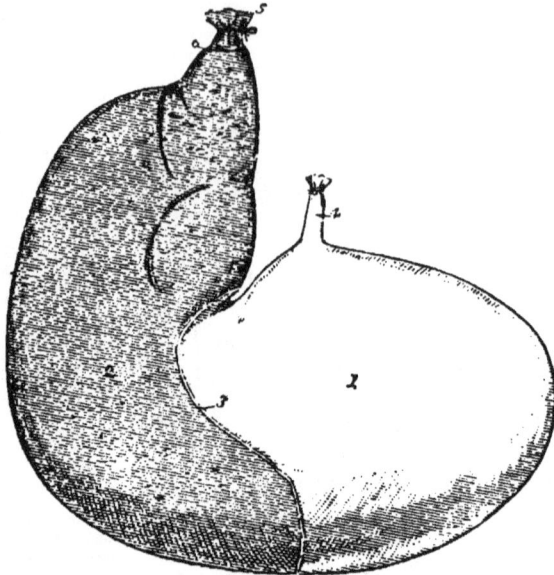

FIG. 39.

STOMACH, EVERTED AND INFLATED.

1. Left (cardiac) sac with its *cuticular* mucous lining; 2. Right (pyloric) sac with its *villous* mucous lining; 3. Cuticular ridge; 4. Termination of œsophagus; 5. Initial part of duodenum; 6. Pyloric ring.

In the interior of the duodenum, about six inches from the pylorus, the openings of the bile and pancreatic ducts will be found. The orifices of the bile duct and duct of Wirsung are placed together on the concave side of the bowel, and are surrounded in common by a ring-like valvular fold of mucous membrane. The opening of the accessory pancreatic duct is placed opposite to these.

STRUCTURE OF THE LIVER.

Lay the organ with its diaphragmatic surface downwards. Find the portal vein, hepatic artery, and bile duct, at the portal fissure, and trace them for a little distance into the liver. Invert the organ, and observe the course of the anterior vena cava in the anterior fissure, and the mouths of the hepatic veins which there discharge themselves into the cava.

Tunics or *Capsules* of the liver. These are two in number: 1. A *peritoneal coat*, giving the free surface of the organ its smooth and glistening characters. 2. A *tunica propria*, or *fibrous coat*, placed beneath the preceding. All over the surface of the liver it sends inwards delicate processes that join the interlobular connective-tissue, and at the portal fissure it furnishes a sheath that accompanies the portal vein, hepatic artery, and bile duct into the liver. This sheath is the *capsule of Glisson.*

Lobules of the Liver.—When a fresh-cut surface of the liver is examined, it shows a system of lines mapping it out into areas about the size of a pin's head. These areas are sec-
tions of the lobules of the liver, which are united together by interlobular connective-tissue. This interlobular connective-tissue is much more abundantly developed in the pig, and, consequently, in that animal the lobulation of the liver substance is much more evident. A lobule may be viewed as having a framework of blood-vessels, in which are set the liver-cells. Between the adjacent cells the rootlets of the bile passages begin, and there are possibly also branches of nerves and lymphatic vessels.

The liver is supplied with blood by two vessels. The first and much the larger of the two is the portal vein, the other is the hepatic artery.

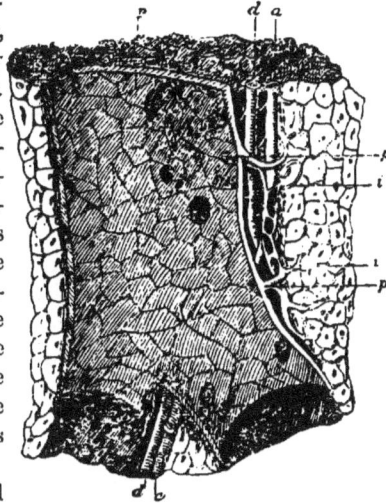

FIG. 40.

LONGITUDINAL SECTION OF A PORTAL CANAL, CONTAINING A PORTAL VEIN, HEPATIC ARTERY, AND HEPATIC DUCT, FROM THE PIG (after *Kiernan*). ABOUT 5 DIAMETERS.

P. Branch of vena portæ, situated in a portal canal, formed amongst the hepatic lobules of the liver; *p. p.* Larger branches of portal vein, giving off smaller ones (*i. i.*), named interlobular veins; there are also seen within the large portal vein numerous orifices of interlobular veins arising directly from it; *a.* Hepatic artery; *d.* Biliary duct.

The PORTAL VEIN collects its blood from the stomach, intestines, spleen, and pancreas. Entering the liver at the portal fissure, this vein comports itself like an artery,

in that it reduces itself by division and subdivision to branches that come progressively smaller until they terminate in a set of capillaries. In their course through the liver, the larger branches of the vein run in tunnels of the liver substance—the portal canals—which contain also branches of the hepatic artery and bile ducts, and are lined by Glisson's capsule. The smaller branches of the portal vein are distributed in the interlobular connective-tissue, where, at the circumference of each lobule, they form an *interlobular plexus.* From this plexus capillary vessels penetrate the lobule, and form within it the *intralobular plexus.* The capillaries of this last plexus converge towards the axis of the lobule, and there empty themselves into what is termed the *central vein* of the lobule. This is the initial vessel of the hepatic system of veins, and at the base of the lobule it joins a larger vessel—

the *sublobular vein*. By the union of these sublobular veins through-out the liver, the larger *hepatic venous trunks* are formed; and these, as already seen, enter the posterior vena cava in the anterior fissure of the liver.

The HEPATIC ARTERY is a branch of the cœliac axis. It enters the liver with the portal vein, and ramifies with it. It has three sets of branches: (1) *capsular branches*, to the tunica propria; (2) *vaginal branches*, to Glisson's capsule and the vessels within it; and (3) *inter-lobular branches*, whose capil-laries pass into the lobule, where they help to form the intralobular plexus, and enter the central vein. The capil-

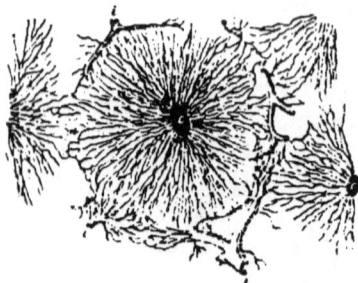

FIG. 41.

TRANSVERSE SECTION THROUGH THE HEPATIC LOBULES (*Turner*).

i, i, i. Interlobular veins ending in the intralobular capillaries; *c, c.* Central veins joined by the intra-lobular capillaries. At *a, a.* the capillaries of one lobule communicate with those adjacent to it.

laries of the vaginal and capsular branches terminate in veins that join the portal vessels.

The *Liver Cells.*—These are granular nucleated masses of protoplasm, often containing fat particles. They are arranged in columns between the strands of the intralobular plexus of capillaries.

The *Bile Passages* begin within the lobule as a network of fine canals —the *bile capillaries*—tunnelled at the lines of apposition of the liver cells. At the periphery of the lobule these become continuous with interlobular bile ducts having a proper wall and a simple columnar epithelial lining. The interlobular bile ducts unite to form the larger ducts that accompany the blood-vessels in the portal canals, and these finally form the main bile duct, which passes in the gastro-hepatic omentum to perforate the wall of the duodenum.

STRUCTURE OF THE SPLEEN.

The spleen, like the liver, possesses two coats, viz., an outer *serous* or *peritoneal coat*, and a deeper *fibrous tunic*, or *tunica propria*. The latter is composed of white fibrous tissue with a considerable admixture of elastic and non-striped muscular fibres. It detaches from its inner sur-face a multitude of *trabeculæ*, which by their anastomosis form a fibrous framework in the interior of the organ. The interspaces of this frame-work are occupied by a grumous material—the *splenic pulp*. If the cut surface of the spleen be washed beneath a tap, the pulp may be removed and the fibrous trabeculæ rendered very evident.

The *Splenic Artery*, a division of the cœliac axis, is a very large vessel. Its branches enter at the concave border of the spleen, and carry with them sheaths derived from the fibrous tunic. These branches reduce themselves by division, and the smaller branches are remarkable in having the outer coat formed of lymphoid tissue   Here and there this lymphoid tissue forms distinct swellings developed either uniformly around the arteries, or more or less to one side. These are the *Malpighian bodies* of the spleen. The arteries terminate in tufts of capillary vessels in the pulp. They are believed to have incomplete walls, allowing their contents to escape and form the pulp. The rootlets of the splenic vein begin in the same manner, having incomplete walls through which their lumen is continuous with the spaces lodging the pulp. Gradually their walls become thicker and complete, and adjacent veins uniting on their course towards the anterior border form the large *splenic vein*, which is one of the main branches of the portal vein.

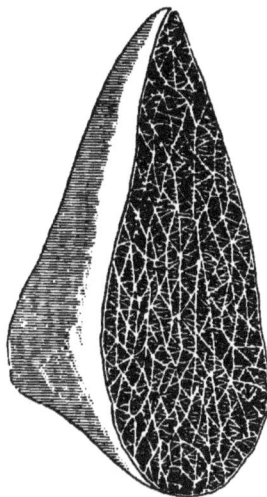

FIG. 42.

CUT SURFACE OF HORSE'S SPLEEN, TRABECULAR FRAMEWORK.

The *Splenic Pulp* possesses a supporting network of retiform connective-tissue; and the meshes of this network are set with many lymphoid cells like the colourless corpuscles of the blood, and with red blood corpuscles, normal or in different stages of disintegration

STRUCTURE OF THE PANCREAS.

The pancreas is a compound tubular or racemose gland. It is composed of lobules held together by a connective-tissue framework. When the main ducts of the gland are traced backwards into the gland, they are found to be formed by the union of smaller ducts, and so on until the smallest ducts are reached. These begin in the *alveoli*, which are lined by secretory epithelium.

STRUCTURE OF THE KIDNEY.

The kidney is invested by a *fibrous capsule*. In health this can without difficulty be stripped off the kidney substance, to which it is connected only by delicate processes and vessels. If a horizontal section be made from the convex border to the hilus of the kidney, the organ will be seen to possess a cavity towards the hilus, termed the *pelvis*, and to consist of two different kinds of tissue—the *cortical* and the *medullary substance* of the kidney.

The renal pelvis is a curved cavity, its extremities being termed the arms. On its outer side there is a horizontal ridge—the *renal crest*—on which the uriniferous tubules open, and on its inner side it is continuous by a funnel-shaped opening with the lumen of the ureter.

The cortical substance forms a layer beneath the capsule; the medulla is disposed around the pelvis and is internal to the cortical substance. The cortex is about twice as thick as the medulla, but the two layers meet along a sinuous line, and slightly interpenetrate one another. It will be noticed that the two layers contrast with one another in the following respects :—The cortex is of a deep red colour, it is granular, friable, and studded with numerous small shining points—the *Malpighian bodies*. The medulla, on the other hand, is pale red, striated, and fibrous-looking, less friable than the cortex, and without any Malpighian bodies.

Uriniferous tubules.—The largest tubes, or *papillary ducts*, open on the crest of the pelvis. If such a tube be traced, it will be found to pass outwards through the medulla, having a straight course, and branching dichotomously. The smaller tubes resulting from this division are called the *collecting tubes;* and, still preserving their rectilinear course, they enter the cortex in bundles termed the *pyramids of Ferrein.* At the surface of these pyramidal bundles, the straight tubes curve outwards in the cortex, and become dilated and tortuous, forming the *intermediary* or *junctional tubules.* Each of these is succeeded by a narrow straight tubule, which descends from the cortex to the medulla, where it forms a bend, or loop, and runs up again into the cortex. There is thus formed the *looped tube of Henle,* which is shaped like the letter U. Having re-entered the cortex, Henle's tube becomes dilated and tortuous, constituting the *convoluted tube,* which becomes constricted and then expands into a bladder-like dilatation—*Bowman's capsule.* Bowman's capsule surrounds a clue-like tuft of capillary vessels called the *glomerulus,* and the whole constitutes a *Malpighian body.* It is more natural, but less simple at first, to regard the tube as beginning not at the crest of the pelvis, but at Bowman's capsule. The student should mentally work it out in that direction for himself. The uriniferous tubules consist of a basement membrane with an epithelial lining. In the convoluted and intermediary tubes the cells are irregularly columnar, but their outlines are obscure; in the descending limb of Henle's tube (nearest the capsule of Bowman) the cells are flattened; and elsewhere the cells lining the tubes are cubical or columnar.

The RENAL VESSELS. The *renal artery* divides into a number of branches which penetrate the kidney near the hilus. Reaching the boundary line between the cortex and medulla, the arteries divide and anastomose to form a series of arches from which both cortical and medullary vessels arise.

The *cortical* or *interlobular arteries* are larger and more numerous than those for the medulla. They pass directly outwards towards the surface of the kidney, giving off lateral branches—the *vasa afferentia*—to Bowman's capsule, and terminal branches to the fibrous coat of the kidney. Each *vas afferens* pierces Bowman's capsule, and resolves itself into the *glomerulus,* or capillary tuft. From this again the blood is led out of Bowman's capsule by the *vas efferens.* The *vasa efferentia* again resolve themselves into capillaries, and these form a network among the convoluted tubes. From this intertubular capillary network, small veins arise and pass to join the *interlobular veins,* running alongside the arteries. These interlobular veins begin at the surface of the kidney by the convergence of a number of minute veins from the capsule—forming the *stellate veins.* The interlobular veins join venous arches disposed in

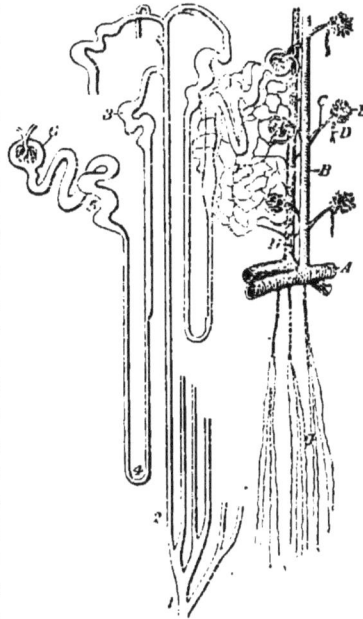

FIG. 43.

VESSELS OF THE KIDNEYS, AND URINIFEROUS TUBULES (modified from *Turner*).

1. Papillary duct; 2. Collecting tube; 3. Intermediary tube; 4. Looped tube of Henle; 5. Convoluted tube; 6. Bowman's capsule; A. Segment of artery forming renal arch; B. Interlobular artery C. Afferent vessel of glomerulus; D. Efferent vessel of the same; E. Glomerulus; F. Plexus formed by vasa efferentia; G. Arteriolæ rectæ; H. Interlobular vein.

the boundary layer between cortex and medulla, and from these arise the larger branches that finally unite to form the large *renal vein* at the hilus.

The medulla is less vascular than the cortex. Springing from the arterial arches in the boundary layer are branches that break up into pencils of long straight arterioles—the *arteriolæ rectæ.* These pass with a rectilinear course between the straight tubules of the medulla, and break up into a wide-meshed capillary network around and between these tubules. Veins having a straight course like the arteries run in company with them, and join the venous arches in the boundary layer.

Connective-tissue of the Kidney.—This exists very sparingly between the tubes in the cortex, but more abundantly in the medulla.

NAME OF MUSCLE.	ORIGIN.	INSERTION.	SOURCE OF NERVE.
Serratus posticus	Vertebral spines, 11th dorsal to 2nd lumbar	Ribs, posterior borders and outer surfaces of eight or nine last	Dorsal nerves.
Serratus anticus	Vertebral spines, 2nd or 3rd to 13th	Ribs, 5th to 13th, anterior borders and outer surfaces	Dorsal nerves.
Transversalis costarum	Lumbar vertebræ, first two, transverse processes; and ribs, anterior borders	Ribs, 1st to 14th, posterior edges; and 7th cervical vertebra, transverse process	Dorsal nerves.
Longissimus dorsi	Ilium, sacral surface; lumbar and dorsal spines (or supraspinous ligament)	Ribs; lumbar vertebræ, transverse and articular processes; dorsal vertebræ, transverse processes; and cervical vertebræ, last four transverse processes, and four spines in front of last	Cervical, dorsal, and lumbar nerves.
Retractor costæ	Lumbar vertebræ, first two or three transverse processes	Last rib, posterior edge	Last dorsal and first lumbar nerve.
Levatores costarum	Dorsal vertebræ, transverse processes	Ribs, outer surfaces	Dorsal nerves.
Semispinalis of the back and loins	Sacrum, lateral lip; lumbar vertebræ, articular tubercles; and dorsal vertebræ, transverse processes	Lumbar and dorsal vertebræ, spines	Dorsal and lumbar nerves.
Lateralis sterni	1st rib, outer surface	Sternum, lateral surface	Intercostal nerves.
External intercostals	Ribs (except last)	Ribs (except first)	Intercostal nerves.
Internal intercostals	Ribs and cartilages (except first)	Ribs and cartilages (except last)	Intercostal nerves.
Triangularis sterni	Sternum, edge of thoracic surface	Costal cartilages, 2nd to 8th; and aponeurosis over internal intercostal muscles	Intercostal nerves.
Obliquus abdominis externus	Ribs, last fourteen, outer surfaces; and tendon of latissimus dorsi	Linea alba; prepubic tendon; ilium, external angle; (and Poupart's ligament)	Last ten intercostal, last dorsal, and first two lumbar nerves.
Obliquus abdominis internus	Ilium, external angle; and Poupart's ligament	Prepubic tendon, linea alba, and four or five last costal cartilages	Last ten intercostal, last dorsal, and first two lumbar nerves.
Rectus abdominis	Sternum, lower face; and costal cartilages, 5th to 9th	Pubis (by prepubic tendon)	Last ten intercostal, last dorsal, and first two lumbar nerves.
Transversalis abdominis	Ribs, last ten, lower extremities or cartilages; and lumbar vertebræ transverse processes	Ensiform cartilage, and linea alba	Last ten intercostal, last dorsal, and first two lumbar nerves.

Psoas magnus } see page 92. Iliacus		
Psoas parvus .	{ Vertebræ, bodies of lumbar and last three or four dorsal.	Ilium, ilio-pectineal eminence . } Lumbar nerves.
Quadratus lumborum .	Sacro-iliac ligament .	{ Lumbar vertebræ, transverse processes; and ribs, last three .
Intertransverse muscles of loins .	Lumbar vertebræ, transverse processes .	Lumbar vertebræ, transverse processes .

CHAPTER XI.

DISSECTION OF THE PELVIS.

UNDER this heading there will be described not only the pelvic cavity and its contents, but also the tail and the hip-joint.

Directions.—The dissection of the abdomen having been completed, the vertebral column should be sawn across or disarticulated about the middle of the lumbar region. If the directions given on page 69 have been attended to, the dissector of the pelvis should find the hip-joint intact, with the femur sawn across below the small trochanter, as in Fig. 48. The muscles or portions of muscles left around the hip-joint should be carefully removed, and the ligaments of the joint are to be dissected, noticing in the first place, however, its movements.

THE HIP-JOINT AND THE LIGAMENTS OF THE PELVIS.

The HIP-JOINT belongs to the class of enarthrodial or ball-and-socket joints.

The bones that enter into its formation are the femur and the os innominatum, the former furnishing a rounded hemispherical *head*, and the latter a cup-like cavity—the *acetabulum*, or the *cotyloid cavity.*

MOVEMENTS. If the stump of the femur be grasped, it will be found to have a great freedom of movement. Thus, it can be *flexed, extended, abducted, adducted, circumducted*, and *rotated.* In *flexion* the femur is carried forwards so as to diminish the angle formed by that bone and the ilium. For the definition of the other terms see page 42. In the horse the hip-joint admits of a greater range of movement than any other joint of the limbs. The movement of abduction, however, is less free than it is in the other domestic animals, being, as will presently be seen, restricted by the pubio-femoral ligament.

The joint possesses four ligaments, viz., capsular, cotyloid, pubio-femoral, and round ligaments.

The CAPSULAR LIGAMENT has the form of a double-mouthed sac, attached, on the one hand, to the rim of the cotyloid cavity, and to the cotyloid ligament, and, on the other hand, to the periphery of the articular head of the femur. It is strengthened in front by an oblique band representing the ilio-femoral ligament of man. Its inner face is lined by the synovial membrane of the joint, while its outer face is supported by the following muscles:—the deep gluteus above, the obturator externus below, the rectus femoris and the rectus parvus in front, and the gemelli behind. The ligament should be incised to show the synovial membrane, after which it may be removed entirely.

The SYNOVIAL MEMBRANE forms a complete internal lining to the capsular ligament, and also invests the pubio-femoral and round ligaments in the interior of the joint.

The COTYLOID LIGAMENT is a ring of fibro-cartilage fixed at the margin of the cotyloid cavity, which it serves to deepen for the reception of the femoral *head*. On the inner side of the joint, where the notch interrupts the rim of the cotyloid cavity, the ligament bridges over the gap, and to this portion of the ring the term *transverse ligament* is sometimes applied. This portion of the ligament, thus, converts the notch into a foramen, through which the pubio-femoral ligament enters the joint.

The PUBIO-FEMORAL LIGAMENT. This ligament derives its fibres from the prepubic tendon of the abdominal muscles, the right and left ligaments intercrossing their fibres in front of the pubes. It is directed outwards and backwards, resting in a groove on the inferior surface of the pubis, and perforating the origin of the pectineus muscle. At the notch on the inner side of the cotyloid ligament, it enters the hip-joint by passing above (in the natural position) the so-called transverse ligament, and it terminates in the depression on the head of the femur. The ligament, being attached across the middle in front, is put upon the stretch when the limb is abducted, and therefore restricts that movement.

The ROUND LIGAMENT (*interarticular ligament*, or *ligamentum teres*). This short and strong ligament is fixed above to the non-articular depression at the bottom of the cotyloid cavity, and below to the excavation on the head of the femur, being confounded at the latter point with the pubio-femoral ligament. It will be best displayed by cutting the transverse ligament and abducting the femur.

Direction.—It is convenient to dissect at this stage the sacro-sciatic ligament, as it is necessary to remove it in order to display the pelvic contents. Along with it, there will be described two other ligaments— the superior and inferior ilio-sacral ligaments.

The SACRO-SCIATIC LIGAMENT (Plate 16, and Fig. 48). This is a large membranous ligament which forms the greater part of the lateral boundary of the pelvis. It is irregularly four-sided in form. Its upper edge, which is pierced by the ischiatic artery, is fixed to the lateral lip of the sacrum, and to the rudimentary transverse processes of the first one or two coccygeal bones; its lower edge is attached to the superior ischiatic spine and to the tuber ischii, and between these points it forms the upper boundary of the small sacro-sciatic foramen ; its anterior edge is short, and forms the posterior boundary of the great sacro-sciatic foramen; its posterior edge, much more extensive than the anterior, is thin, ill-defined, and united to the coccygeal origin of the semimembranosus. Its outer surface is crossed by the great sacro-sciatic nerve, and is covered by the biceps femoris and semitendinosus muscles, which in part arise from it. Its inner surface is lined anteriorly by peritoneum, and is related posteriorly to the compressor coccygis and retractor ani muscles, some of whose fibres take origin from it. The internal pudic nerve and vessels cross this surface, or they may be partly embedded in the texture of the ligament.

The *Great Sacro-sciatic Foramen* is an elliptical opening in the lateral wall of the pelvis, its anterior boundary being formed by the ischiatic edge of the ilium, and its posterior by the sacro-sciatic ligament. It transmits the gluteal nerves and vessels, and the great sciatic nerve.

The *Small Sacro-sciatic Foramen* is an interval in the lower and posterior part of the lateral wall of the pelvis. Its upper edge is formed by the sacro-sciatic ligament; its lower by the smooth and rounded external border of the ischium, between the tuber and the superior ischiatic spine. By this opening the common tendon of the obturator internus and pyriformis emerge from the pelvis, and the nerves to these muscles pass in.

The SUPERIOR ILIO-SACRAL LIGAMENT (Fig. 48) is cord-like, and passes between the internal angle of the ilium (the angle of the croup) and the summits of the sacral spines.

The INFERIOR ILIO-SACRAL LIGAMENT (Fig. 48) is membranous and triangular in form. Its anterior edge is fixed to the upper part of the ischiatic border of the ilium; its lower edge is attached to the lateral lip of the sacrum; its posterior or upper edge is ill-defined, being continuous with the fascia investing the muscles of the tail.

THE CAVITY OF THE PELVIS.

Directions.—Fix the pelvis on a table, with the inlet looking upwards. Sponge out the cavity and distend the bladder with air or some preservative fluid, tying the urethra to prevent its escape.

The pelvis is not distinct from the abdominal cavity, but is merely a backward continuation of it. It is, in fact, that portion of the general cavity of the belly which is posterior to the bony circle formed by the sacrum, pubes, and ilio-pectineal lines. The plane of separation between the abdominal cavity proper and the pelvic cavity, is termed the inlet of the pelvis; the posterior extremity of the pelvic cavity is termed its outlet.

The *inlet* or *brim* of the pelvis is circumscribed by the promontory of the sacrum above, by the anterior margin of the pubic bones below, and by the ilio-pectineal line on each side. It looks downwards and forwards, and it is considerably larger in the mare than the horse. In form it is nearly circular.

The *outlet* of the pelvis is circumscribed by the first one or two coccygeal bones above, by the posterior edges of the ischial bones below, and by the posterior edge of the sacro-sciatic ligament on each side. In outline it is ovoid, with the broad end below ; and it looks backwards and upwards, being nearly parallel to the inlet.

The *Cavity of the Pelvis* is the irregularly tubular passage between the inlet and the outlet. Its transverse section approaches the circular in front, but changes gradually to the oval as it is taken more posteriorly. For convenience of description, however, it may be said to have a roof, a floor, and two lateral walls. The roof is formed by the inferior surface of the sacrum and first one or two coccygeal bones. The floor is formed by the pubic and ischial bones. Each lateral wall is formed for a short space in front by the pelvic surface of the shaft of the ilium, and for the rest of its extent by the sacro-sciatic ligament.

Contents of the Cavity.—These vary with the sex. In both sexes it contains the rectum, the urinary bladder, and the termination of the ureters, and numerous important vessels and nerves. In the male it lodges, besides these, the vasa deferentia (in part), the seminal vesicles, the prostate, Cowper's glands, the ejaculatory ducts, and the prostatic and membranous portions of the urethra. In the female it lodges the posterior part of the uterus, the vagina, and the vulva.

The PERITONEUM. The serous lining of the abdominal cavity is continued into the pelvis, whose walls and contents it in part covers. Thus, if it be followed backwards along the roof of the cavity, it will be seen to cover the lower face of the sacrum about as far as its 4th segment, but at that point it is reflected on to the rectum. Again, if the peritoneum be traced over the pelvic brim at the pubes, it will be found to cover the floor of the pelvis for a short distance, and then to become reflected on to the bladder. In the same way, along a curved line on the side of the pelvis between these two points, the peritoneum leaves the pelvic wall and passes on to the viscera. Since this reflection, however, takes place anterior to the posterior extremity, or outlet, of the cavity, it results that the pelvic viscera get at most only a partial covering of peritoneum. Thus, the rectum for a length of from four to six inches in front of the anus, the posterior extremity of the vesiculæ seminales, and (in the collapsed state) nearly the half of the upper face of the bladder, and three-fourths of its lower face are without a serous covering.

In the mare, in the same manner, the posterior part of the vagina and the whole of the vulva are without a serous covering.

The peritoneum in passing on to the viscera forms certain folds, or ligaments. Thus, it forms below and on each side of the urinary bladder a double fold, the inferior and lateral ligaments of the organ (Plate

44). The *inferior ligament* is a mesial fold attaching the bladder to the pubic symphysis, and to the middle line of the abdominal wall in front of the pubic brim. The *lateral ligaments* pass between the sides of the bladder and the lateral walls of the pelvis, and in the adult the free (anterior) edge of each contains the cord-like remains of the umbilical artery.

Again, the peritoneum, in descending from the roof of the cavity to envelop the first part of the rectum, forms a suspensory fold—the *meso-rectum*, which is continuous in front with the colic mesentery.

On each side the ureter and the vas deferens project narrow bands of peritoneum, and the right and left vasa deferentia where they lie above the bladder are connected by a triangular serous fold which contains between its layers the prostatic vesicle.

In the mare there are formed in an analogous manner the uterine ligaments described at page 303.

Directions.—The pelvis should now be either laid on its side, or suspended in the natural position and at a convenient height. A side view of its contents is to be exposed by the following steps :—With the saw cut through the shaft of the ilium close above the cotyloid cavity. Make another section through the same bone immediately external to the sacro-iliac articulation. Remove the intermediate piece of bone, at the same time separating the peritoneum from its inner aspect. Find the internal pudic artery in the position shown in Plate 16. It will be either internal to the sacro-sciatic ligament or in its texture. Trace it forwards and backwards. It is accompanied by a satellite vein, and where the two vessels pass above the small sacro-sciatic foramen they are crossed outwardly by the internal pudic nerve. This having been found should be followed upwards. Without injury to the nerve and vessels, the sacro-sciatic ligament may then be removed, taking care of the compressor coccygis and retractor ani muscles, which lie internal to the posterior part of the ligament.

The INTERNAL PUDIC ARTERY (Plates 46 and 47) is a branch of the internal iliac, arising at the last lumbar vertebra. Entering the pelvis, it descends obliquely downwards and backwards across the side of the cavity, lying on the inner surface of the sacro-sciatic ligament or within its texture (Plate 16). At the small sacro-sciatic foramen it passes with an inward and backward direction, terminating in a manner that varies with the sex.

In the male it turns round the ischial arch and reaches the perinæum, where it penetrates the urethral bulb. Besides slender hæmorrhoidal and perineal branches, it gives off the vesico-prostatic artery.

The *vesico-prostatic artery* arises about the neck of the bladder, and supplies the prostate, the vesicula seminalis, the posterior part of the bladder, and the terminal part of the vas deferens.

In the female the internal pudic terminates in hæmorrhoidal, vulvar, and bulbous branches; and, instead of the vesico-prostatic, it gives off the *vaginal artery*, which is expended in the bladder, vagina, and cervix uteri, anastomosing with branches of the uterine artery.

The UMBILICAL or HYPOGASTRIC ARTERY. In the adult (Plate 46) this is a comparatively small vessel arising from the internal pudic near its root. It is pervious only in the first few inches of its course, giving off a few twigs to the bladder, and being then continued as a solid cord at the free edge of the lateral ligament of the bladder. In the fœtus, however, it is of great size, and carries the fœtal blood to the placenta to be purified.

The INTERNAL PUDIC VEIN runs in company with the artery. It receives branches corresponding to those of the artery, and terminates in the internal iliac vein.

The PUDIC NERVE is derived from the 3rd sacral. Descending on the inner surface of the sacro-sciatic ligament, it crosses the internal pudic vessels superficially at the small sacro-sciatic foramen. Here it turns slightly inwards, and disappears beneath the ischio-urethral muscle. Having gained the lower face of the urethra, it turns round the ischial arch, and is continued as the dorsal nerve of the penis. Before leaving the pelvis, it detaches a *perinæo-anal branch*, which gives twigs to the muscles of the urethra and penis, and hæmorrhoidal branches that pass upwards on the rectum to reach the anus, some of them appearing to terminate in the sphincter. These latter branches are crossed by descending branches from the hæmorrhoidal nerve.

The lower posterior gluteal nerve (Plate 16) gives fibres to both the trunk of the pudic nerve and its perinæo-anal branch, and in some cases the latter derives the majority of its fibres from this source.

In the female the pudic nerve terminates in branches to the labia, clitoris, and constrictor muscles of the vulva.

The HÆMORRHOIDAL NERVE is derived mainly from the 4th sacral. It descends on the inner face of the sacro-sciatic ligament, and (for a short distance) the compressor coccygis muscle. It supplies a twig to that muscle, and then penetrates it, or emerges between it and the retractor ani. It then divides into branches for the retractor and sphincter muscles of the anus, and for the skin of the perinæum.

The RETRACTOR ANI. This muscle is described with the perinæum (page 276), but it is here exposed in the whole of its extent.

The COMPRESSOR COCCYGIS (Fig. 48) *arises* from the inner surface of the sacro-sciatic ligament, over the superior ischiatic spine. Passing backwards and upwards, it is inserted into the last sacral and first two coccygeal vertebræ. By its inner face it is related to the rectum, except close to its insertion, where the edge of the depressor of the tail intervenes.

Action.—Acting with the opposite muscle, it forcibly depresses the tail, compressing it over the perinæum. Acting singly, it inclines the tail to that side.

Directions.—The preceding two muscles should be entirely removed. Above the rectum there will be found the terminal portion of the posterior mesenteric artery; and on its side, the pelvic plexus of nerves.

The POSTERIOR MESENTERIC ARTERY (Plate 46) is a branch of the abdominal aorta. Its terminal portion enters the pelvic cavity between the layers of the meso-rectum; and passing backwards above the bowel, it terminates above the anus. In its backward course it detaches numerous branches to the wall of the rectum.

The POSTERIOR MESENTERIC VEIN runs in company with the artery. Its initial portion is formed at the posterior part of the rectum, by the union of hæmorrhoidal veins, which communicate with like branches of the internal pudic vein. In the abdominal cavity it concurs in the formation of the portal vein.

The PELVIC PLEXUS of the sympathetic nerve. This is an intricate network of nerves, placed on the side of the rectum, and distributing branches to the pelvic viscera. It receives in front the offsets from the posterior mesenteric plexus, and above it is joined by branches from the inferior sacral nerves. In both sexes it distributes branches to the rectum and bladder; and, besides, it supplies branches to the prostate, vesicula seminalis, and vas deferens in the male, and to the vagina and uterus in the female.

The RECTUM (Plate 46) is the terminal segment of the large intestines. At the entrance to the pelvis it is directly continuous with the small colon, and it terminates at the anus. Its initial portion resembles the small colon in being puckered and of comparatively small calibre. Its terminal portion, on the other hand, is dilated and sac-like, forming a large pouch in which the fæces collect.

In the male it is related inferiorly to the bladder, vesiculæ seminales, vasa deferentia, prostate gland, and pelvic part of the urethra. In the female it is related on the same aspect to the vulva, vagina, and uterus.

Structure.—The wall of the rectum resembles that of the large intestine in general (page 309), possessing serous, muscular, submucous, and mucous layers. As already seen, its peritoneal investment is incomplete, its terminal portion being destitute of peritoneum, and connected by loose areolar tissue to contiguous organs. In front of the anus the longitudinal muscular fibres of the bowel form on each side a band that passes upwards to be inserted into the coccygeal vertebræ. This, which is termed the *suspensory ligament of the rectum,* forms a prominence at the root of the tail. At the anus the last of the circular muscular fibres form what is termed the internal sphincter. Developed in connection with the termination of the rectum are two striped muscles—the sphincter ani externus and the retractor ani. These are described at page 276. In the male the retractor muscles of the penis (page 276) form a kind of sling for the rectum in front of the anus; and similar cords of involuntary muscular tissue unite below the rectum at the same point in the female, and terminate in the vulva.

The URINARY BLADDER (Plates 46 and 47) is the reservoir for the accumulation of the urine. The secretory action of the kidneys is constant; and the urine, passing along the ureter, accumulates in the bladder, to be expelled at intervals. As now seen in its distended condition, the bladder is not wholly contained within the pelvic cavity, but projects a little beyond the pubic brim. When empty, however, it lies entirely within the cavity, resting on the concave upper surface of the pubic bones. In form the distended viscus is ovoid. The broad end, which is free and directed forwards, is termed the *fundus ;* the narrow end has the opposite direction, and becomes continuous by a constricted *neck* with the urethra; the sides of the bladder are related to the pelvic walls; and the upper surface is related to the rectum, vasa deferentia, and vesiculæ seminales in the male, and to the vagina and uterus in the female. It is maintained in position by the peritoneum, which gives it only a partial covering, and by its continuity with the urethra. As already noticed, the peritoneum in passing on to it forms the folds called the middle and lateral ligaments of the organ.

The URETERS (Plates 46 and 47). Each tube having crossed the inlet of the pelvis, passes across its lateral wall, sustained by a narrow band of peritoneum. Finally, it is reflected inwards to perforate the upper wall of the bladder, a little in advance of its neck.

Directions.—Should the subject be a mare, the dissector must now turn to page 351 *et seq.*, where the urethra and reproductive organs of the female are described.

The URETHRA in the male (Plate 47). This is a long tube, extending from the neck of the bladder to the free extremity of the penis. The first few inches of the tube are contained within the pelvis, between the rectum and the ischiatic symphysis; for the rest of its extent it is extra-pelvic, and amalgamated with the penis except at its termination, where it projects as a short tube from the glans penis. The intra-pelvic division of the tube is divided into the *prostatic* and *membranous* portions; the extra-pelvic division is called also the *spongy* portion. The *prostatic portion* includes the first inch or two of the tube behind the neck of the bladder, and it is embraced by the prostrate gland. The *membranous portion* comprises the next two or three inches, extending as far as the ischial arch, where, at a very acute angle, it becomes continuous with the spongy portion. It is at this angle that the point of the catheter is likely to be arrested.

MUSCLES. The membranous part of the urethra has connected with it two muscles. The first of these, termed *Wilson's muscle*, or the *constrictor urethræ*, envelops the tube behind the prostrate gland, from which, indeed, it is not well defined. Its muscular fibres, of a pale red colour, comprise two sets, which extend across the urethra on its upper and lower faces respectively, and embrace the tube like an elliptical sphincter. The most posterior fibres of the muscle pass over Cowper's glands. The other, termed the *ischio-urethral muscle*, consists on each side of a band whose fibres arise from the ischial arch, and pass to the urethra beneath Cowper's gland, blending with Wilson's muscle. Like the preceding, it is composed of pale red muscular fibres.

Action.—These muscles are constrictors of the membranous urethra, and aid in the ejection of urine and semen.

The *spongy portion* of the urethra, with its muscles—the transversus perinæi and accelerator urinæ—has been already described as a constituent part of the penis (page 284).

The PROSTATE GLAND (Plates 46 and 47) embraces the neck of the bladder and the initial part of the urethra. It consists of a middle and two lateral lobes ; and in structure it is glandular, with a considerable admixture of striped muscular tissue. Its glandular texture consists of branching excretory tubes and acini, both having a columnar lining. Its ducts, as will be seen at a later stage, perforate the urethral wall, to which it is adherent.

COWPER'S GLANDS (Plates 46 and 47). Each of these is placed at the side of the membranous urethra, just in front of the ischial arch. They are round, reddish-yellow, and (in the stallion) about the size of a hazel nut. They have the racemose type of structure, and their ducts perforate the adjacent wall of the urethra.

The VASA DEFERENTIA (Plates 46 and 47). These are the excretory ducts of the testicles. As already seen, each is one of the constituents of the spermatic cord. Appearing at the internal abdominal ring, as a tube about the thickness of a goose-quill, it is reflected backwards to enter the pelvis. Crossing the direction of the ureter, it places itself on the upper surface of the bladder, and expands to four or five times its previous calibre, forming what is called the *bulbous portion* of the vas deferens. It then passes backwards beneath the vesicula seminalis ; and contracting again, it terminates under the prostate, by uniting outwardly with the neck of the vesicula to form a short tube termed the ejaculatory duct. Where the vasa deferentia lie above the bladder, they are connected together by a peritoneal fold between whose layers there is contained the *vesicula prostatica*, or *uterus masculinus*. This is a short tube with a blind anterior end, and opening by its posterior extremity into the urethra. It is the homologue of the uterus and vagina of the female.

The VESICULÆ SEMINALES (Plates 46 and 47). These bodies are placed between the rectum and the posterior part of the upper face of the bladder. Each is a small ovoid sac, like a miniature bladder. The anterior end of the sac is rounded and free; the posterior end contracts, and unites with the vas deferens to form the ejaculatory duct. Only the anterior half of the vesicula is covered by peritoneum, which in passing between the two bodies forms a small triangular serous fold.

The COMMON EJACULATORY DUCTS. Each of these is a short tube formed under cover of the prostate, by the union, at a very acute angle, of the neck of the vesicula seminalis with the vas deferens. Its opening into the roof of the urethra will be presently exposed.

Directions.—Carefully raise the fundus of the bladder, and cut its peritoneal and connective-tissue adhesions to the sides and floor of the pelvis. Free, in the same way, the membranous urethra at the ischial arch; and cut the crus penis and its erector muscle from the tuber ischii. This will enable the dissector to remove from the pelvis the organs just described, while maintaining their mutual relations. Lay the bladder on a table with its upper or rectal aspect downwards, and open it by a mesial incision on its lower face. Carry the incision backwards into the urethra, so as to open the whole extent of its prostatic and membranous portions. Care must be taken that the incision in both bladder and urethra is on the inferior face.

STRUCTURE OF THE BLADDER. This comprises four coats :—

1. The *Serous* or *Peritoneal Coat.* This, as already seen, is an incomplete investment.

2. The *Muscular Coat* is composed of bundles of non-striped fibres arranged in all directions. Compared with its condition in many other animals, this coat is very thin; and its fasciculi in the distended bladder seem hardly to form a continuous layer. At the neck of the bladder some of the fibres have a circular disposition, forming the *sphincter vesicæ.*

3. The *Submucous Coat* is composed of vascular areolar connective-tissue, and it loosely unites the muscular and mucous coats.

4. The *Mucous Coat.* This forms a complete internal lining for the bladder, and in the empty viscus it is thrown into folds, or *rugæ.* Observe the slit-like orifices of the ureters, near one another and a little anterior to the urethral orifice (Fig. 44). Pass a probe or bristle into one of them, and notice that the ureter perforates the wall very obliquely—an arrangement which has a valvular action in preventing the regurgitation of urine from the distended bladder. Between the uretral and urethral orifices in the human subject is a triangular area—the *trigone*—over which the mucous membrane is smooth even in the contracted bladder. In the horse, however, this area is wrinkled like the rest of the surface. The epithelium of the mucous membrane is stratified and transitional.

STRUCTURE OF THE URETHRA. The spongy portion has been described with the penis. The prostatic and membranous portions have a *mucous lining*, external to which is a *muscular coat* of non-striped fibres. Observe the following points in connection with the interior of the intra-pelvic part of the urethra (Fig. 44). On the middle line of the roof of the tube, close behind the communication with the bladder, there is a mucous eminence—the *colliculus seminalis*, or *verumontanum*. In the gelding this is often small, and sometimes hardly recognisable, but in the stallion it is sometimes a considerable eminence, like the tip of the little finger. At each side of this projection is the orifice of the ejaculatory duct. These orifices in the stallion are sufficiently large to permit of the tip of the little finger being insinuated into them. This should be remembered, as the point of catheter, if not guided along the floor of the urethra, might easily pass into one of them. At the summit of the colliculus, and therefore on the middle line, is a very minute opening—the orifice of the uterus masculinus. Insert a fine bristle into it, and guide it on into the tube. On the wall of the urethra at each side of the colliculus, observe an irregular series of minute orifices which belong to the ducts of the prostate gland. Behind these on each side, notice another series of small openings with a linear arrangement. These are the orifices of the ducts of Cowper's glands. Insert bristles into a few of each set of openings, and guide them on into the respective glands. Close to the neck of the bladder the epithelium of the urethra is of the same character as in the bladder, but behind that point it is simple and columnar.

STRUCTURE OF THE VESICULÆ SEMINALES. The walls of these are composed of *fibrous, fibro-muscular*, and *mucous* layers; and contain many tubular glands, which discharge their secretion into the cavity, where it mixes with the semen. The bulbous portion of each vas deferens has the same structure.

FIG. 44.

BLADDER AND INTRAPELVIC PORTION OF URETHRA OPENED FROM BELOW (*Leyh*).

1. Vas deferens; 1'. Bulbous part of the same; 2. Peritoneal fold joining the vasa deferentia; 3. Bladder; 4. Vesicula seminalis; 5. Orifices of ureters; 6. Prostate; 7. Verumontanum with orifices of ejaculatory ducts; 8. Orifice of prostatic vesicle; 9. Cowper's gland; 10. Orifices of ducts of prostate; 11. Orifices of ducts of Cowper's gland; 12. Corpus cavernosum; 13. Corpus spongiosum with urethra in its centre.

Directions.—The student must now return to the pelvis, at the roof of which he is to dissect the lumbo-sacral plexus of nerves, and the branches of the internal iliac artery (Plate 48). Thereafter he is to examine the pyriformis and obturator internus muscles.

The LUMBO-SACRAL PLEXUS (Plate 48) is composed of the anastomosing nerve trunks for the supply of the hind limb. It is formed by the inferior primary branches of the last three lumbar (4th, 5th, and 6th) and first two sacral nerves, and it receives also a fasciculus from the corresponding branch of the 3rd lumbar nerve. Each of these roots emerges from the intervertebral foramen behind the vertebra after which it is named; thus, the root from the 6th lumbar nerve emerges by the intervertebral foramen behind the 6th lumbar vertebra, the 1st sacral root by the first inferior sacral foramen, and so on. The branches of the plexus, taken in order from before to behind, are as follows :—

1. ILIACO-MUSCULAR BRANCHES, for the psoas and iliacus muscles. Two of these are seen in Plate 48, one coming from the anterior root of the plexus, and the other from the anterior crural nerve.

2. The ANTERIOR or GREAT CRURAL NERVE. In point of size, this is the second nerve of the plexus. It derives its fibres from the first two roots of the plexus (4th and 5th lumbar), and from the fasciculus furnished by the 3rd lumbar nerve.

3. The OBTURATOR NERVE derives its fibres from the 4th and 5th lumbar roots of the plexus. It descends in company with the obturator vessels, resting on the pelvic surface of the ilium. Under cover of the obturator internus muscle, it passes through the obturator foramen and reaches the thigh.

The 5th lumbar root, having given a branch to aid in the formation of the anterior crural, and another to the obturator nerve, is continued obliquely backwards between the internal iliac artery and the spine, to join a broad nervous fasciculus to which the remaining roots of the plexus (6th lumbar and first two sacral) contribute the whole of their fibres. The remaining branches of the plexus are divisions of this fasciculus.

4. The ANTERIOR GLUTEAL NERVES. Three or four in number, these leave the pelvis and reach the hip by passing through the forepart of the great sciatic opening, with the gluteal vessels.

5. The GREAT SCIATIC NERVE, the largest in the body, passes out into the hip through the great sciatic foramen, behind the preceding.

6. The POSTERIOR GLUTEAL NERVES, distinguished as superior and inferior, pass out behind the great sciatic.

The 3RD SACRAL NERVE. The inferior primary branch of this nerve is continued as the internal pudic nerve, after giving a bundle of fibres to aid in the formation of the hæmorrhoidal nerve.

The 4TH SACRAL NERVE receives the before-mentioned branch from the 3rd nerve, and is continued as the hæmorrhoidal nerve.

The 5TH SACRAL NERVE gives a backward twig to the 1st coccygeal nerve, and is then expended in the skin of the anus and root of the tail.

As in other regions of the spine, each of the inferior primary branches just considered communicates with the contiguous ganglion of the sympathetic cord, by one or more branches detached at the intervertebral foramen; and the sacral nerves send each a filament to the pelvic plexus.

The SYMPATHETIC GANGLIATED CORD in the sacral region. This is the direct backward continuation of the lumbar cord. It is placed on the inferior surface of the sacrum, internal to the inferior sacral foramina, the lateral sacral artery intervening between it and the inferior primary branches of the sacral nerves at their points of emergence. It possesses a ganglion opposite each of the first three sacral foramina; and, as before said, it communicates by filaments passing between these ganglia and the corresponding spinal nerves. The emergent branches of these ganglia are very slender, and pass to the cellular tissue beneath the sacrum, or to the contiguous blood-vessels. The cord terminates at the last ganglion, either abruptly, or by a filament passing on to the middle coccygeal artery.

The INTERNAL ILIAC ARTERY (Plate 48). This is one of the terminal branches of the posterior aorta. Beginning at the intervertebral disc between the 5th and 6th lumbar vertebræ, it passes downwards and backwards across the articulation between the last lumbar transverse process and the sacrum, and then across the sacro-iliac articulation; and at the upper part of the ilio-pectineal line, a little above the eminence of the same name, it divides into the iliaco-muscular and obturator arteries. The vessel is covered by the peritoneum, and in the first inch or two of its course it is separated from the external iliac artery by the common iliac vein. The collateral branches of the internal iliac, taken in the order of their point of detachment, are as follows:—

1. The second last of the series of lumbar arteries arises from the internal iliac at its root. It behaves like the lumbar branches of the aorta. Its upper division, much the larger of the two, passes upwards through the intervertebral foramen between the 5th and 6th lumbar vertebræ.

2. The INTERNAL PUDIC ARTERY. This is a considerable vessel having its origin at the last lumbar vertebra. Entering the pelvis, it descends at the ischiatic edge of the ilium, and then passes backwards in the texture of the sacro-sciatic ligament, or on its inner face.

3. The LATERAL SACRAL ARTERY leaves the parent trunk at the sacro-lumbar articulation, and passes backwards on the lower face of the sacrum, beneath or at the inner side of the inferior sacral foramina. A little behind the middle of the sacrum it divides into the *ischiatic* and *lateral coccygeal* arteries. The former, much the larger of the two, passes out through the edge of the sacro-sciatic ligament to reach the hip (Plate 16); the latter continues the direction of the lateral sacral artery to the tail. The inferior division of the 3rd sacral nerve appears in the angle of separation between these two arteries. The collateral branches of the lateral sacral artery are :—(1) Branches entering the

intervertebral foramen between the last lumbar vertebra and the sacrum (last lumbar artery), and the first two or three inferior sacral foramina. Each of these enters the spinal canal, furnishes there a spinal branch, and then emerges by the corresponding superior foramen, and is distributed to the overlying muscles and skin. (2) The *middle coccygeal* artery is an unpaired vessel, variable as to its origin, but generally, as in Plate 48, furnished by the right lateral sacral artery. It passes inwards to the middle line, and is continued backwards to the tail.

4. The ILIO-LUMBAR ARTERY. This artery is in series with the lumbar arteries, representing, as it were, the abdominal or inferior branch of the last lumbar artery. Arising from the outer side of the parent trunk, it passes outwards across the sacro-iliac joint, giving branches to the iliacus and psoas magnus muscles. Its terminal twigs may reach the gluteus maximus or the tensor vaginæ femoris.

5. The GLUTEAL ARTERY, a large vessel, arises at the edge of the sacrum, and passes out into the hip by the great sacro-sciatic foramen, dividing into a number of branches as it escapes (Plate 16).

The ILIACO-FEMORAL ARTERY, one of the terminal branches of the internal iliac, passes downwards and outwards between the shaft of the ilium and the iliacus muscles, to reach the outer aspect of the thigh. It supplies the nutrient artery of the ilium.

The OBTURATOR ARTERY, the other terminal branch of the internal iliac, passes downwards and backwards on the pelvic surface of the ilium, at the anterior edge of the pyriformis muscle. Under cover of the obturator internus muscle, it passes through the obturator foramen and reaches the thigh. It is accompanied by a satellite vein, and by the obturator nerve, which is placed anterior to the vessels. The tendon of the psoas parvus muscle is inserted in the angle of separation between this and the preceding artery.

The INTERNAL ILIAC VEIN collects the blood from the satellite veins of the foregoing arteries. It unites with the external iliac vein, forming the common iliac vein.

The OBTURATOR INTERNUS and the PYRIFORMIS. For a description of these muscles turn to page 68.

turn to page 68.

REPRODUCTIVE ORGANS IN THE FEMALE.

Comprised under this heading there are : the ovaries, the Fallopian tubes, the uterus, the vagina, and the vulva. The ovaries, the Fallopian tubes, and the uterus (in part) are abdominal organs, and their mode of suspension in that cavity has already been noticed. Their more complete examination can now be undertaken along with the dissection of the purely pelvic parts of the same apparatus, and at the same time it is convenient to examine the female urethra.

The OVARIES, as already seen, are situated in the lumbar region of the

abdominal cavity (see page 303). Each ovary is about half the size of
the testicle—the corresponding organ of the male. In form it is ovoid,
with a distinct depression on its upper surface—the *hilus*. At the hilus
the nerves and vessels of the organ enter from the broad ligament of the
uterus, and in its neighbourhood the expanded end of the Fallopian tube
is attached by one of its fimbriæ to the surface of the ovary. From the
posterior extremity of the ovary a cord of non-striped muscular tissue—
the ligament of the ovary—passes to the uterine cornu. The lateral
surfaces, the inferior border, and the anterior end of the ovary are
rounded and free.

STRUCTURE OF THE OVARY. This comprises (1) an epithelial covering
on the surface of the organ, (2) a fibrous framework, or stroma, and
(3) Graafian follicles.

1. The *Germinal Epithelium*.—This is a single layer of short columnar
cells with granular contents. In veterinary text-books the surface of
the ovary is described as having a serous covering derived from the
broad ligaments. The cells of this surface covering, however, are in
marked contrast to the cells of the broad ligament, which have the
ordinary flattened and transparent endothelial characters. The term
germinal is applied to this layer because the ova, or *germ-cells*, are
separated from it in the fœtal ovary.

2. The *Stroma* is composed of fibrous connective-tissue with some bundles
of non-striped muscular tissue. The blood vessels of the ovary ramify
in it, and it surrounds the Graafian follicles. Around the hilus it is
most vascular and open in texture, and this portion of the stroma is
sometimes termed the *zona vasculosa* or the *medullary substance*, in contradistinction to the peripheral *cortical substance*. A layer of condensed
stroma without any Graafian follicles lies beneath the surface epithelium,
and is sometimes termed the *tunica albuginea* of the ovary.

3. The *Graafian Follicles*, or *Ovi-sacs*.—These are vesicular bodies for
the maturation and extrusion of the ova. A large-sized follicle possesses
the following parts :—

a. The wall of the follicle, composed of an inner delicate *tunica
propria*, and an outer layer—the *tunica fibrosa*—derived from the
surrounding stroma.

b. The *Membrana Granulosa*.—This forms an epithelial lining to the
wall of the follicle, and consists of several layers of cells. At one point
these epithelial cells are heaped up to form the *cumulus* or *discus
proligerus*, the cells of which surround the *ovum*.

c. The *Liquor Folliculi*.—This is a fluid which fills up the remainder of
the cavity of the follicle.

The *Ovum* is a typical animal cell. It consists of an outer envelope—the
zona pellucida; protoplasmic cell-contents—the *vitellus* or *yelk;* a nucleus—
the *germinal vesicle;* and, within the nucleus, a nucleolus—the *germinal spot*.

The Graafian follicles vary greatly in size. The smallest are imbedded in the cortical part of the ovary. These are of microscopic size, and differ from the larger follicles in having only a single layer of cells in the membrana granulosa, and in having no liquor folliculi. Follicles of intermediate size are placed more deeply in the ovary, and differ from the largest chiefly in the small amount of liquor that they contain. These differences of size represent different stages of development of the follicles, the largest being the most mature. When mature, a follicle occupies a considerable space in the substance of the ovary in the neighbourhood of the hilus. Finally it bursts through the surface of the

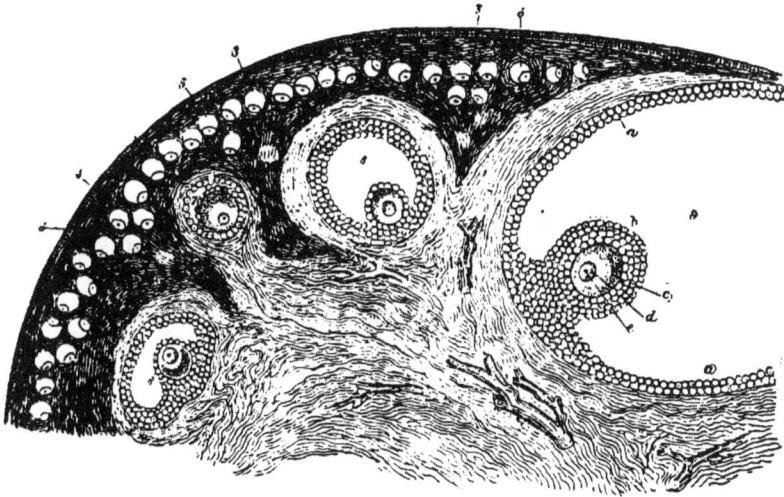

FIG. 45.

SECTION OF CAT'S OVARY, MAGNIFIED (from *Schrön*).

1. Outer covering of the ovary ; 2. Fibrous stroma ; 3. Superficial layer of fibro-nuclear substance; 3'. Deeper parts of the same ; 4. Blood-vessels ; 5. Ovi-sacs forming a layer near the surface ; 6. One or two of the ovi-sacs sinking deeper and beginning to enlarge ; 7. One of the ovi-sacs farther developed, now enclosed by a prolongation of the fibrous stroma, and consisting of a small Graafian follicle, within which is situated the ovum covered by the cells of the discus proligerus ; 8. A follicle farther advanced ; 8'. Another which is irregularly compressed ; 9. the greater part of the largest follicle, in which the following parts are seen ; *a*. Cells of the tunica granulosa lining the follicle ; *b*. The reflected portion named discus proligerus ; *c*. Vitellus or yelk part of the ovum, surrounded by the zona pellucida ; *d*. germinal vesicle ; *e*. Germinal spot.

ovary, and the ovum, along with the liquor folliculi and part of the membrana granulosa, escapes and is caught by the expanded extremity of the Fallopian tube. The follicle then collapses, while it becomes in part filled with blood from the vessels opened by the rupture of its wall. The rupture then heals, and the follicle becomes converted into a

2 A

yellowish body—the *corpus luteum*. In the early stage of a corpus luteum the cells of the membrana granulosa proliferate, while capillaries extend into it from the wall of the follicle. Later on the blood-clot in the centre becomes decolorised, and the granulosa cells become fatty; and finally the corpus luteum shrinks and disappears.

The PAROVARIUM, or the ORGAN OF ROSENMÜLLER. This is a minute body situated in the broad ligament, between the ovary and the Fallopian tube. It consists of a number of short convoluted tubules opening into a longitudinal tube, the latter representing the *canal of Gærtner* in the cow. The parovarium is the homologue of the epididymis of the male.

The FALLOPIAN TUBES, or OVIDUCTS. The Fallopian tube is the duct for the conveyance of the ova from the ovary to the uterus. In its

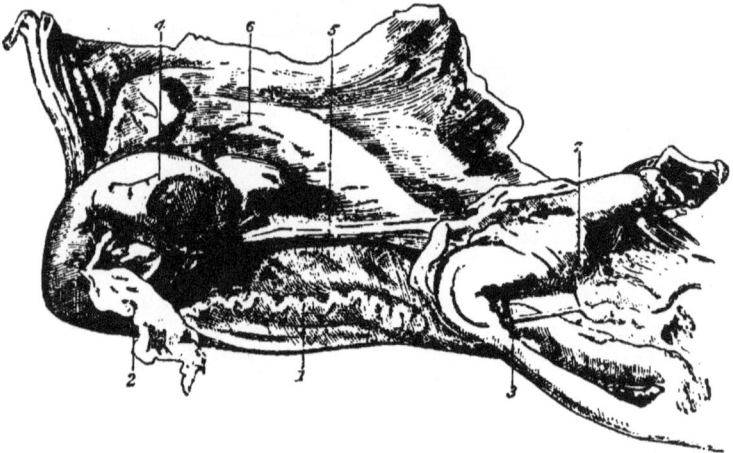

FIG. 46.

RIGHT OVARY AND FALLOPIAN TUBE.

1. Fallopian tube; 2. Abdominal opening (fimbriated extremity) of the same; 3. A probe introduced into the uterine opening of the tube; 4. Ovary; 5. Ligament of the ovary; 6. Broad ligament of the uterus; 7. Tip of uterine cornu laid open.

course between these two organs the tube passes in a flexuous manner at the anterior border of the broad ligament. The ovarian extremity of the tube opens on the surface of an expansion whose rim is cut into a few short fringe-like processes—the *fimbriæ*. Inwardly the rim of this expansion is fixed to the surface of the ovary near the hilus. The upper surface of the expansion is covered by a mucous membrane with delicate rugæ that converge from its rim to its centre, where it shows the orifice of the tube—the *ostium abdominale*. The under surface of the expansion is smooth and covered by peritoneum. The uterine extremity of the tube opens into the extremity of the uterine horn by a minute orifice—the *ostium uterinum.*

Although the Fallopian tube bears to the ovary the relationship of an excretory duct, in that it conveys away the ova, it differs from all other excretory ducts in not having its lumen closely continuous with the interior of the gland whose secretion it conveys. Moreover, this discontinuity between the Fallopian tube and the ovary establishes an indirect communication between the sac of the peritoneum and the surface of the body, and brings about the single exception to the rule that serous membranes form perfectly close sacs.

STRUCTURE OF THE TUBE. The wall of the oviduct comprises the following layers, enumerated from without inwards, viz., (1) an outer *serous coat,* derived from the broad ligament ; (2) a coat of non-striped *muscular tissue,* arranged as an outer longitudinal and an inner circular set of fibres; (3) a *submucous coat* of vascular connective-tissue ; (4) a *mucous coat,* having a ciliated columnar epithelium.

The lumen of the tube is narrowest at its uterine extremity and widest at the ovary.

The UTERUS, or womb, is the organ that receives the ovum, retains it during its development (provided it has been fertilised), and, finally, expels it at the expiration of the full term of pregnancy. In situation the organ is partly abdominal, and partly pelvic, and its mode of suspension by the broad ligaments has already been observed in connection with the peritoneum (page 303).

The organ is single in its posterior portion, and bifid in front.

The anterior bifurcations of the organ are termed its *cornua* or *horns.* At its anterior extremity each horn is pointed, and receives the uterine opening of the Fallopian tube. From this point the calibre of the horn gradually increases to its posterior end, where it opens into the body of the organ. Each horn shows a concave upper border at which the broad ligament reaches it, while its lower border is convex and free. The cornua are entirely abdominal in position, and are related to the intestines.

The posterior single portion of the uterus comprises the *body,* and the *neck,* or *cervix ;* but this division is not apparent on the exterior.

The *body,* placed in front, presents two faces, two borders, and two extremities. The upper face is slightly flattened and related to the rectum ; the lower face, also flattened, is related to the intestines in front, and to the bladder behind ; the borders, right and left, show the insertions of the broad ligaments; the anterior extremity, or *fundus,* is the widest part of the body, and it is joined at each angle by the cornu ; the posterior extremity is continuous with the cervix. The body of the uterus is partly abdominal and partly pelvic in situation.

The *cervix* is the extreme posterior part of uterus. It is directly continuous with the body in front; and its posterior extremity, as will be seen when the organ is laid open, projects into the anterior extremity of the vagina.

The VAGINA is a tubular organ which connects the uterus and the vulva. It is lodged entirely within the pelvis, being related to the rectum above, to the bladder and urethra below, and to the ureters and pelvic walls laterally. Its mode of connection with the two cavities that it connects will be examined later on. Its average length is about nine or ten inches.

The VULVA is the passage that continues the vagina backwards, and opens on the surface of the body beneath the anus.

The tube of the vulva is about five inches in length. It is united by cellular tissue to the rectum above, and to the pelvic floor below, while on each side it is related to the retractor ani muscle. Below and laterally it is covered by a layer of striped muscular tissue —the *anterior constrictor of the vulva.* The fibres of this muscle after embracing the tube of the vulva are lost on the sides of the rectum.

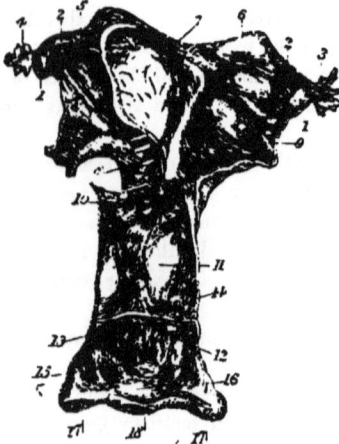

FIG. 47.

GENERATIVE ORGANS OF THE MARE, VIEWED FROM ABOVE.

1, 1. Ovaries ; 2, 2. Fallopian tubes ; 3. Fimbriated extremity of the tube, outer face ; 4. The same, inner face, showing the abdominal orifice ; 5. Ligament of the ovary ; 6. Right cornu, intact ; 7. Left cornu, laid open ; 8. Body of the uterus ; 9. Broad ligament ; 10. Os uteri (externum) ; 11. Interior of the vagina ; 12. Meatus urinarius, with its valve 13 ; 14. Mucous fold, a vestige of the hymen ; 15. Interior of the vulva ; 16. Clitoris ; 17, 17. Labia of the vulva ; 18. Inferior commissure of the vulva.

The external opening of the vulva has the form of a vertical slit, and it is bounded at the sides by the *labia*, which meet above and below to form the *commissures.* The *superior commissure* is acute, and separated from the anus by a narrow interval. The *inferior commissure* is rounded, and immediately within it the clitoris is lodged. The labia are covered externally by skin, which is thin, almost destitute of hairs, and generally black-pigmented ; inwardly they are lined by mucous membrane ; and at their sharp edges these cutaneous and mucous coverings meet. If the cutaneous covering of the labia be removed, the *posterior constrictor of the vulva* will be exposed. This is a red muscle corresponding to the compressor bulbi of human anatomy. Its fibres are elliptically disposed around the extremity of the vulva, being confounded with the sphincter ani above, while inferiorly some of the fibres are attached to the base of the clitoris, and others are attached to the inner surface of the skin below the inferior commissure. When the muscle contracts, it constricts the orifice of the vulva. Its lower fibres may frequently be observed to contract after micturition, depressing the inferior commissure and exposing the clitoris, which is simultaneously erected.

The CLITORIS. This small erectile body is the homologue of the male penis *minus* the urethra. It is lodged within the inferior commissure of the vulva, and presents a base, or attached extremity, a body, and a free extremity. The base is bifid, and attached to the ischial arch by the branches, or *crura*, each crus being covered by a rudimentary *erector clitoridis* muscle—the homologue of the erector penis. The body of the clitoris, which is from two to three inches in length, projects backwards and upwards, and is composed of right and left halves like the *corpora cavernosa* of the penis. The free extremity is formed by a rudimentary *glans*, which is provided with a mucous cap analogous to the prepuce. The clitoris is composed of erectile tissue resembling that of the penis.

The VESTIBULAR BULB. This will be exposed by the removal of the posterior constrictor muscle. It is an erectile body composed of right and left halves, each of which is placed at the side of the vulvar cavity (the vestibule), between the posterior constrictor and the mucous membrane. Inferiorly the two halves of the organ are in communication with one another, and with the erectile tissue of the clitoris, and superiorly each terminates at the side of the vulva by a rounded end. The bulb is the homologue of the corpus spongiosum of the penis.

Directions.—The pelvic viscera must now be removed to allow an examination of the structure and interior of the organs just considered. This is to be effected by cutting the meso-rectum and the peritoneal ligaments of the bladder, carrying the knife above the anus and below the inferior commissure of the vulva, and destroying the vascular and connective-tissue attachments of the various organs to the pelvic walls. The entire generative apparatus will thus be removed along with the urinary bladder and the rectum. The latter organ should be dissected from the vagina and vulva (for its structure see page 344), and the other viscera examined *seriatim*. The canal of the vulva and vagina is to be exposed by a mesial incision on the upper wall of these organs.

THE CANAL OF THE VULVA. This, as already stated, is a tubular passage about five inches in length. When removed from the body and inflated, it assumes a large calibre, but ordinarily its walls are in contact. Tracing the canal in an order inverse to that followed

in the previous description of parts, it may be said to begin on the
surface of the body at the vertical slit already described, and to
terminate in front by joining the tube of the vagina.   In the
adult animal there is little to mark the line of separation between the
two passages, but in the young animal a membranous septum—the
hymen—stretches between the two.   This is occasionally seen also in
the adult mare, and more frequently a few warty projections—the
carunculæ myrtiformes—which are the shrunken remains of the hymen,
stud the line of junction ; but very often the canal of the vulva passes
without interruption into that of the vagina.   The vulva is lined by a
mucous membrane of a rosy, vascular tint.   It possesses numerous
mucous follicles ; and its free surface is formed by a stratified squamous
epithelium, which, towards the external opening, is often pigmented in
spots.

The MEATUS URINARIUS.   The urethra opens on the middle line of
the floor of the vulva immediately behind its point of continuity with
the vagina.   The opening is surmounted by a large mucous fold—
the *valve of the meatus urinarius.*   This valve has its free edge
directed backwards, and it serves to direct the flow of urine towards
the exterior.   Its presence must be remembered in passing the
female catheter, the point of which should be made to press on the floor
of the vulva as it is directed onwards.   The meatus is of large size
when compared with the same orifice in the male, since it readily admits
two fingers.

Directions.—Reverse the natural position of parts, laying the uterus,
vagina, and vulva with their upper surfaces downwards, and open the
bladder by a mesial incision on its lower (in the natural position) face.
For an account of the structure of the bladder turn to page 347.

The URETHRA of the female is very much shorter, but considerably
wider, than the corresponding tube of the male.   Beginning as a funnel-
like prolongation of the neck of the bladder, it passes backwards on the
middle line of the lower face of the vagina, in whose wall it is partially
imbedded ; and after a course of two or three inches it perforates the
lower wall of the vulva, and opens by the meatus already described.
The calibre of the tube is in correspondence with the large size of the
meatus ; and with slight stretching it will accommodate three fingers.
The wall of the urethra is composed of connective-tissue, and non-
striped muscular fibres circularly disposed ; and it is lined internally
by a longitudinally folded mucous membrane with a stratified squamous
epithelium.

STRUCTURE AND INTERIOR OF THE VAGINA.   The tube of the vagina
is about nine or ten inches in length.   Posteriorly it joins the vulva,
and anteriorly it embraces the cervix uteri.   The connection between
the cavities of the vagina and uterus is, thus, not by simple continuity,

but the vaginal wall is carried forwards, so as to cause the os uteri to project freely into the forepart of the vaginal canal.

The wall of the vagina comprises the following layers :—

1. A *Serous Coat.*—This is only a partial covering, the posterior part of the organ being without a peritoneal investment. In the hinder part of the tube the place of the peritoneum is taken by connective-tissue uniting it to surrounding parts. This connective-tissue is loose and areolar towards the rectum ; but between the vagina and the bladder it is closer, and forms a more intimate bond between the two organs.

2. A *Muscular Coat.*—This is composed of non-striped muscular tissue, continuous in front with the muscular coat of the uterus. Posteriorly the muscular tissue is reddish in tint, and passes into the anterior constrictor of the vulva. The fibres are arranged both longitudinally and circularly.

3. A *Mucous Coat.* This lines the tube inwardly, and it is longitudinally folded. It possesses numerous mucous glands, and its epithelium is stratified and squamous. It is of a pinkish, vascular tint, like the mucous lining of the vulva.

Directions.—Lay open one of the horns of the uterus in its whole extent, and carry the incision along the body and cervix to the os.

STRUCTURE AND INTERIOR OF THE UTERUS. The interior of the uterus comprises the cavities of the body and cervix, and those of the horns.

The *Cavity of the Cervix* begins posteriorly at the orifice of the tap-like projection already noticed at the forepart of the vaginal canal. This orifice is termed the *os uteri externum*, or, shortly, the *os uteri*. Ordinarily the orifice is closed, and forms a circular depression from which the folds of mucous membrane radiate outwards, and curve round the circular lip of the os, to be carried to the vaginal wall.

In front the canal of the cervix passes gradually into the wider cavity of the body. (In the human subject the connection between the canal of the cervix and the cavity of the body is abrupt, constituting the *os uteri internum*.)

The *Cavity of the Body* is triangular in form, with the base in front. At its posterior angle it passes into the canal of the cervix, and at each antero-lateral angle it is joined by the cavity of a horn.

The *Cavities of the Horns* are conical and curved. Each is widest at its base, where it joins the cavity of the body ; and it tapers to its anterior extremity, in the centre of which it presents a small tubercle perforated by the uterine orifice of the Fallopian tube.

The wall of the uterus comprises serous, muscular, and mucous layers :—

1. The *Serous Coat* is peritoneum, continuous with the layers of the broad ligaments. It completely envelops the organ.

2. The *Muscular Coat* is composed of non-striped fibres arranged as

an external longitudinal, and an inner circular set. To compensate for the expansion of the uterine wall during pregnancy, and to provide a force to expel the fœtus at parturition, there is, during pregnancy, both an increase in the size, and an addition to the number, of these muscular fibres.

3. The *Mucous Coat* forms a complete lining to the uterus. It is smooth, of a pale pink colour, and thrown into longitudinal wrinkles. The epithelium is simple, columnar, and ciliated, except in the posterior part of the cervix, where it is stratified and squamous, as in the vagina. In the cornua and body the mucous membrane is set with numerous *utricular* glands. The mouths of these glands open on the surface of the membrane, while their blind ends lie against the muscular coat. They lie obliquely in the membrane, and are branched at their deep ends. They are lined by a single layer of columnar ciliated cells.

The mucous membrane of the cervix contains numerous mucous follicles, and the peculiar *ovula Nabothi*, which appear to be mucous glands distended into a vesicular form by their own clear secretion. In pregnancy these cervical glands secrete the mucous plug that closes the os uteri.

Directions.—The student must now return to the dissection of parts remaining in the pelvis, beginning with the lumbo-sacral plexus (page 349).

THE TAIL (FIG. 48).

Directions.—Saw through the ilium that is still intact, making the section across the bone at the great sciatic foramen. By cutting the sacro-sciatic ligament on the same side, the sacro-coccygeal part of the spine, with the sacro-iliac joints, will be isolated. Dissect away the inferior ilio-sacral ligament, and remove the skin from the tail.

The skin of the tail differs from that of the body in general in the greater length of its hairs. On its under surface, however, extending backwards from its root, there is a triangular area without hairs. Along the under surface of the tail, and especially in front, the skin is thin; but on its upper aspect and sides it is thick, and intimately adherent to the subjacent fascia.

The muscles of the tail are enveloped by a strong *coccygeal fascia* which is continuous in front with the inferior ilio-sacral ligament. The isolation of the muscles can be readily effected near the root of the tail, but towards its tip they tend to blend with each other. In each half of the tail there are three muscles, viz., one above, one below, and one at the side. There are also three arteries—one on the middle line below, and one between the inferior and lateral muscles on each side. On each side there are two sets of nerves, one of which accompanies the lateral artery, while the other is on the upper aspect of the bones,

between the lateral and superior muscles. [Besides the three muscles
now to be described, there is the compressor coccygis already dissected
(page 343).]

The ERECTOR COCCYGIS (sacro-coccygeus superior).   This muscle *arises*
from the sides and summits of the sacral spines, and it is *inserted* by
successive short tendons to the upper aspect of the coccygeal vertebræ.

Action.—Acting with its fellow, to elevate the tail directly ; acting
alone, to elevate the tail and incline it laterally.

The CURVATOR COCCYGIS (sacro-coccygeus lateralis).   This muscle
seems to continue backwards the semispinalis of the loins.   It *arises*
from the last two lumbar spines and from the spines of the sacrum,
and it is *inserted* into the lateral aspect of the coccygeal bones.

Action.—To bend the tail to the side of the acting muscle.

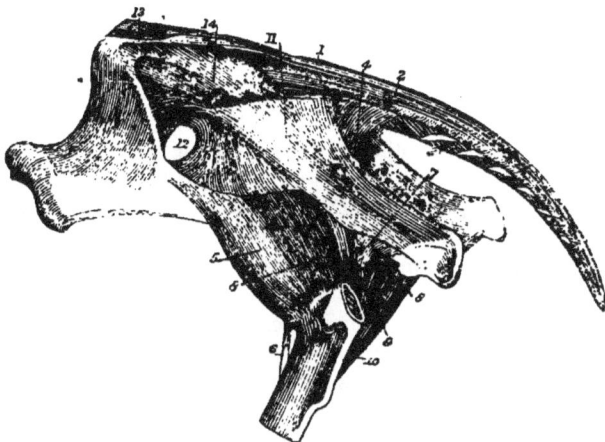

FIG. 48.

MUSCLES OF THE TAIL, DEEP MUSCLES OF THE HIP, AND PELVIC LIGAMENTS (*Chauveau*).

1. Erector coccygis ; 2. Curvator coccygis ; 3. Depressor coccygis ; 4. Compressor coccygis ;
5. Deep gluteus ; 6. Rectus parvus ; 7. Common tendon of obturator internus and pyriformis ;
8. Gemelli ; 9. Accessory fasciculus of the same ; 10. Quadratus femoris ; 11. Sacro-sciatic ligament ;
12. Great sacro-sciatic foramen ; 13. Superior ilio-sacral ligament ; 14. Inferior ilio-sacral ligament.

The DEPRESSOR COCCYGIS.   Anteriorly this muscle consists of an
outer and an inner portion, which Leyh describes as separate muscles.
It *arises* from the lower face of the sacrum, beginning about the 3rd
foramen.   The slips of the inner portion are *inserted* into the first six
coccygeal vertebræ, while the outer portion extends to the extremity of
the tail, and is provided with strong tendons of insertion.

Action.—It inclines the tail laterally or depresses it, according as it
acts alone or with the opposite muscle.

Between the preceding two muscles a number of semi-independent fleshy fasciculi connect adjacent coccygeal bones. Leyh describes these separately as the *intertransversales caudæ*.

At the root of the tail, between the right and left depressors, the retractor muscles of the penis take origin from the 1st and 2nd or 2nd and 3rd coccygeal bones; and behind these the so-called suspensory ligaments of the rectum are inserted (Plate 46).

The MIDDLE COCCYGEAL ARTERY (Plate 48). This is the largest artery of the tail.* It is an unpaired vessel, and in the great majority of cases it is a collateral branch detached from the inner side of the lateral sacral artery towards the middle of the sacrum. Sometimes it is detached in the same way from the left lateral sacral artery. Passing backwards and inwards on the lower surface of the sacrum, it places itself on the middle line, and extends in that position throughout the tail, lying under the coccygeal vertebræ, and between the right and left depressor muscles. In its backward course it gradually reduces itself by giving off lateral branches.

The LATERAL COCCYGEAL ARTERY (Plate 48). Each artery (right or left) is one of the terminal branches of the lateral sacral artery (the ischiatic artery being the other branch). Having its origin towards the middle of the sacrum, it passes backwards in the tail, crossing the sides of the coccygeal bones, between the depressor and curvator muscles, the former muscle separating it from the middle artery. It becomes smaller by the detachment of numerous collateral twigs, the largest of which pass upwards. Leyh designates this vessel the infero-lateral coccygeal artery, describing as the supero-lateral coccygeal artery what is, apparently, an unusually large branch of the first.

VEINS. The foregoing arteries are accompanied by veins of the same names.

COCCYGEAL NERVES. There are five or six pairs of coccygeal nerves, and they are numbered according to the bones behind which they turn outwards, the first issuing behind the first coccygeal vertebra, and so on with the others. The first of them has a loop of communication with the last sacral. As they turn outwards, they divide into an upper and a lower branch corresponding to the superior and inferior primary branches of the spinal nerves in other regions. The branches of each of these sets are directed backwards, detaching slender filaments, and then applying themselves to the next nerve of the same set. In this way there are formed in each half of the tail two composite nerves, one of which accompanies the lateral artery, while the other runs on the upper aspect of the tail between the erector and curvator coccygis. These cords are expended in branches to the muscles and skin of the tail.

---

* Leyh describes and figures this artery as being smaller than the lateral coccygeal, but that, certainly, is not usually the case.

SACRAL NERVES. On the upper aspect of the sacrum the superior primary branches of the sacral nerves will be found at their points of emergence from the spinal canal, the first four issuing by the superior sacral foramina, and the last by the foramen between the sacrum and the first bone of the coccyx. These nerves are much smaller than the corresponding inferior primary branches; and after giving twigs to the muscles on the side of the spine, they pass upwards to the skin of the croup. Slender branches of the lateral sacral artery issue from the spinal canal in company with them.

JOINTS AND LIGAMENTS OF THE SACRUM AND COCCYX.

The sacral portion of the spine in the adult animal does not present any joints between its constituent pieces, which are fused by anchylosis. The lumbar supraspinous ligament is prolonged on the summits of the sacral spines. This region, however, furnishes the important joint between the vertebral column and the skeleton of the hind limb—the sacro-iliac articulation.

The SACRO-ILIAC ARTICULATION. The bony surfaces that concur to form this are the auricular facet on the lateral aspect of the sacrum, and the corresponding facet on the pelvic surface of the ilium. The *movements* permitted in the joint are scarcely appreciable, as the student may prove by grasping the sacrum and the part of the ilium left in connection with it. Since this joint is the bond of connection between the skeleton of the trunk and that of the hind limb, in which, in locomotion, the main propulsive efforts are originated, it is necessary that but slight movement should be permitted, as otherwise these efforts would not be transmitted with precision to the trunk. The stability of the joint is effected mainly by one ligament—the sacro-iliac, and to a less degree by the superior and inferior ilio-sacral ligaments and the sacro-sciatic ligament already described (page 339).

The *Sacro-iliac Ligament.*—This ligament is composed of strong fibres passing between the sacrum and ilium, in close relation to the joint. It consists of an upper and a lower half, corresponding respectively to the anterior and posterior sacro-iliac ligaments of human anatomy. The former is much the stronger of the two; and the necessity for its strength is apparent when one reflects that whatever weight is placed on the back and loins of the horse, tends to drive the sacrum downwards from its connection with the iliac bones, and that this tendency is rather favoured than otherwise by the form of the articular surfaces, which offer an arrangement comparable to an inverted arch.

The bones should be disarticulated to show the articular surfaces. The joint is provided with a rudimentary *synovial membrane*.

SACRO-COCCYGEAL and INTER-COCCYGEAL ARTICULATIONS. Ordinarily these are movable joints, the articular surfaces being the opposed extremities of the rudimentary vertebral bodies. These are connected by small intervertebral discs, which are shaped like a biconcave lens, since the bodies of the vertebræ are here convex on both extremities.

The bones are also invested by a fibrous sheath, which may be supposed to represent the superior and inferior common ligaments of the back and loins.

Movements.—The biconvex form of the vertebral centra, and the suppression of the different processes in this region give a great range and freedom of movement to the tail, which, provided with its appendage of hairs, is admirably fitted to protect the hind quarters of the animal from the attacks of insects. It is interesting to notice the absence of the panniculus carnosus over the area within which the tail is serviceable for this purpose. In animals above middle age it is not uncommon to find the sacro-coccygeal, and even the first intercoccygeal joint, anchylosed.

INDEX.

———✻———

MUSCLES—*continued.*
Hyoideus transversus, 199.
Hyo-pharyngeus, 205.

Iliacus, 63, 326.
Inferior oblique of eye, 210.
Infraspinatus, 17.
Internal intercostal, 98.
,, pterygoid, 185.
Interossei, 32, 80.
Intertransversales of loins, 326.
,, ,, neck, 155.
Ischio-urethral, 346.

Lateralis sterni, 97.
Latissimus dorsi, 9, 14, 94.
Levator anguli scapulæ, 8.
Levatores costarum, 97.
Levator labii superioris alæque nasi, 177.
Levator labii superioris proprius, 178.
Levator menti, 180.
,, palati, 204.
,, palpebræ superioris, 175.
Longissimus dorsi, 96.
Longus colli, 156.
Lumbricales, 31, 80.

Masseter, 180.
Mastoido-auricularis, 162.
Mastoido-humeralis, 10, 152, 166.
Middle gluteus, 65.
,, hyo-glossus, 199.
Mylo-hyoideus, 171, 198.

Obliquus abdominis externus, 289.
,, ,, internus, 291.
,, capitis inferior, 169.
,, ,, superior, 169.
Obturator externus, 61.
,, internus, 67.
Occipito-styloid, 166.
Orbicularis oris, 179.
,, palpebrarum, 175.

Palato-glossus, 199.
Palato-pharyngeus, 203.
Panniculus, 8, 94, 144, 170, 177, 287.
Parieto-auricularis externus, 160.
,, ,, internus, 162.
Parotido-auricularis, 160.
Pectineus, 60.
Peroneus, 75, 81.
Popliteus, 72.
Posterior deep pectoral, 3.
Posterior superficial pectoral, 2.
Psoas magnus, 63, 325.
,, parvus, 326.
Pterygo-pharyngeus, 205.
Pyriformis, 67.

Quadratus femoris, 61.
,, lumborum, 326.

Recti oculi, 209.
Rectus abdominis, 292.
,, capitis anticus major, 155.

MUSCLES—*continued.*
Rectus capitis anticus minor, 208.
,, ,, lateralis, 208.
,, ,, posticus major, 169.
,, ,, posticus minor, 169.
,, femoris, 69.
,, parvus, 69.
Retractor ani, 276.
,, costæ, 96.
,, oculi, 209.
,, penis, 276.
Rhomboideus, 10.

Sartorius, 57.
Scalenus, 156.
Scapulo-humeralis gracilis, 16.
Scapulo-ulnaris, 14.
Scuto-auricularis externus, 162.
,, ,, internus, 162.
Semimembranosus, 60.
Semispinalis of back and loins, 97.
,, colli, 155.
Semitendinosus, 65.
Serratus anticus, 95.
,, magnus, 7, 95.
,, posticus, 95.
Short extensor of foot, 80.
Small hyo-glossus, 199.
,, stylo-pharyngeus, 205.
Soleus, 71.
Sphincter ani, 276.
,, pupillæ, 262.
Splenius, 153.
Stapedius, 268.
Sterno-maxillaris, 145, 166.
Sterno-thyro-hyoideus, 146.
Stylo-glossus, 198.
Stylo-hyoideus, 166, 190.
Stylo-maxillaris, 166.
Stylo-pharyngeus, 205.
Subscapularis, 15.
Subscapulo-hyoideus, 146.
Superficial flexor of digit (fore limb), 23, 34.
Superficial flexor of digit (hind limb), 72, 81.
Superficial gluteus, 64.
Superior oblique of eye, 210.
Supraspinatus, 17.

Temporalis, 185.
Tensor palati, 203.
,, tympani, 268.
,, vaginæ femoris, 64.
Teres major, 14.
,, minor, 17.
Thyro-arytenoid, 229.
Thyro-hyoid, 228.
Thyro-pharyngeus, 205.
Trachelo-mastoid, 153.
Transversalis abdominis, 293.
,, costarum, 96.
Transversus perinæi, 276.
Trapezius, 9.
Triangularis sterni, 120.
Triceps extensor cubiti, 15.

Vastus externus, 69.
,, internus, 62.

NERVES—*continued.*
Small superficial petrosal, 189.
Spheno-palatine, 214, 220.
Spinal, 138.
,, accessory, 140, 146, 151, 194, 255.
Splanchnic, 111, 318.
Staphyline, 204, 214.
Subcutaneous thoracic, 7, 288.
Subzygomatic, 168, 183, 188.
Superior dental, 213.
,, laryngeal, 194, 230.
,, maxillary, 213, 254.
Supra-orbital, 176, 211.
Suprascapular, 13.
Sympathetic cord, cervical, 149, 195.
,, ,, dorsal, 111, 118.
,, ,, lumbar, 325.
,, ,, sacral, 350.

Tenth cranial, 108, 116, 194, 255.
Third cranial, 176, 212, 254.
Trifacial or trigeminal, 254.
Trochlear, 212, 254.
Twelfth, 193, 198, 256

Ulnar, 13, 21.

Vagus, 108, 116, 194, 255.
Vertebral, 157.
Vidian, 214.
Nipple, 285.
Nodulus Arantii, 127.
Nostrils, 176.

Œsophagus, 115, 150.
Olfactory bulbs, 244.
,, cells, 219.
Olivary body, 239.
Omentum, gastro-hepatic, 312.
,, gastro-splenic, 301, 312.
,, great, 300, 312.
Optic thalami, 251.
,, tracts, 243, 252.
Orbicular process, 268.
Organ of Corti, 272.
,, Jacobson, 219.
,, Rosenmüller, 354.
Osseous spiral lamina, 271.
Os uteri, 359.
Otic ganglion, 189.
Otoliths, 270.
Ovaries, 351.
Oviducts, 354.
Ovula Nabothi, 360.
Ovum, 352.

Pacchionian bodies, 234.
Palate, hard, 200.
,, soft, 202.
Palpebral tendon, 175.
Papillæ of tongue, 196.
Papilla optica, 264.
Pancreas, 313, 333.
Parotid gland, 164.
Parovarium 354.
Pelvis, 338.
Peduncles of cerebellum, 241.
,, pineal gland, 251.

Pedunculated hydatid of morgagni, 279.
Penis, 281.
Pericardium, 105.
Perilymph, 270.
Perinæum, 274.
Periople, 36.
Perioplic ring, 38.
Peritoneum, 298.
,, pockets of, 303.
Pes anserinus, 182.
Peyer's patches, 309.
Pharynx, 204.
Pia mater, cranial, 236.
,, spinal, 138.
Pillars of soft palate, 202.
Pineal gland, 251.
Pituitary gland, 243.
,, membrane, 219.
Plantar cushion, 38.
Pleuræ, 101.
PLEXUSES OF NERVES—
Anterior mesenteric, 307.
Aortic, 325.

Carotid, 238.
Cavernous, 238.
Cœliac, 318.

Pelvic, 344.
Posterior mesenteric, 307.

Renal, 320.

Solar, 318.
Spermatic, 325.
Suprarenal, 318.
Pomum Adami, 225.
Pons Tarini, 243.
,, Varolii, 240.
Portal fissure, 314.
Poupart's ligament, 289.
Prepuce, 280.
Prostate, 346.
Puncta lachrymalia, 174.
Pupil, 261.
Pylorus, 311.
Pyramidal body, 40.
Pyramids of Ferrein, 334.
,, medulla, 239.

Quadrilateral space, 244.

Receptaculum chyli, 324.
Rectum, 297, 344.
Reissner's membrane, 271.
Renal crest, 334.
Rete testis, 280.
Retina, 264.
Right lymphatic duct, 119.
Rivinius, ducts of, 197.
Root of lung, 105.
Roots of cranial nerves, 253.
,, spinal nerves, 140.
Rostrum, 248.

Saccule, 270.
Scala intermedia, 272.
,, tympani, 271.
,, vestibuli, 272.

VEINS—*continued.*
Coronary plexus, 42.

Digital, of fore limb, 30.
    ,,   hind limb, 79.
Dorsal, 114, 119, 155.

External iliac, 323.
   ,,   pudic, 291.

Femoral, 59, 61.

Gastric, 317.
Gastro-omental, 317.
Great vena azygos, 119.

Hæmorrhoidal, 344.
Hepatic, 323, 332.

Inferior dental, 186.
Intercostal, 99, 119.
Internal iliac, 323, 351.
   ,,   maxillary, 168, 186, 188.
   ,,   pudic, 276, 343.
   ,,   subcutaneous of fore-arm, 20.
   ,,   saphena, 57, 70.
   ,,   thoracic, 114, 119, 122.

Jugular, 144, 167.

Laminal plexus, 42.
Lateral coccygeal, 362.
Lingual, 186, 198.
Lumbar, 99, 119, 323.

Maxillo-muscular, 168, 182.
Median, 20.
Metacarpal, 30.
Metatarsal, 79.
Middle coccygeal, 362.

Obturator, 62.
Occipital, 170, 193.
Of Galen, 251.
Ophthalmic, 211, 213.

Palatine, 202, 213.
Phrenic, 323.
Popliteal, 75.
Portal, 317, 331.

VEINS—*continued.*
Posterior abdominal, 292.
   ,,   auricular, 163, 168.
   ,,   gastric, 317.
   ,,   mesenteric, 306, 317, 344.
   ,,   radial, 22.
   ,,   tibial, 75.
   ,,   vena cava, 119, 323.
Pterygoid, 186.
Pulmonary, 105.

Renal, 320.

Small vena azygos, 114.
Solar plexus of foot, 42.
Spermatic, 279, 323.
Spinal, 141.
Spheno-palatine, 213, 220.
Splenic, 317.
Spur, 287.
Staphyline, 204.
Subcutaneous abdominal, 287.
   ,,   thoracic, 287.

Vasa vorticosa, 263.
Velum interpositum, 251.
   ,,   pendulum palati, 202.
Ventricle of brain, fourth, 241.
   ,,   ,,   lateral, 249.
   ,,   ,,   third, 252.
Ventricle of heart, left, 128.
   ,,   ,,   right, 126.
Ventricles of larynx, 231.
Ventricular grooves, 106.
Vermiform lobe, 240.
Verumontanum, 348.
Vesiculæ seminales, 346, 348.
Vesicula prostatica, 346.
Vestibular bulb, 357.
Vestibule of ear, 269.
Villi of intestine, 308.
Vitreous humour, 266.
Vocal cord, 227.
Vulva, 356, 357.

Wall of hoof, 36.
Wharton's duct, 190, 197.
Wirsung's duct, 313.

Zonula of Zinn, 265.

www.ingramcontent.com/pod-product-compliance
Lightning Source LLC
Chambersburg PA
CBHW021353210326
41599CB00011B/857